Lecture Notes in Computational Science and Engineering

73

Editors

Timothy J. Barth
Michael Griebel
David E. Keyes
Risto M. Nieminen
Dirk Roose
Tamar Schlick

For further volumes:
http://www.springer.com/series/3527

Hans-Joachim Bungartz · Miriam Mehl
Michael Schäfer
Editors

Fluid Structure Interaction II

Modelling, Simulation, Optimization

 Springer

Editors
Hans-Joachim Bungartz
Miriam Mehl
Technische Universität München
Institut für Informatik
Boltzmannstr. 3
85748 Garching
Germany
bungartz@in.tum.de
mehl@in.tum.de

Michael Schäfer
Technische Universität Darmstadt
Numerische Berechnungsverfahren
Dolivostr. 15
64293 Darmstadt
Germany
schaefer@fnb.tu-darmstadt.de

ISSN 1439-7358
ISBN 978-3-642-26523-5   978-3-642-14206-2 (eBook)
DOI: 10.1007/978-3-642-14206-2
Springer Heidelberg Dordrecht London New York

Mathematics Subject Classification Numbers (2010): 65-06, 65K10, 65Mxx, 65Nxx, 65Y05, 74F10, 74Sxx, 76D05, 76Fxx

Cover design: deblik, Berlin

Printed on acid-free paper

Springer is part of Springer Science + Business Media (www.springer.com)

Preface

Modern Computational Science and Engineering (CSE) is confronted with several challenges of "multi-type": Multi-physics problems involve more than one physical effect; multi-scale models involve different scales with respect to space or time; multi-level methods are needed to efficiently tackle large linear systems; multi-core architectures require a new access to parallelism; and much of the research in CSE requires the collaboration of experts from several disciplines – it is multi-disciplinary. Concerning the first issue, multi-physics problems such as fluid-structure interactions (FSI), i. e. the interplay of some moveable or deformable structure with an internal or surrounding flow field, are one of the most relevant and most intensely studied coupled problems. Despite this high attention, FSI are still not completely understood, and there is an obvious lack of reliable, robust, and efficient computational methods.

Furthermore, there is a somewhat astonishing discrepancy between, on the one hand, how complex specific scenarios have already been successfully simulated (think of airbags or parachutes, e. g.) and, on the other hand, how big the problems are that occur when those codes shall be used for different problems. Hence, there hasn't been any widely accepted numerical benchmark for FSI before this volume's predecessor LNCSE **53** in 2006. Also experimental validation has turned out to be far from trivial: either the experimental setting is too complicated for the numerical tools, or the numerical setting is not feasible for experiments; if, finally, both experiments and numerical simulations can deal with a certain scenario, the effects intended to study often do not show up. All this shows that there are still challenging questions in FSI research, ranging from modelling via numerical treatment up to implementation and software tools – and only their ensemble provides a key to deeper insight in FSI.

The present volume contains selected contributions from the "First International Workshop on Computational Engineering – special topic Fluid-Structure Interactions" held in Herrsching, Germany, in October 2009. This three-day workshop was jointly organized by three initiatives funded by the German Research Foundation (Deutsche Forschungsgemeinschaft, DFG) – the "International Graduate School of Computational Engineering" in Darmstadt, the "International Graduate School of Science and Engineering" in Munich, and the Research Unit 493 "Fluid-Structure Interaction: Modelling, Simulation, Optimization" (FOR 493). FOR 493 was established by the DFG in 2003. In this framework, researchers from seven German universities working in the fields of mathematics, informatics, mechanical engineering, chemical engineering, or civil engineering joined their forces to push forward the state-of-the-art of fundamental computational research in FSI. Designed as a forum for latest results in computational engineering in general and FSI in particular, the workshop in Herrsching brought together leading experts from all over the world, allowing for three highly interesting days of tutorials, invited lectures, and minisymposia, and, now, resulting in the fifteen papers collected in this volume – which is the second book on FSI published by our Research Unit FOR 493.

We would like to thank the editors of Springer's Lecture Notes in Computational Science and Engineering (LNCSE) for admitting our collection to this series for the second time, as well as Springer Verlag and, in particular, Dr. Martin Peters, for their most valuable support from the first idea to the final layout. Furthermore, we are deeply obliged to Michael Lieb, who did a great job in compiling the single contributions to a finally harmonic ensemble. Last, but not least, we want to express our thanks to DFG for more than six years of ongoing funding. Without this support, neither many of the results presented in the contributions of this book nor this volume itself would have become reality.

Munich and Darmstadt   *Hans-Joachim Bungartz*
May 2010   *Miriam Mehl*
                                                                              Michael Schäfer

Contents

Multi-Level Accelerated Sub-Iterations for Fluid-Structure Interaction

A.H. van Zuijlen and H. Bijl

Abstract Computational fluid-structure interaction is most commonly performed using a partitioned approach. For strongly coupled problems sub-iterations are required, increasing computational time as flow and structure have to be resolved multiple times every time step. Many sub-iteration techniques exist that improve robustness and convergence, although still flow and structure problems have to be solved a number of times every time step. In this paper we apply a multilevel acceleration technique, which is based on the presumed existing multigrid solver for the flow domain, to a two-dimensional strongly coupled laminar and turbulent problem and investigate the combination of multilevel acceleration with the Aitken underrelaxtion technique. It is found that the value for the under-relaxation parameter is not significantly different when performing sub-iterations purely on the coarse level or purely on the fine level. Therefore coarse and fine level sub-iterations are used alternately, where it is found that performing 3 or 4 coarse level sub-iterations followed by 1 fine level sub-iteration resulted in the highest gain in efficiency. Although the total number of sub-iterations increases slightly by 30%, the number of fine grid iterations can be decreased by as much as 65–70%.

1 Introduction

In many engineering applications, fluid-structure interaction phenomena play a key role in the dynamic stability of a structure (e.g. aircraft, wind-turbines, suspension bridges, etc.). A fast and accurate computation of the dynamic interaction between

A.H. van Zuijlen
Delft University of Technology, Kluyverweg 1, 2629HS, Delft, The Netherlands
e-mail: A.H.vanZuijlen@tudelft.nl

H. Bijl
Delft University of Technology, Kluyverweg 1, 2629HS, Delft, The Netherlands
e-mail: H.Bijl@tudelft.nl

H.-J. Bungartz et al. (eds.), *Fluid Structure Interaction II*, Lecture Notes
in Computational Science and Engineering 73, DOI 10.1007/978-3-642-14206-2_1,
© Springer-Verlag Berlin Heidelberg 2010

1

flow and structure is therefore of the utmost importance. For fluid-structure interaction often a partitioned approach is chosen to resolve the coupled problem, as it allows reusing existing flow and structure solvers and independent development and optimization of the codes. The drawback of a partitioned approach over a monolithic approach is that the coupling between the flow and structure domain needs additional attention and for strongly coupled physical problems, sub-iterations are required. Sub-iterating increases the computational expense as flow and structure have to be resolved multiple times each time step. In order to reduce computational costs, we propose a multilevel algorithm for reducing the costs of particularly the flow solver.

Basic sub-iteration techniques include block-Gauss-Seidel iterations, which may suffer instability or fixed under-relaxation methods, which are robust, but at the price of slower convergence. In literature several methods can be found for performing sub-iterations in an efficient and robust fashion. Especially when the flow is incompressible, robustness of the sub-iteration technique is an issue [3]. One of the most popular methods is the Aitken under-relaxation method [6], that tunes the under-relaxation parameter to obtain faster convergence. Methods that require more implementation effort are e.g. interface GMRES [5] or reduced order modeling [9]. They have the advantage over the Aitken method that they have faster convergence. Recently a quasi Newton method has been proposed [4] and applied to fluid-structure interaction coupling [2] that has a similar implementation complexity as the Aitken method but with superior convergence.

Whichever method is chosen, a (small) number of sub-iterations still has to be performed. In these sub-iterations solving the fluid dynamic equations is generally the most computationally expensive. In previous research [11] we investigated the possibility of performing sub-iterations initially on a coarse level and showed that a basic coarse level block-Gauss-Seidel sub-iteration would have much the same convergence as sub-iterating on the fine level.

In this paper we apply this model to a more challenging two-dimensional strongly coupled laminar [8] and turbulent problem and investigate the combination of multilevel acceleration with Aitken under-relaxation. We investigate the effectiveness of performing initial sub-iterations on a coarse level and alternating coarse and fine level sub-iterations.

2 Coupled problem

In this paper a fluid-structure interaction problem is addressed which is based on the benchmark problem by Turek [8]. The original problem consists of an incompressible fluid around a circular cylinder with a flexible trailing flap. The coupled problem consists of a fluid domain Ω_f, see Fig. 1, which is here modeled as a compressible fluid, and a structure domain Ω_s which is modeled as a linear elastic body. The boundaries of the domain are given by Γ_s, Γ_f, which result in boundary conditions for the structure and fluid dynamics respectively. In the partitioned approach

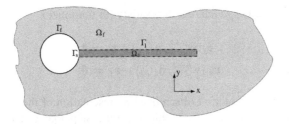

Fig. 1 Computational domain.

the interface boundary between the fluid and the structure domains Γ_I is denoted by two boundaries Γ_{sf} and Γ_{fs} which close the structure and fluid domains so that each domain can be treated separately from the other. The coupling between flow and structure is introduced in the boundary conditions that are imposed on Γ_{sf} and Γ_{fs} and that should yield continuity of displacement (of the interface) and stresses. Since the flap has the highest flexibility in y-direction and the shear stresses mainly act in x-direction, we simplify the continuity of stresses to a continuity in pressure so that the conditions at the interface are

$$d_{\Gamma_{fs}} = d_{\Gamma_{sf}}, \tag{1}$$
$$p_{\Gamma_{sf}} = p_{\Gamma_{fs}}, \tag{2}$$

where d denotes the displacement of the interface boundary and p the pressure. At the moment it is still assumed that the spatial coupling is continuous and the temporal coupling instantaneous. Since the domains have been split and the coupling is performed using boundary conditions, readily available flow and structure solvers can be used to discretize and resolve their own dynamics on their own domains. Therefore, we do not address the specific spatial discretization of the solvers and simply write

$$\frac{d\mathbf{w}_s}{dt} + \mathbf{D}_s(\mathbf{w}_s, \mathbf{p}_{\Gamma_{sf}}) = \mathbf{S}_s, \tag{3}$$

$$\frac{d\mathbf{w}_f}{dt} + \mathbf{D}_f(\mathbf{w}_f, \mathbf{d}_{\Gamma_{fs}}) = \mathbf{S}_f, \tag{4}$$

wherein $\mathbf{w}_s$ and $\mathbf{w}_f$ are the discrete state vectors for the structure state and fluid state respectively. They contain e.g. the structural displacement or the fluid density. The spatial discretization of the governing equation is simplified by the operator $\mathbf{D}$, which depends both on the state $\mathbf{w}$ and on the fluid-structure interface conditions $\mathbf{p}_{\Gamma_{sf}}$ (the discrete pressure acting on the structure) and $\mathbf{d}_{\Gamma_{fs}}$ (the displacement of the fluid domain boundary). The right hand side may contain terms $\mathbf{S}$ that may arise from boundary conditions on Γ_s and Γ_f. Equations (3) and (4) are in semi-discrete form. We assume that the time integration is performed by the same implicit scheme in both domains. The structure and flow solver programs can then be described as solution techniques that can find solutions $\mathbf{w}_s^{n+1}$ and $\mathbf{w}_f^{n+1}$ under the boundary conditions $\mathbf{p}_{\Gamma_{sf}}$ and $\mathbf{d}_{\Gamma_{fs}}$ such that they satisfy (or minimize)

$$\mathbf{r}_s(\mathbf{w}_s^{n+1}, \mathbf{p}_{\Gamma_{sf}}) - \mathbf{s}_s = \mathbf{0}, \tag{5}$$

$$\mathbf{r}_f(\mathbf{w}_f^{n+1}, \mathbf{d}_{\Gamma_{fs}}) - \mathbf{s}_f = \mathbf{0}, \tag{6}$$

wherein $\mathbf{r}$ the residual function (discretized representation of the governing equations), $\mathbf{s}$ a constant source term within the time step that can depend on e.g. previous solutions or boundary conditions, $\mathbf{p}_{\Gamma}$ the discrete pressures in the boundary nodes and $\mathbf{d}_{\Gamma}$ the discrete displacements of the boundary nodes. The subscript s, f denotes that the discrete quantities belong to the structure and fluid domains respectively. A Computational Structure Dynamics (CSD) package is capable of finding a $\mathbf{w}_s^{n+1}$ such that (5) is satisfied for a given pressure load $\mathbf{p}_{\Gamma_{sf}}$. A Computational Fluid Dynamics (CFD) package is able to find a $\mathbf{w}_f^{n+1}$ such that (6) is satisfied for a given boundary displacement $\mathbf{d}_{\Gamma_{fs}}$. A fully implicit (or fully coupled) solution would require

$$\mathbf{r}_s(\mathbf{w}_s^{n+1}, \mathbf{p}_{\Gamma_{sf}}^{n+1}) - \mathbf{s}_s = \mathbf{0}, \tag{7}$$

$$\mathbf{r}_f(\mathbf{w}_f^{n+1}, \mathbf{d}_{\Gamma_{fs}}^{n+1}) - \mathbf{s}_f = \mathbf{0}, \tag{8}$$

wherein the superscript $^{n+1}$ denotes the discrete solution at the new time level t_{n+1}. The coupling between (7) and (8) now poses a problem in a partitioned approach as the pressure acting on the structure interface $\mathbf{p}_{\Gamma_{sf}}^{n+1}$ depends on the fluid state $\mathbf{w}_f^{n+1}$ and the displacement of the fluid boundary $\mathbf{d}_{\Gamma_{fs}}^{n+1}$ depends on the structure state $\mathbf{w}_s^{n+1}$. Both the spatial coupling (transferring data from the flow to the structure mesh and vice versa) and the temporal coupling (obtaining an implicitly coupled solution) are addressed in the next sections.

2.1 Spatial coupling

The coupling between flow and structure takes place at the fluid-structure boundary. In the continuous case, this boundary Γ_I would be identical for both fluid and structure domains, however, at the discrete level, the boundary Γ_{sf} and Γ_{fs} do not have to be matching and gaps or overlaps may occur. In [1] several interpolation techniques have been studied for transfer of displacements and pressures at the fluid-structure interface. It was concluded that a consistent mesh interpolation was preferred over the conservative interpolation method. For the consistent interpolation, two separate interpolations are defined for the transfer from fluid to structure mesh and from structure mesh to fluid mesh. The interpolation methods do not have to be the same kind, e.g. a radial basis function interpolation can be used for transfer of displacements from the structure to the flow mesh and a simple nearest neighbor algorithm can be used to transfer pressures from the flow mesh to the structure mesh.

For the conservative approach on the other hand one can choose an interpolation method for transferring displacements from the structure to the flow, but one has to use the transposed of the interpolation to transfer forces from the flow to the structure. First the coupling of displacements is performed by transferring displacements from the discrete structure boundary to the discrete flow boundary through an interpolation $\mathscr{I}_{fs}$

$$\mathbf{d}_{\Gamma_{fs}} = \mathscr{I}_{fs}(\mathbf{d}_{\Gamma_{sf}}). \tag{9}$$

The displacements at the structure boundary follow directly from the structure state vector $\mathbf{d}_{\Gamma_s} = g(\mathbf{w}_s)$. When the structure state vector $\mathbf{w}_s$ already contains the displacements of the boundary nodes as part of its degrees-of-freedom, g could simply be represented by a Boolean matrix that extracts only the boundary displacement from the structure state vector.

The second part of the coupling is the transfer of pressure loads from the flow to the structure by an interpolation $\mathscr{I}_{sf}$

$$\mathbf{p}_{\Gamma_s f} = \mathscr{I}_{sf}(\mathbf{p}_{\Gamma_{fs}}), \tag{10}$$

and the pressure at the fluid boundary follows directly from the fluid state vector $\mathbf{p}_{\Gamma_f} = f(\mathbf{w}_f)$. This time the function f may be more complicated as the fluid state $\mathbf{w}_f$ can be defined in cell centers (for a cell-centered finite volume solver), whereas the pressures on the interface may be defined in face centers or vertex locations. In that case an interpolation from cell centered values to boundary values is also taking place in the function f. Additionally, the fluid state vector may only contain the conservative variables and not the primitive variable p, in which case f also computes the pressure from the conservative variables.

2.2 Temporal coupling

In partitioned fluid-structure interaction, obtaining the coupled solution described by (7) and (8) would require sub-iterating, e.g. when a sequential algorithm is used

$$\mathbf{r}_s(\mathbf{w}_s^i, \hat{\mathbf{p}}_{\Gamma_{sf}}^i) - \mathbf{s}_s = \mathbf{0}, \tag{11}$$

$$\mathbf{r}_f(\mathbf{w}_f^i, \mathscr{I}_{fs}(g(\mathbf{w}_s^i))) - \mathbf{s}_f = \mathbf{0}, \tag{12}$$

wherein the superscript i denotes the i-th sub-iteration and $\hat{\mathbf{p}}_{\Gamma_{sf}}^i$ is the *estimation* of the fluid pressure acting on the structure for the i-th iteration. The simplest choice for the estimation is

$$\hat{\mathbf{p}}_{\Gamma_{sf}}^i = \mathscr{I}_{sf}(g(\mathbf{w}_f^{i-1})), \tag{13}$$

which results in a block-Gauss-Seidel type of iteration, but which is not guaranteed to be stable. To increase robustness under-relaxation can be applied, but generally at the expense of slower convergence rate. In this paper we focus on the widely applied

Aitken method [6], which applies an adaptive under-relaxation to the estimation for the next time step

$$\hat{\mathbf{p}}_{\Gamma_{sf}}^{i+1} = \hat{\mathbf{p}}_{\Gamma_{sf}}^{i} + \theta^{i+1}(\mathbf{p}_{\Gamma_{sf}}^{i} - \hat{\mathbf{p}}_{\Gamma_{sf}}^{i}), \tag{14}$$

for which the under-relaxation parameter θ^{i+1} is obtained from

$$\theta^{i+1} = \theta^{i} \left(1 - \frac{(\Delta\mathbf{e}^{i})^{T}(\mathbf{e}^{i})}{(\Delta\mathbf{e}^{i})^{T}(\Delta\mathbf{e}^{i})} \right), \tag{15}$$

with $\mathbf{e}^{i} = \mathbf{p}_{\Gamma_{sf}}^{i} - \hat{\mathbf{p}}_{\Gamma_{sf}}^{i}$ the error between the estimated and the resulting pressure after solving (11) and (12) for iteration i and $\Delta\mathbf{e}^{i} = \mathbf{e}^{i} - \mathbf{e}^{i-1}$. For the first under-relaxation step a θ has to be chosen as $\mathbf{e}^{i-1}$ is not available yet. One can use last known value and at the very start of the computation any (sufficiently small) value can be taken.

3 Two-level acceleration

In order to reduce the computational time for the (initial) sub-iterations, the two-level acceleration scheme solves for a correction of the solution on a coarsened mesh. The way the method works is best demonstrated on a linear system first.

3.1 Linear fine grid operator

Therefore, the coupled system (7) and (8) is written as a linear system

$$L_h \begin{pmatrix} \mathbf{w}_{h,s} \\ \mathbf{w}_{h,f} \end{pmatrix} - \begin{pmatrix} \mathbf{s}_{h,s} \\ \mathbf{s}_{h,f} \end{pmatrix} = \mathbf{0}, \tag{16}$$

wherein L_h is a linear operator (constant matrix) which consists of

$$L_h = \begin{pmatrix} A_{h,s} & A_{h,sf} I_{h,sf} f_h \\ A_{h,fs} I_{h,fs} g_h & A_{h,f} \end{pmatrix}, \tag{17}$$

which contains the discretization of the governing equations in the structure $A_{h,s}$ and fluid $A_{h,f}$ domains and the discretization of the boundary conditions $A_{h,sf}$ and $A_{h,fs}$. The coupling also depends on the (linear) interpolation algorithms at the fluid-structure interface ($I_{h,sf}$ and $I_{h,fs}$) and the interpolation functions f_h and g_h. Suppose the approximation of the coupled solution is denoted by $(\hat{\mathbf{w}}_{h,s} \ \hat{\mathbf{w}}_{h,f})^{T}$, we obtain a residual $\mathbf{r}_h$

$$L_h \begin{pmatrix} \hat{\mathbf{w}}_{h,s} \\ \hat{\mathbf{w}}_{h,f} \end{pmatrix} - \begin{pmatrix} \mathbf{s}_{h,s} \\ \mathbf{s}_{h,f} \end{pmatrix} = \hat{\mathbf{r}}_h. \tag{18}$$

Subtraction of (18) from (16) yields

$$L_h \begin{pmatrix} \mathbf{w}_{h,s} - \hat{\mathbf{w}}_{h,s} \\ \mathbf{w}_{h,f} - \hat{\mathbf{w}}_{h,f} \end{pmatrix} = -\hat{\mathbf{r}}_h, \tag{19}$$

which can written as

$$L_h \boldsymbol{\varepsilon}_h = -\hat{\mathbf{r}}_h, \tag{20}$$

wherein the correction $\boldsymbol{\varepsilon}_h$ is defined as

$$\boldsymbol{\varepsilon}_h = \begin{pmatrix} \mathbf{w}_{h,s} - \hat{\mathbf{w}}_{h,s} \\ \mathbf{w}_{h,f} - \hat{\mathbf{w}}_{h,f} \end{pmatrix}. \tag{21}$$

Instead of solving the original coupled equation (16), the coupled system (19) term can be solved to obtain the correction term $\boldsymbol{\varepsilon}_h$. This obviously involves solving a coupled system of the same complexity as the original equation. Following the standard multigrid approach, the correction term is computed on a coarse mesh which reduces computational costs.

3.2 Linear coarse grid operator

The coarse level system is obtained by restriction R and prolongation P operators commonly used in the multigrid algorithm. Prolongation and restriction are only applied to flow and structure domains individually, e.g.

$$R = \begin{pmatrix} R_s & 0 \\ 0 & R_f \end{pmatrix}, \tag{22}$$

$$P = \begin{pmatrix} P_s & 0 \\ 0 & P_f \end{pmatrix}, \tag{23}$$

so that there does not exist a coarse level variable that consists of a combination of a fine grid flow variable and fine grid structure variable. Left multiplication of (19) with R and denoting the correction on the fine mesh as a prolongation of the correction on the coarse mesh $\boldsymbol{\varepsilon}_h = P\boldsymbol{\varepsilon}_H$ yields

$$RL_h P\boldsymbol{\varepsilon}_H = -R\hat{\mathbf{r}}_h, \tag{24}$$

wherein the coarse grid operator $L_H = RL_h P$ is

$$L_H = \begin{pmatrix} R_s A_{h,s} P_s & R_s A_{h,sf} I_{h,sf} f_h P_f \\ R_f A_{h,fs} I_{h,fs} g_h P_s & R_f A_{h,f} P_f \end{pmatrix}. \tag{25}$$

Several remarks can be made with respect to the coarse grid coupled system to be solved in (24): first of all the right-hand-side term is directly related to the residual from the fine mesh, and therefore, when the fine grid solution is converged, the coarse grid correction reduces to zero as well. Secondly, the coarse level operator is often constructed by simply making the discretization of the governing equations on a coarse grid, instead of constructing it from the fine grid operator by restriction and prolongation, e.g. use $A_{H,f}$ instead of $R A_{h,f} P$. The discretization of $A_{H,f}$ does not necessarily have to be done by the same discretization, e.g. the fine grid discretization could be a second order central discretization, whereas the coarse grid discretization could be based on a first order upwind scheme. Thirdly, the coarse grid coupling terms (off-diagonal blocks in (25)), may be interpreted as follows: the coupling term $R_s A_{h,sf} I_{h,sf} f_h P_f$ denotes that when we have a coarse grid correction term in the flow domain $\varepsilon_{H,f}$, that this term is prolonged to the fine mesh level $\varepsilon_{h,f}$ and its influence on the boundary value for the pressure $\delta \mathbf{p}_{h,\Gamma_{fs}}$ is computed through f_h. This change in the boundary pressure field is interpolated to the fine grid level of the structure with the $I_{h,sf}$ operator to obtain $\delta \mathbf{p}_{h,\Gamma_{sf}}$ and finally, its influence $A_{h,sf} \delta \mathbf{p}_{h,\Gamma_{sf}}$ on the coarse structure level is obtained by restriction to the coarse level. This shows that the coupling at the fluid structure interface (both for the mesh interpolation as the functions f and g takes place on the *fine* grid level. Alternatively one could make a new coupling on the *coarse* level $A_{H,sf} I_{H,sf} f_H$, which requires the definition of an interpolation between the coarse level meshes $\mathscr{I}_{H,sf}$ and f_H. The two options, performing the coarse level coupling through the *fine* grid level and performing the coarse level coupling on the *coarse* grid level itself, are schematically depicted in Fig. 2(a) and Fig. 2(b) respectively. The definition of the coarse grid operator (25) reflects the coupling as depicted in Fig. 2(a), whereas Fig. 2(b) is represented by the coarse grid operator

$$\tilde{L}_H = \begin{pmatrix} R_s A_{h,s} P_s & A_{H,sf} I_{H,sf} f_H \\ A_{H,fs} I_{H,fs} g_H & R_f A_{h,f} P_f \end{pmatrix}. \tag{26}$$

In this paper only the first option is explored in further detail as the second option may have a drawback in the accuracy of the coupling when the interface data is not constructed accurately using f_H as mentioned in Sect. 4.1.1. In any case, the resulting coarse level system is a coupled system again. Resolving it would require again a partitioned or monolithic approach.

3.3 Coarse level sub-iterations

For the same reason as that the fine level was resolved in a partitioned fashion, the coarse level system is also resolved in a partitioned fashion, e.g. for a standard Gauss-Seidel iteration

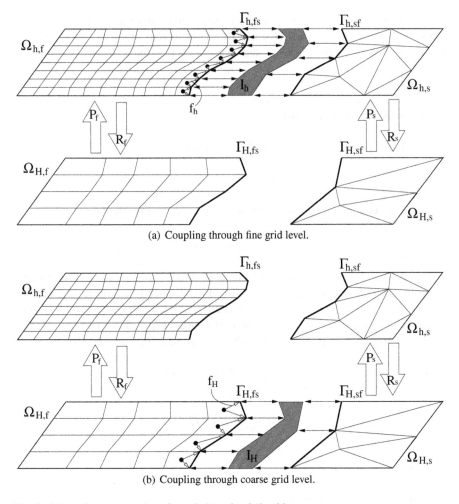

(a) Coupling through fine grid level.

(b) Coupling through coarse grid level.

Fig. 2 Schematic representation of coupled two-level algorithm.

$$
\begin{pmatrix} R_s A_{h,s} P_s & 0 \\ R_f A_{h,fs} I_{h,fs} g_h P_s & R_f A_{h,f} P_f \end{pmatrix} \begin{pmatrix} \boldsymbol{\varepsilon}_{H,s} \\ \boldsymbol{\varepsilon}_{H,f} \end{pmatrix}^i
$$
$$
= -R\hat{\mathbf{r}}_h^0 - \begin{pmatrix} 0 & R_s A_{h,sf} I_{h,sf} f_h P_f \\ 0 & 0 \end{pmatrix} \begin{pmatrix} \boldsymbol{\varepsilon}_{H,s} \\ \boldsymbol{\varepsilon}_{H,f} \end{pmatrix}^{i-1}, \quad (27)
$$

or the equivalent

$$
\begin{pmatrix} R_s A_{h,s} P_s & 0 \\ R_f A_{h,fs} I_{h,fs} g_h P_s & R_f A_{h,f} P_f \end{pmatrix} \begin{pmatrix} \delta\boldsymbol{\varepsilon}_{H,s} \\ \delta\boldsymbol{\varepsilon}_{H,f} \end{pmatrix}^i = -R\hat{\mathbf{r}}_h^{i-1}, \quad (28)
$$

wherein $\delta\varepsilon_H^i = \varepsilon_H^i - \varepsilon_H^{i-1}$, the change in correction term for the next sub-iteration and $\hat{\mathbf{r}}_h^{i-1}$ the fine grid residual obtained for the estimated fine grid solution $\hat{\mathbf{w}}_h^{i-1}$. The two options for subiterating on the coarse level (27) and (28) are explained in more detail.

Coarse grid sub-iteration options

The original equation to solve on the fine grid reads

$$L_h \mathbf{w}_h + \mathbf{s}_h = \mathbf{0}. \tag{29}$$

Assume a first estimation at iteration 0 $\mathbf{w}_h^0$, so that the first residual becomes

$$\mathbf{r}_h^0 = L_h \mathbf{w}_h^0 + \mathbf{s}_h. \tag{30}$$

The coarse grid operator is obtained by restriction and prolongation $L_H = RL_h P$ and split into two matrices for the partitioned iterations on the coarse grid $L_H = L_{H,L} + L_{H,R}$, where $L_{H,L}$ remains on the left-hand-side of the equation and $L_{H,R}$ is taken to the right-hand-side. Using the initialization $\varepsilon_H^0 = \mathbf{0}$, the first iteration on the coarse grid following (27) provides the first coarse grid correction term

$$L_{H,L}\varepsilon_H^1 = -R\mathbf{r}_h^0 - L_{H,R}\mathbf{0}, \tag{31}$$

or

$$\varepsilon_H^1 = -L_{H,L}^{-1} R\mathbf{r}_h^0. \tag{32}$$

Iterating only on coarse grid

A second iteration according to (27) gives

$$L_{H,L}\varepsilon_H^2 = -R\mathbf{r}_h^0 - L_{H,R}\varepsilon_H^1, \tag{33}$$

or

$$\begin{aligned}
\varepsilon_H^2 &= -L_{H,L}^{-1}\left[R\mathbf{r}_h^0 + L_{H,R}\varepsilon_H^1\right], \\
&= \varepsilon_H^1 - L_{H,L}^{-1}L_{H,R}\varepsilon_H^1,
\end{aligned} \tag{34}$$

so that after 2 coarse grid iterations the correction solution on the fine grid yields

$$\mathbf{w}_h^2 = \mathbf{w}_h^0 + P\varepsilon_H^2 = \mathbf{w}_h^0 + P\varepsilon_H^1 - PL_{H,L}^{-1}L_{H,R}\varepsilon_H^1. \tag{35}$$

Iterating through fine grid

Starting from the first correction term (32), the solution on the fine grid is updated

$$\mathbf{w}_h^1 = \mathbf{w}_h^0 + P\varepsilon_H^1. \tag{36}$$

For the new estimation, the residual of the system is obtained

$$\begin{aligned}
\mathbf{r}_h^1 &= L_h\mathbf{w}_h^1 + s_h \\
&= L_h\left[\mathbf{w}_h^0 + P\varepsilon_H^1\right] + s_h \\
&= \mathbf{r}_h^0 + L_h P\varepsilon_H^1,
\end{aligned} \tag{37}$$

and the correction term of the previous iteration on the coarse grid is *not required* when one solves for the second iteration on the coarse grid

$$L_{H,L}\delta\varepsilon_H^2 = -R\mathbf{r}_h^1, \tag{38}$$

since this can be written as

$$\begin{aligned}
\delta\varepsilon_H^2 &= -L_{H,L}^{-1}R\mathbf{r}_h^1 \\
&= -L_{H,L}^{-1}R\left[\mathbf{r}_h^0 + L_h P\varepsilon_H^1\right] \\
&= -L_{H,L}^{-1}R\mathbf{r}_h^0 - L_{H,L}^{-1}RL_h P\varepsilon_H^1,
\end{aligned} \tag{39}$$

and since $L_H = RL_h P$ and using (32) one obtains

$$\begin{aligned}
\delta\varepsilon_H^2 &= \varepsilon_H^1 - L_{H,L}^{-1}L_H\varepsilon_H^1 \\
&= \varepsilon_H^1 - L_{H,L}^{-1}(L_{H,L} + L_{H,R})\varepsilon_H^1 \\
&= \varepsilon_H^1 - \varepsilon_H^1 - L_{H,L}^{-1}L_{H,R}\varepsilon_H^1 \\
&= -L_{H,L}^{-1}L_{H,R}\varepsilon_H^1,
\end{aligned} \tag{40}$$

so that the solution on the fine grid after two coarse grid iterations becomes

$$\mathbf{w}_h^2 = \mathbf{w}_h^1 + P\delta\varepsilon_H^2 = \mathbf{w}_h^0 + P\varepsilon_H^1 - PL_{H,L}^{-1}L_{H,R}\varepsilon_H^1, \tag{41}$$

which shows that the same solution is obtained as with method one (35), although no information needs to be retained at the coarse grid level to perform consecutive iterations.

The advantage of the second option (28) is that there is no need to "remember" the value of the coarse grid correction term to take into account in the right-hand-side of (27). Sub-iterating on the coarse level may be unstable as well when the fine grid sub-iterations are unstable. Therefore, a more advanced sub-iteration technique is required on the coarse level as well. Here we choose the *same* sub-iteration technique (Aitken) as applied for the fine level sub-iterations. Under-relaxation could then be applied to the correction term ε_H, but we choose to apply the under-relaxation to the equivalent fine level variable $\mathbf{p}_{h,\Gamma_{sf}}$ (14). Essentially the coarse grid correction term in the fluid domain is prolonged to the fine grid to update the estimated solution on the fine grid

$$\mathbf{w}_{h,f}^i = \mathbf{w}_{h,f}^{i-1} + P\varepsilon_{h,f}^i. \tag{42}$$

Next, the pressure at the fluid boundary is updated using the function f_h

$$\mathbf{p}_{h,\Gamma_{fs}}^i = f_h \mathbf{w}_{h,f}^i, \tag{43}$$

which allows the transfer of the boundary pressure to the structure mesh by

$$\mathbf{p}_{h,\Gamma_{sf}}^i = I_{sf} \mathbf{p}_{h,\Gamma_{fs}}^i, \tag{44}$$

to which the Aitken under-relaxation is applied (14) to obtain the new estimation of the pressure on the structure mesh

$$\hat{\mathbf{p}}_{h,\Gamma_{sf}}^{i+1} = \hat{\mathbf{p}}_{h,\Gamma_{sf}}^i + \theta^{i+1}(\mathbf{p}_{h,\Gamma_{sf}}^i - \hat{\mathbf{p}}_{h,\Gamma_{sf}}^i), \tag{45}$$

for which the under-relaxation θ is computed by (15). Note that the computation of the Aitken under-relaxation does not differ in any sense from the Aitken under-relaxation applied directly to fine level sub-iterations, i.e. the Aitken algorithm cannot "distinguish" between coarse and fine level sub-iterations. Therefore, coarse and fine level iterations can alternate each other without modification to the Aitken algorithm. The question remains whether the computation of the underrelaxation parameter θ by alternating coarse and fine level iterations affects the stability and convergence properties of the Aitken method.

3.4 Nonlinear coarse level flow sub-iterations

In our applications, the structure requires only a small amount of computational time to solve compared to the flow solver. Therefore, only the fluid domain is coarsened, so that $P_s = R_s = I$. The coarse level sub-iteration (28) is now written as

$$\begin{pmatrix} A_{h,s} & 0 \\ R_f A_{h,fs} I_{h,fs} g_h & R_f A_{h,f} P_f \end{pmatrix} \begin{pmatrix} \delta\varepsilon_{h,s} \\ \delta\varepsilon_{H,f} \end{pmatrix}^i = -\begin{pmatrix} \hat{\mathbf{r}}_{h,s}^{i-1} \\ R\hat{\mathbf{r}}_{h,f}^{i-1}, \end{pmatrix}, \tag{46}$$

which is graphically represented in Fig. 3. So far the system is still assumed to be linear. To find an equivalent when non-linear operators (e.g. A_f) are used, the change in correction is written in terms of the flow and structure states

$$\delta\varepsilon^i = \varepsilon^i - \varepsilon^{i-1} = \mathbf{w}^i - \mathbf{w}^{i-1}, \tag{47}$$

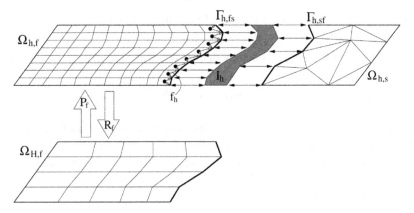

Fig. 3 Schematic representation of coarse grid flow – fine grid structure coupling.

and (46) is written in two consecutive steps

$$A_{h,s}\left(\mathbf{w}_{h,s}^i - \mathbf{w}_{h,s}^{i-1}\right) = -\hat{\mathbf{r}}_{h,s}^{i-1}, \tag{48}$$

$$R_f A_{h,f} P_f \left(\mathbf{w}_{H,f}^i - \mathbf{w}_{H,f}^{i-1}\right) = -R_f \left[\hat{\mathbf{r}}_{h,f}^{i-1} + A_{h,fs} I_{h,fs} g_h \left(\mathbf{w}_{h,s}^i - \mathbf{w}_{h,s}^{i-1}\right)\right], \tag{49}$$

wherein (48) is equivalent to solving the structure with an Aitken prediction for the pressure at the structure interface

$$A_{h,s}\mathbf{w}_{h,s}^i + A_{h,sf}\hat{\mathbf{p}}_{h,\Gamma_{sf}}^i - \mathbf{s}_{h,s} = \mathbf{0}, \tag{50}$$

and (49) is equivalent to a coarse grid solve for the fluid for a residual that consists of the fluid at state $\mathbf{w}_{h,f}^{i-1}$ and the coupling of the structure at state $\mathbf{w}_{h,s}^i$ since using (17) and (18) we obtain

$$\hat{\mathbf{r}}_{h,f}^{i-1} + A_{h,fs} I_{h,fs} g_h \left(\mathbf{w}_{h,s}^i - \mathbf{w}_{h,s}^{i-1}\right) = A_{h,f}\mathbf{w}_{h,f}^{i-1} + A_{h,fs} I_{h,fs} g_h \mathbf{w}_{h,s}^i - \mathbf{s}_{h,f},$$

$$= A_{h,f}\mathbf{w}_{h,f}^{i-1} + A_{h,fs}\mathbf{d}_{h,\Gamma_{fs}}^i - \mathbf{s}_{h,f}, \tag{51}$$

so that (49) becomes

$$R_f A_{h,f} P_f \left(\mathbf{w}_{H,f}^i - \mathbf{w}_{H,f}^{i-1}\right) = -R_f \left[A_{h,f}\mathbf{w}_{h,f}^{i-1} + A_{h,fs}\mathbf{d}_{h,\Gamma_{fs}}^i - \mathbf{s}_{h,f}\right]. \tag{52}$$

Now, in order to perform the coarse grid iterations for nonlinear systems, we substitute the nonlinear operator $\mathbf{r}_{h,s}(\mathbf{w}_{h,s}^i, \hat{\mathbf{p}}_{h,\Gamma_{sf}}^i)$ for the linear terms $A_{h,s}\mathbf{w}_{h,s}^i + A_{sf}\hat{\mathbf{p}}_{h,\Gamma_{sf}}^i$ to obtain

$$\mathbf{r}_{h,s}(\mathbf{w}_{h,s}^i, \hat{\mathbf{p}}_{h,\Gamma_s}^i) - \mathbf{s}_{h,s} = \mathbf{0}, \tag{53}$$

for the structure solve. Next in (52) $r_{h,f}(w^i_{h,f}, d^i_{h,\Gamma_{fs}})$ is substituted for the linear terms $A_{h,f}w^i_{h,f} + A_{h,fs}d^i_{h,\Gamma_{fs}}$ and the coarse grid operator $RA_{h,f}P$ is approximated by the nonlinear operator defined on the coarse grid $r_{H,f}$ to obtain

$$r_{H,f}(w^i_{H,f}, d^i_{H,\Gamma_{fs}}) - r_{H,f}(w^{i-1}_{H,f}, d^i_{H,\Gamma_{fs}}) = -R_f \left[r_{h,f}(w^{i-1}_{h,f}, d^i_{h,\Gamma_{fs}}) - s_{h,f} \right].$$

(54)

Note that on the left-hand-side of (54), the nonlinear coarse grid functions are both evaluated with the boundary displacement $d^i_{H,\Gamma_{fs}}$ such that these terms would cancel on the left-hand-side (in the linear case). Both the coarse level boundary displacement as the fluid solution at iteration $i - 1$ can be obtained through restriction from the fine level

$$w^{i-1}_{H,f} = R_f w^{i-1}_{h,f},$$

(55)

$$d^i_{H,\Gamma_{fs}} = R_f d^i_{h,\Gamma_{fs}}.$$

(56)

The right-hand-side of (54) consists of the restriction of the residual on the fine fluid level which is present for the solution $w^{i-1}_{h,f}$ under the boundary condition $d^i_{h,\Gamma_{fs}}$. In order to obtain the coarse grid solution $w^i_{H,f}$, the flow solver has to solve

$$r_{H,f}(w^i_{H,f}, d^i_{H,\Gamma_{fs}}) - \hat{s}_{H,f} = 0,$$

(57)

wherein $\hat{s}$ is a source term on the coarse grid level equal to

$$\hat{s}_{H,f} = -R_f \left[r_{h,f}(w^{i-1}_{h,f}, d^i_{h,\Gamma_{fs}}) - s_{h,f} \right] + r_{H,f}(w^{i-1}_{H,f}, d^i_{H,\Gamma_{fs}}),$$

(58)

which consists of the restriction of the fine grid residual and the coarse grid residual for the solution $w^{i-1}_{H,f}$. Notice that (57) is of the same form as the original fine grid system (8), except with a different formulation for the source term. Therefore, an existing solver (multigrid) can be used to resolve the coarse grid solution $w^i_{H,f}$. Once the solution on the coarse grid is obtained at a computational expense much smaller than resolving the flow on the fine level, the coarse grid correction term is

$$\varepsilon^i_{H,f} = w^i_{H,f} - w^{i-1}_{H,f},$$

(59)

and the corrected fine grid solution becomes

$$w^i_{h,f} = w^{i-1}_{h,f} + P_f \varepsilon^i_{H,f}.$$

(60)

The impact of $r_{H,f}(w^{i-1}_{H,f}, d^i_{H,\Gamma_{fs}})$ in the source term (58) can be indicated as follows: suppose the fully coupled solution is obtained, so that the fine grid residual is identical to zero. In that case one would not want the coarse grid acceleration to compute a correction other than zero. When the fine grid residual is zero,

$\hat{s}_{H,f} = r_{H,f}(w_{H,f}^{i-1}, d_{H,\Gamma_{fs}}^i)$ and, therefore, (57) yields

$$r_{H,f}(w_{H,f}^i, d_{H,\Gamma_{fs}}^i) - r_{H,f}(w_{H,f}^{i-1}, d_{H,\Gamma_{fs}}^i) = 0, \tag{61}$$

which shows that the obvious solution is $w_{H,f}^i = w_{H,f}^{i-1}$, so that also the correction term $\varepsilon_{H,f} = 0$ and the solution is not adversely affected.

3.5 Two-level accelerated sub-iteration algorithm

The algorithm that is used in this paper for acceleration of the Aitken sub-iterations consists of the following steps:

1. Solve the structure (11) using Aitken for the prediction of the pressure force (14),

$$r_{h,s}(w_{h,s}^i, \hat{p}_{h,\Gamma_s}^i) - s_{h,s} = 0.$$

2. Obtain the fine grid residual for the fluid state still at the *previous* sub-iteration (54)

$$\tilde{r}_{h,f}^{i-1} = r_{h,f}(w_{h,f}^{i-1}, \mathscr{I}_{h,fs}(g(w_s^i))) - s_{h,f}.$$

3. Restrict the fine grid residual and the fine grid solution to the coarse level and compute the coarse level modified source term (58).

$$r_{H,f}(w_{H,f}^i, R\mathscr{I}_{h,fs}(g_h(w_{h,s}^i))) - r_{H,f}(w_{H,f}^{i-1}, R\mathscr{I}_{h,fs}(g_h(w_{h,s}^i))) = -\tilde{r}_{H,f}.$$

4. Solve the coarse level flow problem (57)

$$r_{H,f}(w_{H,f}^i, d_{H,\Gamma_{fs}}^i) - \hat{s}_{H,f} = 0,$$

to obtain the coarse level correction term (59): $\varepsilon_{H,f}^i = w_{H,f}^i - w_{H,f}^{i-1}$
5. Update the fluid state at the fine level by applying the correction by prolongation (P) of $\Delta w_{H,f}^i$ to the fine grid (60)

$$w_{h,f}^i = w_{h,f}^{i-1} + P\Delta w_{H,f}.$$

6. Continue the procedure starting at step 1 until sufficiently converged or the maximum number n_c of coarse level iterations is performed.
7. Perform standard sub-iterations on the fine grid until sufficiently converged or the maximum number n_f of fine level iterations is performed.
8. When not sufficiently converged continue at step 1.

The algorithm allows the subsequent use of coarse and fine grid sub-iterations and is denoted by $CGP_{n_c}C_{n_f}$, e.g. CGP_1C_1 means that 1 coarse level prediction step is performed followed by 1 fine level correction (smoothing) sub-iteration, whereas

$CGP_\infty C_\infty$ (also denoted by CGP-FG) denotes that first all sub-iterations are performed on a coarse level until converged, thereafter all sub-iterations are performed on the fine level. At the moment only two level acceleration has been discussed, but as with multigrid techniques, even coarser levels could be employed.

4 Results

The proposed multilevel acceleration technique is applied to two strongly coupled two-dimensional test problems. The first is a laminar test case and the second a turbulent test case. The flow solver is a unstructured, cell-centered, finite volume solver with a second order central discretization and artificial Jameson dissipation. Coupling of the CFD code to the linear structure code is performed by FLECS [7], a open source flexible coupling shell that allows the coupling of different codes and performs the interpolation between the interface meshes. The interpolation between flow and structure meshes ($\mathscr{I}_{h,sf}$ and $\mathscr{I}_{h,fs}$) is a radial basis function interpolation with a thin-plate-spline function [1]. The structure has its unknowns in the element nodes which are already on the boundary of the elements. Therefore, the function g_h is simply selecting the boundary displacements from the structure state vector. The fluid unknowns are located in the cell centers and, therefore, the function g_h performs an interpolation of the cell centered data to boundary vertex data.

At the moment, the flow solver re-initializes for every sub-iteration, meaning that the coarse mesh levels are regenerated every sub-iteration. The multigrid levels are obtained by agglomeration of fine level cells such that the coarsening ratio is close to the theoretical optimal value of 4 (for two-dimensional applications), see Fig. 4. However, as the mesh is deforming, the lay-out of the coarse grid may change and even the amount of cells on the coarse level may differ from step to step. Therefore, by performing the coupling still through the fine grid level, no information on the coarse level correction term needs to be stored, which would be particularly difficult when the coarse level meshes change in lay-out. Of course, regenerating the whole mesh structure does increase computational costs a lot. For the coarse level solve, most of the computing time is spend on deforming the fluid mesh and regenerating the mesh data structure rather than the iterations for solving the coarse level problem.

4.1 Two-dimensional laminar flow problem

The laminar case is based on [8] and consists of a circular cylinder of diameter 0.1m in a channel with height $H = 0.41$m, length $L = 2.5$m, with an elastic flap behind it of length $l = 0.35$m and thickness $h = 0.02$m, see Fig. 5. The inflow is a parabolic velocity profile (see [8] for details) with a mean velocity of 2 and a maximum velocity of 3m/s. Differently from the original test problem, compressibility

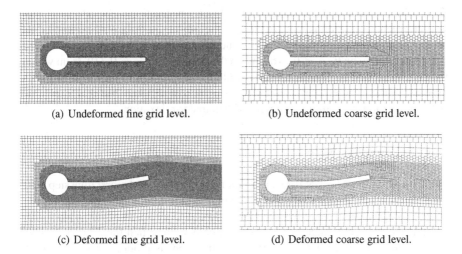

(a) Undeformed fine grid level.
(b) Undeformed coarse grid level.

(c) Deformed fine grid level.
(d) Deformed coarse grid level.

Fig. 4 Fine and coarse level meshes for undeformed and deformed situation.

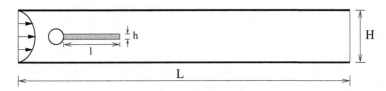

Fig. 5 Two-dimensional laminar testcase.

is allowed, since the solver we use is a compressible Reynolds Averaged Navier-Stokes solver. The reference Mach number based on the mean flow velocity is set to $M_0 = 0.14$. The Reynolds number is based on the mean velocity and cylinder diameter and is $Re = 200$. The structure is modeled as a linear elastic structure with a density equal to the flow density $\rho = 1000$kg/m^3 and a Young's modulus of $E = 5.6 \cdot 10^6$kg/(m.s^2). Time integration is performed by an implicit, third-order accurate, multistage Runge-Kutta scheme with a time step $\Delta t = 0.01$s (for details on time integration see [10]). Each implicit Runge-Kutta stage is sub-iterated until $||\mathbf{p}^i_{\Gamma_s} - \hat{\mathbf{p}}^i_{\Gamma_s}||_2 \leq 10^{-2}$. For this test problem the transient to a periodic state is simulated in 500 time steps. The flow solver uses 3 grid levels for the multigrid (MG) solver. For the coarse level iterations, we use the intermediate MG mesh. The ratio of fine to coarse level mesh cells is 20737 to 5442, which is ≈ 3.8. During the simulations the averaged under-relaxation factor $\bar{\theta}$ is determined. This value is a measure of how strongly coupled the simulation is. A value of $\bar{\theta}$ close to 1 means that hardly any under-relaxation is required, whereas a small value of $\bar{\theta}$ indicates a strong coupling with much under-relaxation. When only Aitken sub-iterations are performed on the fine grid (FG), the averaged value of the under-relaxation $\bar{\theta} = 0.321$, indicating that the problem is strongly coupled and that standard Gauss-Seidel iterations

diverge. In a second simulation, first coarse level sub-iterations are performed until
the convergence criterion is met after which the sub-iterations are all performed on
the fine grid (CGP-FG). The averaged value of the under-relaxation on the coarse
level turned out to be $\bar{\theta}_{CG} = 0.316$, and on the fine level $\bar{\theta}_{FG} = 0.319$, from which
it is concluded that the fine and coarse grid sub-iterations perform "equally" with
respect to Aitken: therefore it would seem possible to alternately use coarse and fine
grid sub-iterations in combination with Aitken under-relaxation.

4.1.1   Sub-iteration behavior

Now we take a closer look at the sub-iterations performed for the first time step. The
initial solution consists of a developed vortex street and a structure that is at rest. In
the first computational step the structure is allowed to move. The Aitken algorithm
is initialized by a first value of $\theta = 0.7$, which is higher than the averaged value.
In Fig. 6 the resulting lift force (force in y direction) is plotted for the multigrid
iterations performed by the flow solver for the first time step. When the flow solver
is sufficiently converged or has done 20 MG iterations, a sub-iteration is carried
out. This coincides with a sharp change in lift as the structure has moved to a new
position. In Fig. 6(a) it can be noted that when only sub-iterating on the fine mesh
(Aitken-FG) the lift converges to around 2500 in the first flow solve, and to around
-7500 for the second flow solve, which shows an increase in error with respect to the
fully coupled solution which turns out to be around 150. The reason is that for the
second flow solve, Aitken is applied with an under-relaxation of $\theta = 0.7$, which is
in fact too high and for the next sub-iteration θ is automatically lowered by Aitken
and the lift turns out at 250 (Fig. 6(b)) which is already much closer to the correct
value. Another striking difference is the value for the lift computed on the coarse
grid level and the fine grid level: this is best seen in Fig. 6(b), where the CGP-FG

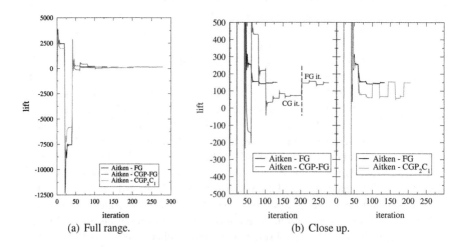

(a) Full range.   (b) Close up.

Fig. 6 History of lift for MG iterations performed by flow solver.

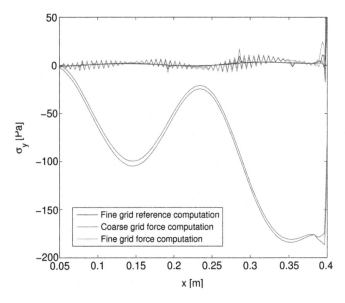

Fig. 7 Computation of interface stresses.

algorithm, first only iterates the flow on the coarse grid level, after which only fine grid iterations are used. The lift on the coarse level seems to converge to 70, whereas the fine grid iterations converge to 150. Also for the CGP_2C_1 algorithm, the coarse and fine level sub-iterations are easily identified because of the large difference in lift. The reason for this discrepancy is that the flow solver, when iterating on the coarse level, computes the boundary forces based on the solution on the coarse level, e.g. $f_H(w_{H,f})$. In Fig. 7 the computation of the stress in y direction over the fluid structure interface is shown. The results of the CGP-FG computation are used. The 'fine grid reference computation' denotes the stress computed after the final fine grid iteration (fully coupled solution). After 12 coarse grid sub-iterations, just before the fine grid sub-iterations start, the computation of the force on the coarse grid and fine grid is shown. It is clear that the computation of the interface stress on the coarse level has a large discrepancy with the reference solution. However, when for the *same* solution, the computation of the interface stress is computed on the fine mesh, the error with respect to the reference is already much smaller. Should the coupling between the coarse level flow and the structure be done directly, this discrepancy will be introduced in the coupling and it may be conceivable that the coarse level sub-iterations converge to a solution, very different from the coupled fine grid solution and, therefore, may not contribute to a faster convergence.

4.1.2 Convergence in interface pressure and displacement

The next property that is investigated is the convergence in interface pressure and displacements for the fine grid and two-level sub-iterations. It was concluded in [11]

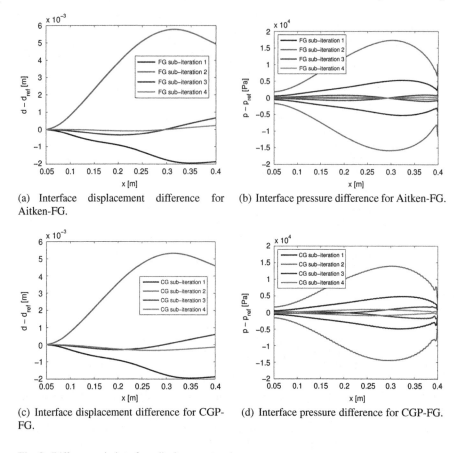

(a) Interface displacement difference for Aitken-FG.

(b) Interface pressure difference for Aitken-FG.

(c) Interface displacement difference for CGP-FG.

(d) Interface pressure difference for CGP-FG.

Fig. 8 Differences in interface displacement and pressure.

that coarse level sub-iterations perform just as well as fine grid sub-iterations when the partitioning error is still large in the lower modes and that the error in the higher modes that cannot be represented on the coarse level remain until a fine grid solver is done. However, that test case was only a one-dimensional piston problem and the fluid-structure interface only consisted of a single point. This time the coupling is stronger and the fluid-structure interface is two-dimensional as well. To verify that the same convergence behavior can be observed the difference in the interface pressure and displacement with respect to the final (fully coupled) solution is shown in Fig. 8. In 8(a) and 8(b) the results for the Aitken-FG sub-iterations are shown. It can be seen that at first the error in pressure and displacement increases, as for the first Aitken sub-iteration the under-relaxation parameter $\theta = 0.7$ which is too high for this strongly coupled problem as explained in the previous section. Also the largest errors in displacement and pressures are caused by low mode shapes, which should enable the multilevel algorithm to efficiently tackle these modes first on the coarse level before resolving the high mode shapes with a more expensive fine level solve.

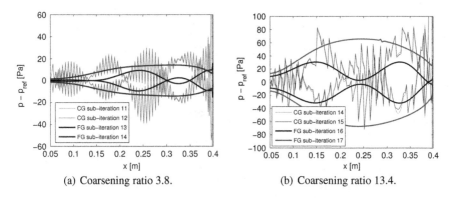

(a) Coarsening ratio 3.8. (b) Coarsening ratio 13.4.

Fig. 9 Convergence in interface pressure when switching from coarse to fine level sub-iterations.

This is shown in Fig. 8(c) and Fig. 8(d) which show the convergence of the interface displacement and pressure for the first four coarse level sub-iterations. The convergence on the coarse level has a close resemblance with the convergence on the fine level, although there seem to be some small (high wave number) oscillations present in the pressure distribution. Another conclusion from [11] was that a fine grid solve effectively reduces the high wave number mode errors that cannot be represented on the coarse level. To show this property for the more two-dimensional test problem, in Fig. 9 the convergence for the pressure at the interface is shown for the last two coarse level sub-iterations in the CGP-FG algorithm and the first two subsequent fine level sub-iterations. In Fig. 9(b) the coarse level is represented by the coarsest multigrid mesh generated by the flow solver, which, in this case, consists of only 1553 cells which yields a coarsening ratio of 13.4 with respect to the fine mesh. In both cases the convergence on the coarse level stagnates as the high wave number error cannot reduce any further. Performing a fine grid solve smooths out the high wave number error and the remaining partitioning error is again smooth and has a large low wave number component. Therefore, in order to avoid the stagnation of convergence on the coarse level, it is beneficial to solve the fine level problem after a few coarse level iterations. In order to increase computational efficiency, the coarse level sub-iterations can be used after the fine level smoothing, as the remaining partitioning error has a (relatively) large low wave number mode component again. This is illustrated in Fig. 10 where for the CGP_3C_1 all the sub-iterations to convergence are shown in blocks of 4 sub-iterations: 3 coarse level iterations, followed by 1 fine level smoothing. Fig. 10(d) only shows one coarse level iteration as the coarse level iteration already reached the required tolerance after 1 iteration. The final fine level iteration shows an error of zero as this result is taken as the reference value. The effectiveness of alternating coarse and fine level sub-iterations is shown. Application to other sub-iteration techniques seems straight forward, especially when these techniques rely on the approximation of the low wave number modes first, e.g. the quasi-Newton subiterations of [2], as these modes can be accurately represented on the coarse level.

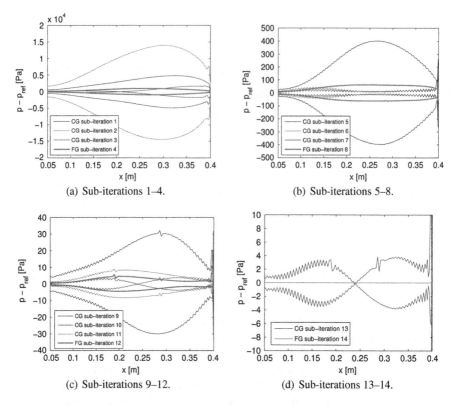

Fig. 10 Convergence in interface pressure for CGP$_3$C$_1$.

4.1.3 Costs of sub-iterations

It must be noted that the first computational step is quite demanding in terms of finding a coupled solution: as the solution starts from a flow and structure state that are completely uncoupled, the initial transient to a periodic motion is quite strong so that the first few time steps require more sub-iterations per time step then during the remainder of the computation. Of course also the somewhat high value of $\theta = 0.7$ for the first sub-iteration increases the partitioning error initially. To compare the computational costs in terms of fine and coarse level sub-iterations, we therefore determine the averaged number of sub-iterations per stage during the whole simulation from initial condition to periodic motion. In Fig.11 the averaged number of sub-iterations per stage in terms of coarse and fine level iterations are shown. The first point in the plot gives the reference number of sub-iterations as these correspond to Aitken sub-iterations on the fine grid (FG) only. The second point (CGP-FG) first only performs sub-iterations on the coarse level, thereafter it continues with sub-iterations only on the fine level. CGP-FG reduces the number of fine grid sub-iterations by roughly 40–50%, but on the other hand increase the

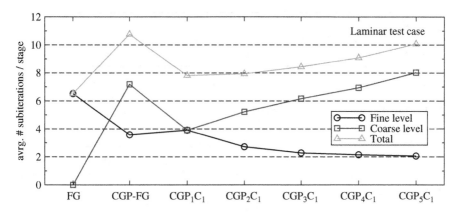

Fig. 11 Averaged number of coarse and fine grid sub-iterations per stage for the laminar case.

total number of sub-iterations more than 60%. Therefore, a lot of computational efficiency gained by performing less fine grid flow solves is lost by the additional coarse level sub-iterations to perform, which also include moving the fluid mesh, transfer of data between flow and structure and solving the structure. Alternating coarse level/fine level sub-iterations, reduces the number of fine level sub-iterations as well as the total number of sub-iterations compared to CGP-FG. Increasing the number of coarse grid sub-iterations before a fine grid sub-iteration reduces the number of required fine grid solves, which seems to converge to ≈2. Increasing the number of coarse grid sub-iterations even further, only increases the total number of sub-iterations required, without an obvious gain in reduction of fine grid sub-iterations. The most effective scheme is found to be CGP_3C_1, reducing the number of fine grid iterations by 65% at the expense of an increase in total number of sub-iterations by 30%.

4.2 Two-dimensional turbulent flow problem

The proposed method is applied to a two-dimensional laminar and a two-dimensional turbulent test case, which, geometrically, differs slightly from the laminar case as the circular cylinder is replaced by a square cylinder, see Fig. 12, with edges of 0.1m to increase the shed vorticity. The Reynolds number is increased to $Re = 1 \cdot 10^5$ and the Spalart-Allmaras turbulence model is used. The Young's modulus of the structure is increased to $E = 22.4 \cdot 10^6 kg/(m.s^2)$. The initial condition is taken after the transient to a periodic coupled solution, and only a single period (≈ 32 time steps) is simulated with a time step of $\Delta t = 0.01$. The ratio of fine to coarse level mesh cells is 56746 to 15151, which is ≈ 3.7. Simulations are run for different settings and Fig. 13 shows the averaged number of sub-iterations required per stage in terms of coarse and fine level iterations. From the computation on only

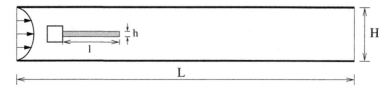

Fig. 12 Two-dimensional turbulent testcase.

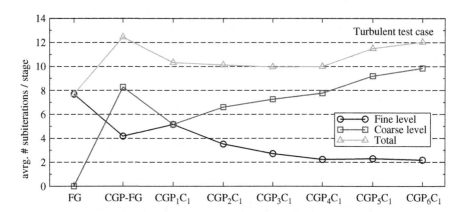

Fig. 13 Averaged number of coarse and fine grid sub-iterations per stage for the turbulent case.

the fine grid (FG) it was found that the average under-relaxation $\bar{\theta} = 0.396$ and that for CGP-FG: $\bar{\theta}_{CG} = 0.373$ and $\bar{\theta}_{FG} = 0.343$. Again these values are in close range of each other, indicating a similar behavior of fine and coarse grid sub-iterations. Much of the same behavior as for the laminar case can be found: performing more coarse grid sub-iterations before a fine grid solve reduces the number of required fine grid solves substantially, although at some point the increase in total number of sub-iterations reduces computational efficiency. The most effective schemes is found to be CGP_4C_1, reducing the number of fine grid iterations by 70% at the expense of an increase in total number of sub-iterations by 30%.

5 Conclusions

In this paper we investigated the combination of multilevel acceleration with the Aitken underrelaxtion technique applied to a two-dimensional strongly coupled laminar and turbulent test problem. Since the value for the under-relaxation parameter is not significantly different for coarse and fine level sub-iterations, they are used alternately, where it is found that performing 3 or 4 coarse level sub-iterations followed by 1 fine level sub-iteration results in the highest gain in efficiency. Although the total number of sub-iterations increases slightly by 30%, the number of fine grid iterations was decreased by as much as 65–70%.

Performing the coupling in the coarse level sub-iterations over the fine mesh, allows alternating fine and coarse level sub-iterations: 1) without modifications to the Aitken algorithm, 2) without the necessity to store coarse level information and 3) with an automatic incorporation of the coupling terms. However, it comes at the price of having to perform a fine grid mesh deformation, even for a coarse level sub-iteration. It is worth while investigating whether a fully coarse level coupling and sub-iteration can be used as well.

Up to this point only the Aitken under-relaxation was used in combination with multilevel acceleration and only few strategies were tested for optimal convergence. Application to other sub-iteration techniques seems straight forward, especially when these techniques rely on the approximation of the low wave number modes first, as these modes can be accurately represented on the coarse level.

References

1. A. de Boer, A.H. van Zuijlen, and H. Bijl. Review of coupling methods for non-matching meshes. *Comput. Methods Appl. Mech. Engrg.*, 196:1515–1525, 2007.
2. J. Degroote, K-J. Bathe, and J. Vierendeels. Performance of a new partitioned procedure versus a monolithic procedure in fluidstructure interaction. *Comput. Struct.*, 87:793–801, 2009.
3. C. Forster, W.A. Wall, and E. Ramm. Artificial added mass instabilities in sequential staggered coupling of nonlinear structures and incompressible viscous flows. *Comput. Methods Appl. Mech. Engrg.*, 196(7):1278–1293, 2007.
4. R. Haelterman, J. Degroote, D. van Heule, and J. Vierendeels. The quasi-newton least squares method: A new and fast secant method analyzed for linear systems. *SIAM J. Numer. Anal.*, 47:2347–2368, 2009.
5. C. Michler, E.H. Brummelen, and R. de Borst. An interface Newton-Krylov solver for fluid-structure interaction. *Int. J. for Num. Meth. in Fluids*, 47(10-11):1189–1195, 2005.
6. D.P. Mok, W.A. Wall, and E. Ramm. Accelerated iterative substructure schemes for insta-tionary fluid-structure interaction. In *First MIT Conference on Computational Fluid and Solid Mechanics*, pages 1325–1328, 2001.
7. Margreet Nool, Erik Jan Lingen, Aukje de Boer, and Hester Bijl. Flecs, a flexible coupling shell application to fluid-structure interaction. In *PARA'06: State-of-the-Art in Scientific and Parallel Computing*, Umeøa, Sweden, 2006.
8. S. Turek, J.: Proposal for Numerical Benchmarking of Fluid-Structure Interaction between an Elastic Object Hron, Schäfer-M. (eds.) Fluid-Structure Interaction Lecture Notes Comput.Šci. Laminar Incompressible Flow. In: Bungartz, H.-J., and pp. 371-385-Springer Heidelberg Eng., LNCSE **53**. 2006.
9. J. Vierendeels, L. Lanoye, J. Degroote, and P. Verdonck. Implicit coupling of partitioned flu-idstructure interaction problems with reduced order models. *Comput. Struct.*, 85:970–976, 2007.
10. A.H. van Zuijlen, A. de Boer, and H. Bijl. Higher order time integration through smooth mesh deformation for 3d fluid-structure interaction simulations. *J. Comput. Phys.*, 224:414–430, 2007.
11. A.H. van Zuijlen, S. Bosscher, and H. Bijl. Two level algorithms for partitioned fluid-structure interaction computations. *Comput. Methods Appl. Mech. Engrg.*, 196:1458–1470, 2007.

A Classification of Interface Treatments for FSI

C.A. Felippa, K.C. Park, and M.R. Ross

Abstract This paper proposes a taxonomy of methods for the treatment of the fluid-structure interface in FSI coupled problems. The top-level classification is based on the presence or absence of Additional Interface Variables (AIV) as well as their type. Associated prototype methods: Direct Force Motion Transfer (DFMT), Mortar and Localized Lagrange Multipliers (LLM) are defined. These are later studied in more detail using a specific FSI benchmark problem used in Ross' 2006 thesis. Desirable attributes of the interfacing methods are stated and commented upon.

1 Introduction

Computational fluid-structure interaction (FSI) emerged in the late 1960. It was primarily driven by linear (or linearized) problems in acoustics, flutter and vibration. Given the speed and memory limitations of computers of that time, early analysis techniques naturally favored analytic or semi-analytic modal formulations in the frequency domain; see e.g. [23]. Structural discretizations were used primarily for modal eigenextraction. Time-domain FSI began to make headway in the 1970s as interest developed in exterior and nonlinear FSI problems, which lie outside the scope of the frequency domain approach. A serious obstacle to sustained growth was the great variety of fluid problems. While computer-based structural analysis was dominated by the Finite Element Method (FEM) since the mid 1960s, no comparable universal treatment was available on the fluid side. For example, an acoustic fluid model is useless for high speed gas dynamics, turbulent or multiphase flow.

C.A. Felippa and K.C. Park
Department of Aerospace Engineering Sciences and Center for Aerospace Structures,
University of Colorado at Boulder, Boulder, Colorado 80309-0429, USA
e-mail: carlos.felippa@colorado.edu

M.R. Ross
Analytical Dynamics Department, Sandia National Laboratories, P. O. Box 5800, MS 0346,
Albuquerque, NM 87185-0346, USA
e-mail: mross@sandia.gov

H.-J. Bungartz et al. (eds.), *Fluid Structure Interaction II*, Lecture Notes
in Computational Science and Engineering 73, DOI 10.1007/978-3-642-14206-2_2,
© Springer-Verlag Berlin Heidelberg 2010

The lack of an universal discretization methodology for the fluid component fostered *specialization* in modeling and computer implementation as technological advances stoked the need for simulations. A key motivation behind the partitioned methods developed in the 1970s and 1980s [13–16, 24–28] was to "divide and conquer" the coupled problem. This division facilitated reuse of FEM and CFD software resources developed since the early 1950s, as well as the expertise and experience accumulated during that time. But combinatorial complexity has remained in two aspects: time marching methods, and treatment of the fluid-solid interface. To find a path through that maze, a classification of those aspects is proposed here. Emphasis is placed on the interface treatment because this aspect has experienced more rapid development in recent years.

The present paper has evolved from M. R. Ross' thesis work [36], which has appeared in a recent article sequence [37, 38]. The comparison of interface treatments outlined in [38] formed the skeleton for a tentative classification, which is expanded upon in the next section.

2 Classification Ingredients

Four ingredients of computational FSI are sketched in Fig. 1: fluid, structure, interface and time stepping. The first three are realized as discrete spatial models. The last one identifies the time domain response method.

As far as the proposed classification is concerned, there is no need to reinvent labels for the fluid and structure models when viewed as *isolated* individual entities, since pertinent terminology is well established by now. Echoing introductory remarks, the structural model will normally be FEM based, whereas the fluid model will be problem dependent; for example a finite volume discretization in aeroelasticity or a boundary-element discretization in acoustics.

Terminology for the other components: interface treatment and time stepping, is not so well established. For the latter, the top-level division into monolithic and partitioned, formally introduced in a 1983 review chapter [26], is gradually gaining acceptance. But interface treatment nomenclature remains volatile since some techniques, especially those introducing Lagrange multipliers, are of more recent coinage. On this account we will focus on that component.

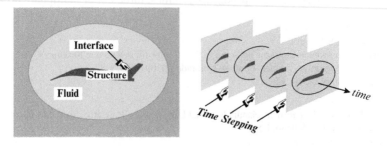

Fig. 1 Two classification ingredients: interface treatment and time-stepping methods.

2.1 Interface Treatment: Top Level Classification Criteria

There is a interface treatment feature that strongly drives computer implementation: *whether additional interface variables (AIV) are present or not*. Accordingly we shall categorize the top level interface treatment into three major classes.

1. **Primal**. There are no AIVs
2. **Dual**. AIVs of dual type (Lagrange multipliers) are introduced.
3. **Primal-dual**. AIVs of both dual and primal type are introduced.

Any of these may be associated with either monolithic or partitioned time-stepping solution procedures, although admittedly some combinations might not be practical. It follows that the foregoing classification may be viewed as "orthogonal" to one based on time stepping, a distinction that simplifies terminology. The orthogonality is illustrated in Fig. 2. The gridlike arrangement places interface treatments horizontally and time stepping solution procedures vertically. That figure also displays important interface treatment realizations, as discussed next.

2.2 Interface Treatment Prototypes

Fig. 3 sketches practical realization schemes (referred to as "prototypes" for brevity) for each of the three major interface treatment categories: primal, dual and primal-dual. It pictures nonmatching meshes since that is often the case in FSI.

The prototype for primal methods is the Direct Force-Motion Transfer, or DFMT, scheme:

$F \Rightarrow S$ Interface fluid pressures are passed as forces to structural nodes,
$F \Leftarrow S$ Structural motions are passed as interface fluid-particle velocities.

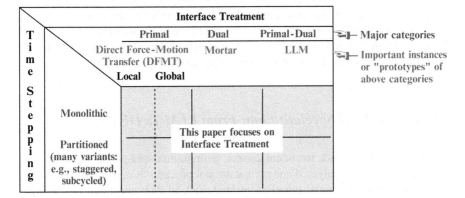

Fig. 2 Two "orthogonal," top-level classification criteria.

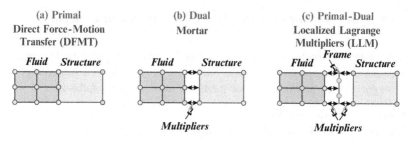

Fig. 3 Sketches for practically important realizations of three interface treatment prototypes.

The gapless sketch of Fig. 3(a) means that transfers are done directly; there are no AIV intermediaries. Two DFMT subclasses are distinguished in Fig. 2 by qualifiers *local* and *global*, which convey interface coupling extent. In a local DFMT scheme, a fluid (structure) interface freedom interacts only with neighboring structure (fluid) interface freedoms. In a global DFMT scheme, the coupling extends beyond that, and possibly to the entire interface.

Over the past decade the Mortar scheme [5] has emerged as prototype for dual methods. Interface freedoms are linked through Lagrange multipliers. Variants arise primarily through the choice of multiplier discretization spaces. Figure 3(b) shows delta-function multipliers collocated at interface nodes. This choice, which is that studied in [36–38], has the advantage of implementation simplicity and immediate physical interpretation. Smoother spaces of distributed multipliers may be used at the cost of additional implementation complexity, as well as need of knowing inter-face shape function details for spatial integration. For structure-structure interaction, as in the contact problem, the idea behind this interfacing method can be traced way back, to a variational principle proposed by Prager for solid mechanics [33].

The Localized Lagrange Multiplier (LLM) scheme, sketched in Fig. 3(c), is a prototype for primal-dual methods. A kinematic "frame" is placed between the fluid and structure. Multipliers are collocated at the interface between the fluid and frame, as well as between structure and frame. The idea, originally introduced for contact problems [30], synthesizes two trends: the FETI domain decomposition methods developed in the early 1990s [8] and, as the background theory evolved, the varia-tional treatment of hybrid finite elements proposed by Atluri [2] for solid mechanics. For the three-decade evolution of this governing functional see [27, 31].

2.3 Hierarchical Specialization: From LLM To DFMT

The three major interface treatment classes: primal, dual and primal-dual, are not impervious islands. Analytical and computational procedures may be used to move over from the most general top class: primal-dual, to dual and finally to primal. This "hierarchical cascade," AIV-reduction process is flowcharted in Fig. 4. Links

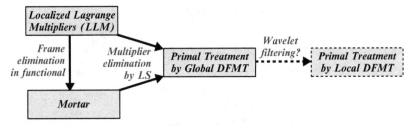

Fig. 4 Reduction "cascade" of additional interface variables (AIV) among interfacing prototypes.

displayed in that figure shed light on interrelations among seemingly disparate schemes that may be found in the FSI literature.

More specifically: the LLM interface functional may be reduced to a Mortar functional method by introducing constraints among the interface variables, as described in Sect. 4.3. Upon discretization of either LLM or Mortar functionals, algebraic elimination of interface variables by least-square (LS) techniques beget *global* DFMT schemes. As of this writing it is not known whether one may subsequently precipitate a *local* DMFT scheme through further localization, although in principle model reduction schemes such as wavelet filtering could be explored along those lines. Accordingly the rightmost link in Fig. 4 bears a question mark.

2.4 Background

The hierarchy displayed in Fig. 4 did not historically develop in the left-to-right "top-down" fashion. It germinated and grew in a more circuitous fashion, mostly going from specific to general. At a fluid-solid "wet" interface one deals with two interacting agents: fluid pressures that coerce a flexible structure into resisting forces, and structural velocities that impel the fluid. Upon discretization, interface pressures can be lumped to structure node forces whereas structure motions are prescribed as velocities on the fluid interface, using appropriate interpolation should meshes mismatch. This transparent physical interpretation is easily understood by engineers. Understandably this simple scheme was the first one to be adopted in FSI computations.

An early DFMT implementation was used in the development of partitioned solution procedures for underwater shock [17, 24], an application sketched in Fig. 5(a). The sketch reflects mid-1980 capabilities, which included hull cavitation modeling [14], later extended to free-surface bulk cavitation to simulate surface ship attacks. The simulation coupled three computational models: a FEM-discretized submerged structure, surrounded by a FEM-discretized, near-field, acoustic bilinear fluid able to cavitate, closed with a BEM-discretized silent boundary implemented through the Doubly Asymptotic Approximation (DAA) of Geers [18–20]. Meshes were usually nonmatching. Interface exchanges were done through the local DFMT scheme

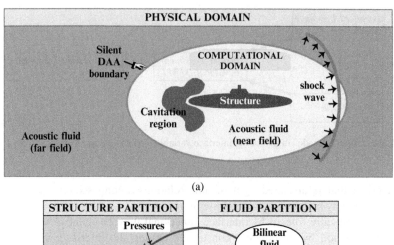

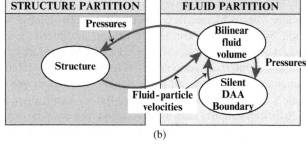

Fig. 5 Early Local DFMT treatment: underwater shock analysis.

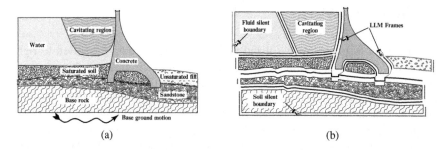

Fig. 6 First LLM application to FSI: dam under seismic action; (a) section of gravity dam, (b) LLM partitioning of computational domain.

flowcharted in Fig. 5(b). We know now that this treatment is not energy conserving. However, for explosion-induced shockwave simulations in which only a few milliseconds of early-time response are of interest to assess survivability, this defect is of little importance.

The opposite in complexity is the primal-dual LLM scheme that evolved from the sources described in Sect. 2.2. It was initially used only for structure-structure interaction in contact problems. The first FSI application, described in [36–38], dealt with a dam under seismic action, as sketched in Fig. 6. The multiple-partition model pictured in Fig. 6 is, however, too complicated for systematic research studies.

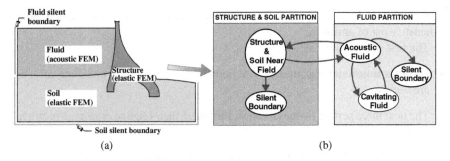

Fig. 7 Two-partition decomposition of dam problem.

A simpler two-partition model of the dam problem is shown in Fig. 7(a). This simplification gets rid of cross points, which may cause serious difficulties with some interface treatments. The interaction diagram is flowcharted in Fig. 7(b).

The main motivation for trying this scheme was program modularity. Interposing the frame conceals the fluid from the structure and vice-versa: *each program component only sees forces* passed as Lagrange multipliers. It *does not need to know details of the opposite mesh*, such as element shape functions. (In commercial codes those may be unknown to users anyway.) Nonmatching meshes, which are important in the contact problems for which LLM was first developed, are conveniently handled. Model reduction of one of the components (for example, replacing a linear structure by a set of vibration or Ritz modes) is eased as such replacements are fully hidden by the frame. Stand-alone testing and staged development are greatly simplified.

What price LLM modularity? The introduction of a large number of AIVs, and the fact that the frame must be discretized through special rules if certain conservation attributes are enforced. The AIV impact can be lessened by hierarchical elimination. Getting rid of the frame reduces LLM to Mortar, which carries less AIVs. Eliminating all AIVs by least-square methods yields global DFMT schemes, a reduction that has special appeal in linear dynamic problems (because it can be preprocessed). Since full AIV elimination typically couples *all* interface variables, interface processing costs remain high with respect to local DFMT schemes. As discussed in following sections, tradeoffs of various nature inevitably emerge at all stages.

3 Desirable Attributes

From a glance at the grid of Fig. 2, it is obvious that combining interface treatments with a profusion of time stepping schemes begets a large number of possible FSI computational methods for a given problem. Which one to pick? To help answer this question one should consider two aspects: which method attributes are desirable, and which priority is assigned to those attributes. Following is a commented, non-exhaustive list of attributes.

The first one is essential:

- **Handles nonmatching meshes**. Methods that only work for matching meshes are useless.

The next three pertain to conservation attributes; their relative priority is problem dependent.

- **Flux transmission consistency**. This may be verified by the Interface Force Patch Test (IFPT) described later.
- **Rigid body motion transmission consistency**. This may be verified by the Interface Rigid Motion Patch Test (IRMPT) described later.
- **Energy conservation**. No artificial energy is gained or lost at interfaces. Testable by duality.

The next set collects beneficial algorithmic attributes.

- **Interface equation stability**. Primarily germane to dual and primal-dual treatments: interface matrices must have correct rank.
- **Time stepping stability**. Applicable to partitioned time stepping: temporal stability of monolithic solution procedure should not be degraded by partitioning. The interface treatment can have an indirect effect on this property, as discussed in Appendix A of [37].
- **Time stepping accuracy**. Similar to above, with "stability" replaced by "accuracy order".
- **Observer independence**. Pertinent to dual and primal-dual methods that require designation of master and slave faces at interfaces. Solution should not depend on choice of master.
- **Correct handling of cross points**. A cross point ("cross line" in 3D) is one at which three or more interfaces meet. Some treatments may fail there if meshes do not match.
- **Interface equations derivable from functional**. Advantages are listed in Sect. 4.1.
- **Interface error monitoring for stepsize control**. Equations obtained from interface functionals befit this attribute. For a discussion of energy error measures of this kind see [38].

Desirability of the following ones is affected by project goals and problem features, as noted in the comments.

- **Fast implementation**. Crucial in time critical projects. Prioritization favors primal methods.
- **Overall computational efficiency**. Qualifier "overall" means to stress here that what matters is the efficiency with which the *full coupled problem is solved*, rather than that attained for individual components. (Often one problem component dominates processing time.)

- **Parallelization friendly**. Obviously linked to previous one.
- **Handles nonlinear problems**. Relative weight of this attribute depends on whether the interface treatment is affected by nonlinearities, as is the case in contact problems.
- **Handles multiscale phenomena**. Of interest for a certain class of problems. For example, turbulent flow or capillar effects.
- **Handles gaps and interpenetration**. Similar to above one in being limited to problems that may exhibit volume mesh overlapping. For example, multiphase flow or gas bubbles.
- **Adapts well to Reduced Order Modeling (ROM) of a component**. Here the information-hiding and data-filtering properties of primal-dual methods are beneficial.
- **Facilitates software reuse**. This applies to treatments in which open source software is available for fluid, structure, or both. For example, ability to use separate mesh generators.
- **Facilitates black box (BB) reuse**. Differs from above in that commercial or proprietary software is used for fluid, structure, or both. As code modifications are precluded, communication must be done via I/O. In addition, details such as element shape functions might be unavailable, which can demerit interfacing methods that rely on such knowledge.

As may be expected, no known FSI methodology meets all these desirable attributes, or even a majority of them. Prioritization as per application and goals is essential. For example, in shock-wave-excited problems correct flux (force) transmission is important but energy conservation is not. Conservation attributes are studied in more detail in following sections.

4 A Specific FSI Problem: Dam Under Seismic Action

As previously noted, the present classification emerged from a study of alternative interface formulations in Ross' thesis [36], summarized in a recent article sequence [37, 38]. The compared formulations were used to simulate seismic actions on two existing water-filled dams: the Koyna gravity dam located in India, and the Morrow Point arch dam located in Colorado, USA. These were analyzed under standard digitized earthquake records available from the COSMOS database [1] using 2D and 3D models, respectively. The Koyna dam, pictured in Fig. 8(a), was more extensively studied since the 2D plane strain model allowed systematic convergence studies, quantified by various error mesaures, to be rapidly carried out using Matlab on a PC. A typical 2D discrete model is pictured in Fig. 8(b). Note that fluid-structure meshes are nonmatching.

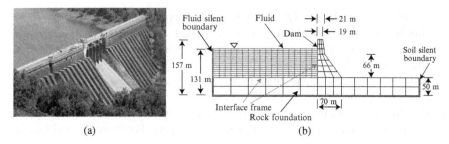

Fig. 8 LLM application to a FSI problem: Koyna gravity dam under seismic action; (a) dam picture, (b) spatial discretization using a plane strain "slice".

4.1 Variationally Based Interface Coupling

All interface coupling methods used in the dam problems are variationally based. The following advantages can be cited for working within a variational framework:

1. Lagrange multipliers that appear in dual and primal-dual treatments are naturally accommodated. Their presence in a variational statement guides decisions as to the selection of multiplier interpolation spaces.
2. If space discretization of the physical problem components produce symmetric equations, symmetry of the master coupled equations is guaranteed. This facilitates the use of more efficient equation solvers as well as coupled system eigensolvers. As regards the latter, maintaining symmetry excludes possible appearance of nonphysical complex roots.
3. Testing for interface energy conservation is simplified.

Even if the interface treatment is not variationally based, as was the case with the local DFMT scheme used in the underwater shock problem of Fig. 5, comparison with DFMT equations derived through variational methods is illuminating.

In the present section the LLM variational framework of [37, 38] is recalled for the reader's convenience. Two other interface methods developed as LLM specializations: the Mortar and global DFMT methods, are discussed in Sects. 5 and 6, respectively.

4.2 Equations of Motion

Below we summarize the coupled equations of motion (EOM) of the LLM discretization shown in Figs. 9 and 10 for the matching and nonmatching mesh case, respectively. Structural damping, silent boundaries and nonlinear fluid-cavitation effects are omitted from the governing EOM for brevity. Only elastic material behavior and small displacement motions are considered. Fluid flow effects are ignored as only the early time response under seismic actions is of interest. Under the foregoing assumptions the coupled EOM may be derived from the total functional

$$\Pi = \Pi_S + \Pi_F + \Pi_B. \tag{1}$$

Here Π_S and Π_F are governing functionals used for the structure and fluid, respectively, considered as separated individual entities. When existing FEM programs are used, the choice of Π_S and Π_F is implicit from whatever models were used therein, and need not be discussed further. The new ingredient is Π_B, called the *interface functional*. This term produces the interface coupling equations studied below.

Separate discretization of the fluid and structure models, as well as LLM treatment of the fluid-structure interface as illustrated in Figs 9 and 10 yields the following semidiscrete matrix EOM in terms of displacements and interface forces:

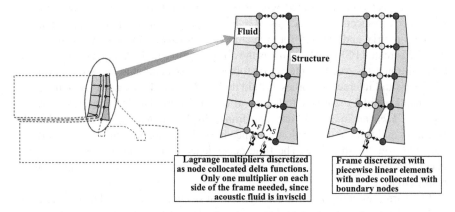

Fig. 9 Zoom on interface elements coupled by a LLM treatment: matching mesh case. Frame nodes coincide with fluid and structure boundary nodes on each side.

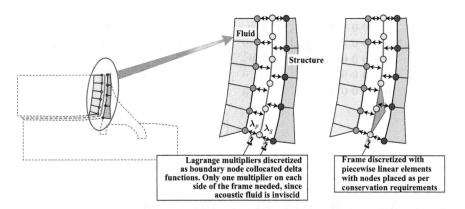

Fig. 10 Zoom on interface elements coupled by a LLM treatment: nonmatching mesh case. Frame nodes do not generally coincide with fluid or structure boundary nodes on each side.

$$
\begin{bmatrix} \mathbf{M}_S & 0 & 0 & 0 & 0 \\ 0 & \mathbf{M}_F & 0 & 0 & 0 \\ 0 & 0 & 0 & 0 & 0 \\ 0 & 0 & 0 & 0 & 0 \\ 0 & 0 & 0 & 0 & 0 \end{bmatrix} \begin{Bmatrix} \ddot{\mathbf{u}}_S \\ \ddot{\mathbf{u}}_F \\ \ddot{\lambda}_S \\ \ddot{\lambda}_F \\ \ddot{\mathbf{u}}_B \end{Bmatrix} + \begin{bmatrix} \mathbf{K}_S & 0 & \mathbf{B}_{Sn} & 0 & 0 \\ 0 & \mathbf{K}_F & 0 & \mathbf{B}_{Fn} & 0 \\ \mathbf{B}_{Sn}^T & 0 & 0 & 0 & -\mathbf{L}_{Sn} \\ 0 & \mathbf{B}_{Fn}^T & 0 & 0 & -\mathbf{L}_{Fn} \\ 0 & 0 & -\mathbf{L}_{Sn}^T & -\mathbf{L}_{Fn}^T & 0 \end{bmatrix} \begin{Bmatrix} \mathbf{u}_S \\ \mathbf{u}_F \\ \lambda_S \\ \lambda_F \\ \mathbf{u}_B \end{Bmatrix}
$$

$$
= \begin{Bmatrix} \mathbf{f}_S \\ \mathbf{f}_F \\ 0 \\ 0 \\ 0 \end{Bmatrix}. \tag{2}
$$

For the structure model, $\mathbf{u}_S$ is the array of structural node displacements. whereas $\mathbf{M}_S$, $\mathbf{K}_S$ and $\mathbf{f}_S$ denote the master mass matrix, stiffness matrix and applied force vector, respectively, associated with $\mathbf{u}_S$. For the fluid model, $\mathbf{u}_F$ is the array of fluid node displacements. whereas $\mathbf{M}_F$, $\mathbf{K}_F$ and $\mathbf{f}_F$ denote the master mass matrix, stiffness matrix and applied force vector, respectively, associated with $\mathbf{u}_F$. Over the LLM-treated FSI interface, $\mathbf{u}_B$ is the array of frame node displacements, λ_S the array of frame-to-structure interaction forces at wet structural nodes, λ_F the array of frame-to-fluid interaction forces at fluid nodes, $\mathbf{B}_{Sn}$ and $\mathbf{B}_{Fn}$ are Boolean matrices that map λ_S and λ_F onto the full set of structural and fluid node forces, respectively, $\mathbf{L}_{Sn}$ and $\mathbf{L}_{Fn}$ are matrices that map frame displacements $\mathbf{u}_B$ to structural node freedoms and fluid node freedoms, respectively. Structure and fluid nodes need not coincide over the interface. The time dependence of the state vectors and forces is omitted from (2) for brevity. A superposed dot denotes time differentiation with respect to time t.

Fluid irrotationality is enforced by the transformation $\mathbf{u}_F = \mathbf{D}_F \psi$, where ψ collects displacement potential degrees of freedom at fluid mesh nodes and $\mathbf{D}_F$ is a generally rectangular transformation matrix. (Since the displacement potential is a scalar field, there is only one ψ freedom per node.) Carrying out a congruential transformation on fluid freedoms yields

$$
\begin{bmatrix} \mathbf{M}_S & 0 & 0 & 0 & 0 \\ 0 & \mathbf{M}_{F\psi} & 0 & 0 & 0 \\ 0 & 0 & 0 & 0 & 0 \\ 0 & 0 & 0 & 0 & 0 \\ 0 & 0 & 0 & 0 & 0 \end{bmatrix} \begin{Bmatrix} \ddot{\mathbf{u}}_S \\ \ddot{\psi} \\ \ddot{\lambda}_S \\ \ddot{\lambda}_F \\ \ddot{\mathbf{u}}_B \end{Bmatrix} + \begin{bmatrix} \mathbf{K}_S & 0 & \mathbf{B}_{Sn} & 0 & 0 \\ 0 & \mathbf{K}_{F\psi} & 0 & \mathbf{B}_{F\psi n} & 0 \\ \mathbf{B}_{Sn}^T & 0 & 0 & 0 & -\mathbf{L}_{Sn} \\ 0 & \mathbf{B}_{F\psi n}^T & 0 & 0 & -\mathbf{L}_{Fn} \\ 0 & 0 & -\mathbf{L}_{Sn}^T & -\mathbf{L}_{Fn}^T & 0 \end{bmatrix} \begin{Bmatrix} \mathbf{u}_S \\ \psi \\ \lambda_S \\ \lambda_F \\ \mathbf{u}_B \end{Bmatrix}
$$

$$
= \begin{Bmatrix} \mathbf{f}_S \\ \mathbf{f}_{F\psi} \\ 0 \\ 0 \\ 0 \end{Bmatrix}, \tag{3}
$$

in which $\mathbf{M}_{F\psi} = \mathbf{D}_F^T \mathbf{M}_F \mathbf{D}_F$, $\mathbf{K}_{F\psi} = \mathbf{D}_F^T \mathbf{K}_F \mathbf{D}_F$, $\mathbf{B}_{F\psi n} = \mathbf{D}_F^T \mathbf{B}_{Fn}$ and $\mathbf{f}_{F\psi} = \mathbf{D}_F^T \mathbf{f}_F$. For detailed tutorial examples on the configuration of these EOM, see Appendix C in [37].

4.3 Interface Functionals

The "wet" fluid-structure boundary is denoted by Γ_B. Since an acoustic fluid model is inviscid, only the normal displacements u_{Fn} and u_{Sn}, as well as normal tractions t_{Fn} and t_{Sn}, appear in the strong interaction conditions: $u_{Fn} = u_{Sn}$ and $t_{Fn} + t_{Sn} = 0$. (Note that $t_{Fn} = -p_B$, in which p_B is the hydrodynamic fluid pressure on Γ_B, positive if compressive.) Two weak forms, originally proposed for elasticity problems by Prager [33], Pian and Tong [31] and Atluri [2], can be stated in terms of the following interface functionals

$$\Pi_B^M[\lambda_{Bn}] = \int_{\Gamma_B} (u_{Fn} - u_{Sn}) \lambda_{Bn} \, d\Gamma,$$

$$\Pi_B^L[\lambda_{Bn}, \lambda_{Fn}, u_{Bn}] = \int_{\Gamma_B} \left\{ (u_{Fn} - u_{Bn}) \lambda_{Fn} + (u_{Sn} - u_{Bn}) \lambda_{Sn} \right\} d\Gamma. \quad (4)$$

Here independently varied fields are identified in square brackets on the left-hand side. In Π_B^M, λ_{Bn} is a global Lagrange multiplier function that connects directly the fluid and structure faces. In Π_B^L, λ_{Fn} and λ_{Sn} are localized Lagrange multiplier functions that link the independently varied normal displacement u_{Bn} of a frame introduced between fluid and structure. The LLM treatment of the interface is based on Π_B^L whereas the variational-based Mortar method outlined later derives from Π_B^M. Substituting

$$u_{Bn} \Rightarrow \tfrac{1}{2}(u_{Fn} + u_{Sn}), \qquad \tfrac{1}{2}(\lambda_{Fn} - \lambda_{Sn}) \Rightarrow \lambda_{Bn} \quad (5)$$

into Π_B^L reduces it to Π_B^M. Note that the replacements in (5) must be adjusted at *cross points* where more than two partitions meet. At such locations special handling is needed because no unique normal exists. Excluding that situation, Π_B^L embodies Π_B^M as special case.

4.4 Discrete Interface Equations

To produce matrix connection equations, functionals (4) are spatially discretized by assuming shape functions for the independently varied fields: either λ_{Bn} in Π_B^M, or $\lambda_{Fn}, \lambda_{Sn}$ and u_{Bn} in Π_B^L. Boundary normal displacements u_{Sn} and u_{Fn} come from elements used for the fluid and structure models, respectively, and are prescribed data for both interface functionals. Restricting attention to the more general Π_B^L, assume $\lambda_{Fn} = \mathbf{N}_{\lambda F} \lambda_F$, $\lambda_{Sn} = \mathbf{N}_{\lambda S} \lambda_S$ and $u_{Bn} = \mathbf{N}_B \mathbf{u}_B$. For the boundary displacements take $u_{Fn} = \mathbf{N}_{FB} \mathbf{u}_F$ and $u_{Sn} = \mathbf{N}_{FB} \mathbf{u}_S$, where $\mathbf{N}_{FB}$ and $\mathbf{N}_{SB}$ are shape functions for fluid and displacement elements, respectively, evaluated on Γ_B and projected over the normal n. Insert these interpolations in Π_B^L, and integrate over Γ_B to get the discretized functional

$$\Pi_B^L[\lambda_F, \lambda_S, \mathbf{u}_B] = \lambda_F^T (\mathbf{B}_F \mathbf{u}_F - \mathbf{L}_F \mathbf{u}_B) + \lambda_S^T (\mathbf{B}_S \mathbf{u}_S - \mathbf{L}_S \mathbf{u}_B), \quad (6)$$

in which the connection matrices are defined as

$$\mathbf{B}_F = \int_{\Gamma_B} \mathbf{N}_{\lambda F}^T \mathbf{N}_{FB} \, d\Gamma, \quad \mathbf{B}_S = \int_{\Gamma_B} \mathbf{N}_{\lambda S}^T \mathbf{N}_{SB} \, d\Gamma,$$

$$\mathbf{L}_F = \int_{\Gamma_B} \mathbf{N}_{\lambda F}^T \mathbf{N}_B \, d\Gamma, \quad \mathbf{L}_S = \int_{\Gamma_B} \mathbf{N}_{\lambda S}^T \mathbf{N}_B \, d\Gamma. \tag{7}$$

Some notational simplifications have been made in the foregoing expressions for brevity: subscript n is dropped throughout whereas Γ_B denotes the *discretized* "wet" interface surface, which may differ from the original one for curved geometries.

The integral evaluations in (7) are significantly simplified by taking $\mathbf{N}_{\lambda F}$ and $\mathbf{N}_{\lambda S}$ to be *delta functions collocated at the fluid and structure interface nodes*, respectively, a configuration illustrated in Fig. 10. (If meshes match, multipliers, frame and boundary nodes coalesce as pictured in Fig. 9.) If so $\mathbf{B}_F$ and $\mathbf{B}_S$ become Boolean matrices that select and normal-project node boundary freedoms from the complete state vectors: $\mathbf{u}_{BF} = \mathbf{B}_F \, \mathbf{u}_F$ and $\mathbf{u}_{BS} = \mathbf{B}_S \, \mathbf{u}_S$. For interpolating frame displacements $\mathbf{u}_B$ we have so far only used *piecewise linear*, C^0-continuous shape functions $\mathbf{N}_B$ with nodes placed according to distribution rules discussed in [29, 30]. Frame interpolation spaces of lower (C^{-1}) continuity remain to be investigated.

Setting the first variation $\delta \Pi_B^L$ to zero yields the three matrix equations

$$\mathbf{u}_{BF} = \mathbf{B}_F \mathbf{u}_F = \mathbf{L}_F \, \mathbf{u}_B, \quad \mathbf{u}_{BS} = \mathbf{B}_S \mathbf{u}_S = \mathbf{L}_S \, \mathbf{u}_B, \quad -\mathbf{L}_F^T \, \boldsymbol{\lambda}_F - \mathbf{L}_S^T \, \boldsymbol{\lambda}_S = \mathbf{0}. \tag{8}$$

When adjoined to the fluid and structure uncoupled EOMs, these appear as the last three matrix equations of the coupled EOM (2) in displacement coordinates. For future use, introduce the matrices

$$\mathbf{Q}_{FF} = \mathbf{L}_F \mathbf{L}_F^T, \quad \mathbf{Q}_{SS} = \mathbf{L}_S \mathbf{L}_S^T, \quad \mathbf{R}_{FF} = \mathbf{L}_F^T \mathbf{L}_F, \quad \mathbf{R}_{SS} = \mathbf{L}_S^T \mathbf{L}_S. \tag{9}$$

Inverses of the $\mathbf{Q}$ matrices appear in ensuing derivations. Since $\mathbf{L}_F$ and $\mathbf{L}_S$ are generally rectangular for nonmatching meshes, one or more of (9) could become singular, in which case ordinary inverses do not exist. In the equations below $\mathbf{A}^{-G}$ denotes the Moore-Penrose generalized inverse of $\mathbf{A}$, also popularly known as the pseudoinverse [34].

4.5 LLM Interface Force-Motion Relations

Solving the last equation of (8) by least-squares methods yields

$$\boldsymbol{\lambda}_S = -\mathbf{T}_{SF} \, \boldsymbol{\lambda}_F, \quad \boldsymbol{\lambda}_F = -\mathbf{T}_{FS} \, \boldsymbol{\lambda}_S, \tag{10}$$

in which $\mathbf{T}_{SF} = \mathbf{Q}_{SS}^{-G} \mathbf{L}_S \mathbf{L}_F^T$ and $\mathbf{T}_{FS} = \mathbf{Q}_{FF}^{-G} \mathbf{L}_F \mathbf{L}_S^T$ are called *force transfer* matrices. Here pseudoinverses should be replaced by ordinary inverses if appropriate. Products $\mathbf{T}_{FS}\mathbf{T}_{SF}$ and $\mathbf{T}_{SF}\mathbf{T}_{FS}$ are orthogonal projectors. Elimination of $\mathbf{u}_B$

from the first two equations of (8): $\mathbf{u}_{BF} = \mathbf{L}_F \mathbf{u}_B$ and $\mathbf{u}_{BS} = \mathbf{L}_S \mathbf{u}_B$, yields the corresponding transformations between boundary displacement vectors:

$$\mathbf{u}_{BS} = \mathbf{U}_{SF} \, \boldsymbol{\lambda}_F, \quad \mathbf{u}_{BF} = \mathbf{U}_{FS} \, \boldsymbol{\lambda}_S, \quad \text{in which} \quad \mathbf{U}_{SF} = \mathbf{T}_{FS}^T, \quad \mathbf{U}_{FS} = \mathbf{T}_{SF}^T. \tag{11}$$

Here $\mathbf{U}_{SF} = (\mathbf{L}_S^T)^{-G} \, \mathbf{R}_{SS} \, \mathbf{R}_{FF}^{-G} \, \mathbf{L}_F^T = \mathbf{T}_{FS}^T$ and $\mathbf{U}_{FS} = (\mathbf{L}_F^T)^{-G} \, \mathbf{R}_{FF} \, \mathbf{R}_{SS}^{-G} \, \mathbf{L}_S^T = \mathbf{T}_{SF}^T$ will be called *motion transfer* matrices. To prove that $\mathbf{U}_{SF} = \mathbf{T}_{SF}^T$ from linear algebra, start from the identity $\mathbf{R}_{SS}^{-G} \, \mathbf{L}_S^T = \mathbf{L}_S^T \, \mathbf{Q}_{SS}^{-G}$, premultiply both sides by $\mathbf{L}_F^T \, (\mathbf{L}_F^T)^{-G} \, \mathbf{R}_{FF}$ and use projector properties. Likewise for $\mathbf{U}_{FS} = \mathbf{T}_{SF}^T$. The transformation duality (11) can be established more directly from work theorems, noting that no energy is gained or lost at the interface, as follows. The complementary virtual work δW_B^* of interface displacements on their conjugate multiplier variations is

$$\delta W_{BS}^* = \mathbf{u}_{BS}^T \, \delta \boldsymbol{\lambda}_S, \quad \delta W_{BF}^* = \mathbf{u}_{BF}^T \, \delta \boldsymbol{\lambda}_F, \quad \delta W_B^* = \delta W_{BF}^* + \delta W_{BS}^* = 0. \tag{12}$$

Setting $\delta \boldsymbol{\lambda}_F = -\mathbf{T}_{FS} \, \delta \boldsymbol{\lambda}_S$ and $\mathbf{u}_{BS}^T = \mathbf{u}_{BF}^T \mathbf{U}_{SF}^T$ in $\delta W_B^* = 0$ gives $\mathbf{u}_{BF}^T (\mathbf{U}_{SF}^T - \mathbf{T}_{FS}) \delta \boldsymbol{\lambda}_S = 0$ for arbitrary $\mathbf{u}_{BF}$ and $\delta \boldsymbol{\lambda}_S$, whence $\mathbf{T}_{FS} = \mathbf{U}_{SF}^T$. Setting $\delta \boldsymbol{\lambda}_S = -\mathbf{T}_{SF} \, \delta \boldsymbol{\lambda}_F$ and $\mathbf{u}_{BF}^T = \mathbf{u}_{BS}^T \mathbf{U}_{FS}^T$ gives $\mathbf{u}_{BS}^T (\mathbf{U}_{FS}^T - \mathbf{T}_{SF}) \delta \boldsymbol{\lambda}_F = 0$ for arbitrary $\mathbf{u}_{BS}$ and $\delta \boldsymbol{\lambda}_F$, whence $\mathbf{T}_{SF} = \mathbf{U}_{FS}^T$.

4.6 LLM Interface Patch Tests

The duality (11) holds for *any* $\mathbf{L}_F$ and $\mathbf{L}_S$, even if those matrices were filled with random numbers. Thus it is independent of the frame discretization choice, or even of whether a frame is present. This generality can be used to advantage for other treatments, e.g. the Mortar and Global DFMT methods derived through AIV elimination. The placement of frame nodes may affect, however, the results of the Interface Patch Test (IPT) as shown by the following example, which is a variant of one studied in [30].

Consider the two coarse nonmatching 2D meshes pictured in Fig. 11(a). The interface Γ_B of height H connects three bilinear rectangular fluid elements of height $\frac{1}{3}H$ to two bilinear rectangular structural elements of height $\frac{1}{2}H$. All elements have uniform thickness h. Of the four frame nodes, two (1 and 4) are placed at both ends and two (2 and 3) symmetrically located at distance αH from the middle fluid boundary node positions, as indicated. Here $-\frac{1}{3} \leq \alpha \leq \frac{1}{6}$ is a dimensionless free parameter; conventionally $\alpha > 0$ if the middle frame nodes lie in the center thirdspan, as drawn in the Fig. 11(a). If $\alpha = \frac{1}{6}$ frame nodes 2 and 3 coalesce and the frame has 3 nodes. If $\alpha = -\frac{1}{3}$ nodes 2 and 3 coalesce with the end nodes and the frame has 2 nodes. The only displacement DOF considered are the interface displacements shown in Fig. 11(b). These are collected in vectors

$$\mathbf{u}_{BS}^T = [u_{S1} \; u_{S2} \; u_{S3}], \quad \mathbf{u}_{BF}^T = [u_{F1} \; u_{F2} \; u_{F3} \; u_{F4}],$$
$$\mathbf{u}_B^T = [u_{B1} \; u_{B2} \; u_{B3} \; u_{B4}]. \tag{13}$$

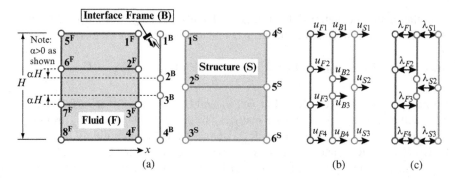

Fig. 11 Example to illustrate interface patch test and Zero Moment Rule (ZMR).

The interface multiplier freedoms, displayed in Fig. 11(c), are collected in vectors

$$\boldsymbol{\lambda}_S^T = [\lambda_{S1} \; \lambda_{S2} \; \lambda_{S3}], \quad \boldsymbol{\lambda}_F^T = [\lambda_{F1} \; \lambda_{F2} \; \lambda_{F3} \; \lambda_{F4}]. \tag{14}$$

The connection and force-transfer matrices for arbitrary α, excluding the node-coalescence cases $\alpha = -\frac{1}{3}$ and $\alpha = \frac{1}{6}$, are

$$\mathbf{L}_F = \begin{bmatrix} 1 & 0 & 0 & 0 \\ \dfrac{3\alpha}{1+3\alpha} & \dfrac{1}{1+3\alpha} & 0 & 0 \\ 0 & 0 & \dfrac{1}{1+3\alpha} & \dfrac{3\alpha}{1+3\alpha} \\ 0 & 0 & 0 & 1 \end{bmatrix}, \quad \mathbf{L}_S = \begin{bmatrix} 1 & 0 & 0 & 0 \\ 0 & 1/2 & 1/2 & 0 \\ 0 & 0 & 0 & 1 \end{bmatrix},$$

$$\mathbf{T}_{SF} = \begin{bmatrix} 1 & \dfrac{3\alpha}{1+3\alpha} & 0 & 0 \\ 0 & \dfrac{1}{1+3\alpha} & \dfrac{1}{1+3\alpha} & 0 \\ 0 & 0 & \dfrac{3\alpha}{1+3\alpha} & 1 \end{bmatrix}, \quad \mathbf{T}_{FS} = \frac{1}{2}\begin{bmatrix} 2 & -3\alpha & 0 \\ 0 & 1+3\alpha & 0 \\ 0 & 1+3\alpha & 0 \\ 0 & -3\alpha & 2 \end{bmatrix}. \tag{15}$$

To check the Interface Force Patch Test (IFPT) one assumes that the 3 fluid elements are under uniform pressure p, positive if compression. If displacement shape functions vary linearly along the edge, as in the case of a bilinear fluid element, the consistent interface fluid node forces are $\tilde{\mathbf{f}}_{BF} = -\frac{1}{6}phH\,[1 \; 2 \; 2 \; 1]^T$. Assuming a similar linear shape function variation over the structural elements and uniform stress $\sigma_{xx} = -p$, others zero, the consistent structural node forces are $\tilde{\mathbf{f}}_{BF} = \frac{1}{4}phH\,[1 \; 2 \; 1]^T$. To apply the IFPT, set $\hat{\boldsymbol{\lambda}}_F = -\mathbf{f}_{BF}, \tilde{\boldsymbol{\lambda}}_S = \tilde{\mathbf{f}}_{BS}$ and use the multiplier transformation equations (10) to compute

$$\boldsymbol{\lambda}_S = \mathbf{T}_{SF}\hat{\boldsymbol{\lambda}}_F = \frac{pHh}{6(1+3\alpha)}\begin{bmatrix} 1+9\alpha \\ 4 \\ 1+9\alpha \end{bmatrix}, \quad \boldsymbol{\lambda}_F = \mathbf{T}_{FS}\hat{\boldsymbol{\lambda}}_S = \frac{pHh}{4}\begin{bmatrix} 1-3\alpha \\ 1+3\alpha \\ 1+3\alpha \\ 1-3\alpha \end{bmatrix}. \tag{16}$$

The IFPT is passed if $\lambda_S = \tilde{\lambda}_S$ and $\lambda_F = \tilde{\lambda}_F$. Clearly this happens if and only if $\alpha = \frac{1}{9}$. This is the only 4-node-frame configuration that satisfies the Zero Moment Rule (ZMR) [29, 30].

The interface rigid-motion patch test (IRMPT) checks whether a linearly varying fluid boundary displacement field is correctly transmitted to the structure and vice-versa. If d denotes uniform translation along x and θ the rotation about z, the correct node displacement values are $\hat{\mathbf{u}}_{BF} = d \begin{bmatrix} 1 & 1 & 1 & 1 \end{bmatrix}^T + \frac{1}{6}\theta H \begin{bmatrix} -3 & -1 & 1 & 3 \end{bmatrix}^T$ and $\hat{\mathbf{u}}_{BS} = d \begin{bmatrix} 1 & 1 & 1 \end{bmatrix}^T + \frac{1}{2}\theta H \begin{bmatrix} -1 & 0 & 1 \end{bmatrix}^T$. Application of the displacement transformations (11) gives a $\mathbf{u}_{BS} = \mathbf{U}_{SF} \hat{\mathbf{u}}_{BF}$ that reproduces $\hat{\mathbf{u}}_{BS}$ exactly for any α. The converse transformation, however, is exact only for $\alpha = \frac{1}{6}$.

5 Mortar Method

Since its inception in 1990 [5] the Mortar method has gained popularity as a interfacing scheme for multiphysics capable of handling non-matched meshes [3, 4, 6, 7, 12, 22, 40]. With growing acceptance the name has come to designate a set of loosely related mesh coupling techniques. The common feature is the use of one and only one "gluing" Lagrange multiplier field that directly links the two sides of the interface (that is, without a kinematic frame). For coupling of an inviscid fluid (in particular, an acoustic fluid) to an elastic structure the gluing field is a scalar, which was called λ_{Bn} in Sect. 4.3.

Mortar interface equations have been usually constructed from Galerkin or other weighted residual methods, see e.g. [22]. This is inevitable for general fluid models, which are not derivable from variational principles. In the present context, however, we will restrict those equations to be based on the functional Π_B^M of (4), a variational framework that preserves symmetry. The only field to be discretized is the scalar multiplier function λ_{Bn} on Γ_B. Physically this is the normal-to-the-interface surface traction, also known as wet wall pressure.

In most of the published literature on Mortar methods, multipliers are distributed functions interpretable as surface tractions. To make a fair comparison with LLM, however, we take the multiplier space for λ_{Bn} to be that of delta functions, as pictured in Fig. 12; which can be viewed as interaction point forces. An important question is, where should those forces be placed? For matching meshes the answer is easy, see Fig. 12(a). For nonmatching meshes, as pictured in Fig. 12(b), the choice is not obvious. One solution is to declare a face as master (typically that pertaining to the finer mesh, as explicitly recommended in [38]), and collocate point forces at the master nodes. This is illustrated in Fig. 12(c) with the fluid face picked as master. A dual-master scheme, e.g. [21], collocates multipliers at all interface nodes as shown in Fig. 12(d); this avoids master vs. slave decisions but may lead to singularity or ill-conditioning. Similar freedom placement decisions may be necessary when using distributed Lagrange multiplier spaces. Regardless of choice, a master mesh must know (in the nonmatching case) about the boundary shape functions of the slave one and the modularity of the LLM treatment is lost.

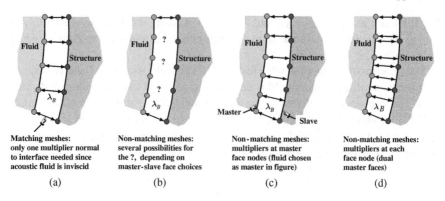

Fig. 12 Mortar treatment with node-collocated delta function multipliers.

5.1 Mortar Equations Of Motion

From now on subscript n will be again omitted for brevity. With the discrete multiplier assumption written as $\lambda_B = N_{\lambda B}\lambda_B$, insertion into Π_B^M gives the discretized functional

$$\Pi_B^M[\lambda_B] = \left(\mathbf{u}_F^T \hat{\mathbf{B}}_F - \mathbf{u}_S^T \hat{\mathbf{B}}_S\right)\lambda_B. \qquad (17)$$

where $\hat{\mathbf{B}}_F = \int_{\Gamma_B} N_{FB}^T N_{\lambda B}$ and $\hat{\mathbf{B}}_S = \int_{\Gamma_B} N_{SB}^T N_{\lambda B}$. Since $N_{\lambda B}$ consists of delta functions, the integrals reduce to collocation on the master face and interpolation on the other. For nonmatching meshes $\hat{\mathbf{B}}_F$ and $\hat{\mathbf{B}}_S$ are not Boolean and so will generally differ from the $\mathbf{B}_F$ and $\mathbf{B}_S$ of the LLM treatment. Setting $\delta\Pi_B^M = 0$ yields the matrix connection equation $\hat{\mathbf{B}}_S^T\mathbf{u}_S = \hat{\mathbf{B}}_F^T\mathbf{u}_F$. For interpretation, let $\mathbf{f}_{BS} = \hat{\mathbf{B}}_S^T\lambda_B$ and $\mathbf{f}_{BF} = -\hat{\mathbf{B}}_F^T\lambda_B$ be the boundary force arrays conjugate to $\mathbf{u}_{BS}$ and $\mathbf{u}_{BF}$, respectively. Hence $\mathbf{f}_{BS} + \mathbf{f}_{BF} = \mathbf{0}$, which expresses discrete interface force equilibrium. Adjoining this to the dynamic equations of the uncoupled fluid and structure gives the matrix equations of motion in terms of structural and fluid displacements. If damping is ignored the EOMs are

$$\begin{bmatrix} \mathbf{M}_S & \mathbf{0} & \mathbf{0} \\ \mathbf{0} & \mathbf{M}_F & \mathbf{0} \\ \mathbf{0} & \mathbf{0} & \mathbf{0} \end{bmatrix}\begin{Bmatrix} \ddot{\mathbf{u}}_S \\ \ddot{\mathbf{u}}_F \\ \ddot{\lambda}_B \end{Bmatrix} + \begin{bmatrix} \mathbf{K}_S & \mathbf{0} & \hat{\mathbf{B}}_S \\ \mathbf{0} & \mathbf{K}_F & -\hat{\mathbf{B}}_F \\ \hat{\mathbf{B}}_S^T & -\hat{\mathbf{B}}_F^T & \mathbf{0} \end{bmatrix}\begin{Bmatrix} \mathbf{u}_S \\ \mathbf{u}_F \\ \lambda_B \end{Bmatrix} = \begin{Bmatrix} \mathbf{f}_S \\ \mathbf{f}_F \\ \mathbf{0} \end{Bmatrix}. \qquad (18)$$

As before fluid irrotationality is enforced by the transformation $\mathbf{u}_F = \mathbf{D}_F\,\psi$, where ψ collects displacement potential degrees of freedom at fluid mesh nodes. Carrying out a congruential transformation on fluid freedoms yields

$$\begin{bmatrix} \mathbf{M}_S & \mathbf{0} & \mathbf{0} \\ \mathbf{0} & \mathbf{M}_{F\psi} & \mathbf{0} \\ \mathbf{0} & \mathbf{0} & \mathbf{0} \end{bmatrix}\begin{Bmatrix} \ddot{\mathbf{u}}_S \\ \ddot{\psi} \\ \ddot{\lambda}_B \end{Bmatrix} + \begin{bmatrix} \mathbf{K}_S & \mathbf{0} & \hat{\mathbf{B}}_S \\ \mathbf{0} & \mathbf{K}_{F\psi} & -\hat{\mathbf{B}}_{F\psi} \\ \hat{\mathbf{B}}_S^T & -\hat{\mathbf{B}}_{F\psi}^T & \mathbf{0} \end{bmatrix}\begin{Bmatrix} \mathbf{u}_S \\ \psi \\ \lambda_B \end{Bmatrix} = \begin{Bmatrix} \mathbf{f}_S \\ \mathbf{f}_{F\psi} \\ \mathbf{0} \end{Bmatrix}, \qquad (19)$$

in which $\mathbf{M}_{F\psi} = \mathbf{D}_F^T \mathbf{M}_F \mathbf{D}_F$, $\mathbf{K}_{F\psi} = \mathbf{D}_F^T \mathbf{K}_F \mathbf{D}_F$, $\hat{\mathbf{B}}_{F\psi n} = \mathbf{D}_F^T \mathbf{B}_{Fn}$ and $\mathbf{f}_{F\psi} = \mathbf{D}_F^T \mathbf{f}_F$. A response analysis can be carried out by the partitioned analysis procedure described in [37]. The numerical stability of that procedure in conjunction with the Newmark time integrator is studied in Appendix A of that reference.

5.2 Mortar Interface Force-Motion Relations

As done in Sect. 4.5 a relationship between the fluid boundary displacements $\mathbf{u}_{BF}$ and the structure boundary displacements $\mathbf{u}_{BS}$ can be found on solving $\hat{\mathbf{B}}_S^T \mathbf{u}_S = \hat{\mathbf{B}}_F^T \mathbf{u}_F$ by least-square methods. Defining $\hat{\mathbf{Q}}_{FF} = \hat{\mathbf{B}}_F \hat{\mathbf{B}}_F^T$ and $\hat{\mathbf{Q}}_{SS} = \hat{\mathbf{B}}_S \hat{\mathbf{B}}_S^T$ we have

$$\mathbf{u}_{BF} = \hat{\mathbf{Q}}_{FF}^{-G} \hat{\mathbf{B}}_F \hat{\mathbf{B}}_S^T \mathbf{u}_{BS} = \hat{\mathbf{T}}_{FS} \mathbf{u}_{BS}, \quad \mathbf{u}_{BS} = \hat{\mathbf{Q}}_{SS}^{-G} \hat{\mathbf{B}}_S \hat{\mathbf{B}}_F^T \mathbf{u}_{BF} = \hat{\mathbf{T}}_{SF} \mathbf{u}_{BF}. \tag{20}$$

The dual force transformations

$$\mathbf{f}_{BS} = \mathbf{T}_{SF} \mathbf{f}_{BF}, \quad \mathbf{f}_{BF} = \mathbf{T}_{FS} \mathbf{f}_{BS}, \tag{21}$$

follow from interface energy conservation as follows. The complementary virtual work δW_B^* over Γ_B is $\delta W_B^* = \mathbf{u}_{BF}^T \delta \mathbf{f}_{BF} + \mathbf{u}_{BS}^T \delta \mathbf{f}_{BS} = \mathbf{u}_{BS}^T [\hat{\mathbf{T}}_{FS}^T \delta \mathbf{f}_{BF} + \delta \mathbf{f}_{BS}] = \mathbf{u}_{BS}^T [\hat{\mathbf{T}}_{FS}^T - \hat{\mathbf{U}}_{SF}] \delta \mathbf{f}_{BF} = \mathbf{u}_{BS}^T [\hat{\mathbf{T}}_{FS}^T - \hat{\mathbf{U}}_{SF}] \hat{\mathbf{B}}_F^T \delta \lambda_B = 0$ for arbitrary $\mathbf{u}_{BS}$ and $\delta \lambda_B$, whence $\hat{\mathbf{U}}_{SF} = -\hat{\mathbf{T}}_{FS}^T$. Similarly $\hat{\mathbf{U}}_{FS} = -\hat{\mathbf{T}}_{SF}^T$.

5.3 Mortar Method Assessment

The Mortar method based on delta-function multipliers (point interaction forces) produces simpler interaction equations than LLM, as a comparison of the coupled EOMs (2)–(3) versus (18)–(19) makes obvious. For nonmatching meshes, forming $\hat{\mathbf{B}}_F$ and $\hat{\mathbf{B}}_S$ involves more work than the Boolean $\mathbf{B}_F$ and $\mathbf{B}_S$ since shape function interpolations are generally needed in the former, but there is no need for the LLM connection matrices $\mathbf{L}_S$ and $\mathbf{L}_F$. This simplicity, however, is counterbalanced by two features:

(1) The interface force patch test may be violated if collocated face displacements are used.
(2) Interaction equations may become singular at cross points (2D) or cross lines (3D), if meshes are nonmatching.

Weakness (1) is illustrated by a 2D example given in [38], in which two iso-P bilinear elements are linked to a 9-node biquadratic iso-P element. See Fig. 5 of that reference. It can be also verified that the IRMTPT is also violated whence motion transfer is incorrect. To pass the patch test it is necessary to pair the multipliers with weighted node displacements.

Weakness (2) has no simple cure within the Mortar context. In fact it does not depend on whether multipliers are distributed or lumped as point forces, because the problem is caused by the lack of unique normal.

6 Direct Force-Motion Transfer Methods

The term *Direct Force-Motion Transfer* or DFMT was introduced in [38] as collective label for a wide class of interfacing methods with historical and practical importance. Their common feature is that no AIV are introduced. Boundary forces and displacements (or velocities) are moved directly from fluid to structure and vice-versa. It is convenient to distinguish two subclasses, qualified here as Local and Global.

6.1 Local DFMT for Blast Simulations

As described in Sect. 2.4, local DFMT methods for FSI are the oldest as well as the easiest to understand because they are based on simple physics. This matrix-free process was used in simulations of submerged structures subject to underwater shock (UWS) using staggered time integration. The initial implementation, studied in [24], coupled a coarse BEM fluid model to a finer structural FEM mesh. BEM to FEM force lumping was based on contributing wet-surface areas. Interface energy conservation was not enforced or even monitored: this is unimportant in shock wave transient simulations that span milliseconds. Despite that shortcoming this DFMT performed well for the chief goal of assessing structural vulnerability.

For the hull-cavitation simulations of [14] a fluid-volume FE mesh was placed between the structure and the BEM model operating as silent boundary; see Fig. 5. Many local DFMT flavors exist. For expediency in [38] they were identified by the code in which implemented. Examples: DFMT-CFA and DFMT-CASE for those used in the UWS code CFA (Cavitation Fluid Analyzer) of [14] and CASE (Cavitation Spectral Elements) of [39], respectively.

6.2 Consistent Interpolation

A similar transfer method was initially used by Farhat and his team [9, 10] for the aeroelasticity problem. It was driven by the need to determine aircraft surface boundary forces from the aerodynamic pressure provided by a gas-dynamics fluid-volume (FV) code. Here the fluid pressure may experience rapid changes in space, especially with Navier-Stokes fluid models. Unlike UWS simulations, the near field fluid mesh is typically finer than the structure mesh to facilitate capture of those spatial gradients. The aeroelasticity DFMT process incorporated refinements: fluid pressures were transferred to structural mesh Gauss points, from which structure node forces are obtained by quadrature. The motion transfer is carefully done by consistent interpolation, as described in more detail below. Application of this method to the example of Fig. 11 gives the following fluid-to-structure force transfer $\mathbf{f}_{BS} = \mathbf{T}_{SF}\,\mathbf{f}_{BF}$ and the structure-to-fluid motion transfer $\mathbf{u}_{BF} = \mathbf{U}_{FS}\,\mathbf{u}_{BS}$:

$$\mathbf{f}_{BS} = \begin{Bmatrix} f_{S1} \\ f_{S2} \\ f_{S3} \end{Bmatrix} = - \begin{bmatrix} 1 & \frac{1}{3} & 0 & 0 \\ 0 & \frac{2}{3} & \frac{2}{3} & 0 \\ 0 & 0 & \frac{1}{3} & 1 \end{bmatrix} \begin{Bmatrix} f_{F1} \\ f_{F2} \\ f_{F3} \\ f_{F4} \end{Bmatrix},$$

$$\mathbf{u}_{BF} = \begin{Bmatrix} u_{F1} \\ u_{F2} \\ u_{F3} \\ u_{F4} \end{Bmatrix} = \begin{bmatrix} 1 & 0 & 0 \\ \frac{1}{4} & \frac{3}{4} & 0 \\ 0 & \frac{3}{4} & \frac{1}{4} \\ 0 & 0 & 1 \end{bmatrix} \begin{Bmatrix} u_{S1} \\ u_{S2} \\ u_{S3} \end{Bmatrix}. \tag{22}$$

It is easily checked that (22) pass both the stress and rigid-motion patch tests. However since $\mathbf{T}_{SF}$ is not the transpose of $-\mathbf{U}_{FS}$, interface energy conservation is not verified. In aeroelastic calculations that may span minutes of real time, correct interface energy balance is important if the principal aim is detecting flutter or divergence onset, so as to establish flight envelopes. Accordingly the original transfer scheme was modified [11, 32] into the Consistent Interpolation (CI) method as follows.

First, motion transfer was carefully implemented to take into account peculiarities of FV fluid codes. Each FV grid point j on the fluid boundary is paired with the closest structural element e^S (or elements if j is equally distant from more than one). The structural element natural coordinates ξ_j of j (or of its projection on the structure element if j is offset from the interface) are determined. (In cell-centered FV schemes grid point j may be offset from the fluid-structure interface Γ_B; in that case the projection on Γ_B is used.) The fluid displacement $\mathbf{u}_{Fj}$ is determined using the structural element shape functions evaluated at ξ_j. A structure-to-fluid displacement transformation matrix $\mathbf{U}_{FS}$ is built by repeating this procedure over all j so that $\mathbf{u}_{FB} = \mathbf{U}_{FS}\,\mathbf{u}_{SB}$. Second, the transformation from fluid to structure forces: $\mathbf{f}_{SB} = \mathbf{T}_{SF}\mathbf{f}_{SB}$, follows from duality: $\mathbf{T}_{FS} = -\mathbf{U}_{SF}^T$. This enforces interface energy conservation. In typical aeroelastic problems the fluid mesh is more refined then the structure. Thus, the CI method interpolates refined mesh values from coarse mesh values, which helps to produce well conditioned coupling matrices.

Application of the CI method to the example of Fig. 11 yields

$$\mathbf{f}_{BS} = \begin{Bmatrix} f_{S1} \\ f_{S2} \\ f_{S3} \end{Bmatrix} = - \begin{bmatrix} 1 & \frac{1}{3} & 0 & 0 \\ 0 & \frac{2}{3} & \frac{2}{3} & 0 \\ 0 & 0 & \frac{1}{3} & 1 \end{bmatrix} \begin{Bmatrix} f_{F1} \\ f_{F2} \\ f_{F3} \\ f_{F4} \end{Bmatrix},$$

$$\mathbf{u}_{BF} = \begin{Bmatrix} u_{F1} \\ u_{F2} \\ u_{F3} \\ u_{F4} \end{Bmatrix} = \begin{bmatrix} 1 & 0 & 0 \\ \frac{1}{3} & \frac{2}{3} & 0 \\ 0 & \frac{2}{3} & \frac{1}{3} \\ 0 & 0 & 1 \end{bmatrix} \begin{Bmatrix} u_{S1} \\ u_{S2} \\ u_{S3} \end{Bmatrix}. \tag{23}$$

To check the interface force patch test (IFPT), apply uniform pressure p over the 3 fluid elements. Then $\mathbf{f}_{BF} = -\frac{1}{6} ph H [1\ 2\ 2\ 1]^T$ and $\mathbf{f}_{BS} = (1/18) ph H$ $[5\ 8\ 5]^T$. Since $\mathbf{f}_{BS} \neq \tilde{\mathbf{f}}_{BS} = \frac{1}{4} ph H [1\ 2\ 1]^T$, the test is not passed. Notice that the transformation matrices in (23) are the same as those provided by LLM treatment in the example of Sect. 4.6 if $\alpha = \frac{1}{6}$. This configuration does not satisfy the Zero Moment Rule (ZMR) stated in [30]. Thus it is not surprising that the IFPT fails.

6.3 Global DFMT Methods

Aother subclass of DFMT is that of *global* DFMT methods. These are not based on local interface physics. They are instead built in two stages. First a multiplier based discretization such as LLM or point-force Mortar is constructed. Interface unknowns are then eliminated by least-square methods as described in Sects. 4.5 and 5.2 to yield the force transfer matrices $\mathbf{T}_{SF}$ and $\mathbf{T}_{FS}$. The motion transfer matrices follow from duality.

The qualifier *global* indicates that transfer matrices are generally fully populated, meaning that each interface DOF is coupled to every other one. By construction interface energy conservation is satisfied *a priori*, but interface patch tests are not necessarily passed, as previous examples make clear. The following Table summarizes method labeling used for the application examples discussed in [38].

Label	Fluid-structure interface treatment
LLM	LLM with step-by-step solution of interface equations
Mortar	Point-force Mortar with step-by-step solution of interface equations
DFMT-CASE	Local DFMT procedure used in CASE spectral code [39]
DFMT-LLM	Global DFMT with LLM-derived transfer matrices (10) and (11)
DFMT-Mortar	Global DFMT with Mortar-derived transfer matrices (20) and (21)

Do LLM and DFMT-LLM produce identical results? Only under very special conditions. For instance, if connection matrices $\mathbf{L}_F$ and $\mathbf{L}_S$ are square and of full rank. Otherwise the least-squares elimination of interface unknowns can be expected to work as a *low pass filter* that projects interface patterns on the column span of the $\mathbf{Q}$ matrices. This will typically mollify the computed response. A similar remark applies to Mortar versus DFMT-Mortar. Whether this kind of filtering is acceptable or desirable can be expected to be problem and goal dependent.

7 Conclusions

The Introduction of the survey article [17] notes that one obstacle to rapid progress in computational multiphysics is combinatorial complexity in method design and implementation. Restricting ourselves to the FSI problem, model-based simulation involves making decisions on several computational ingredients:

S   Structure spatial model

F   Fluid spatial model

FSI Fluid-structure interface treatment

TS Time stepping

Aux Auxiliary components for some problems, especially those involving exterior domains; for example silent boundaries, ALE mesh mover, ...

One difficulty in coming to terms with ingredients **FSI** and **TS** is agreement on terminology, especially as regards the former. (As noted in the Introduction, some convergence on the classification of time stepping methods appears to be emerging.) Although the literature on coupled problem simulation is growing steadily, lack of coherent nomenclature may hinder the impact of individual contributions. It is hoped that the proposed classification of interface treatments in Sect. 2 may help to alleviate textual obstacles to result dissemination. More specifically, it is hoped that the proposed terminology and in particular understanding the "orthogonality" between **FSI** and **TS**, may foster comparisons and correlation of results obtained by separate research teams.

Assuming some consensus on terminology is eventually achieved, another area that deserves attention is the association of desirable attributes with specific coupled problems, project objectives and model choices. From the long list compiled in Sect. 3 it should be obvious that no "universal" method that possesses even a majority of those attributes can be hoped to be found. Consequently, prioritization by linking attributes to methods and objectives is necessary.

To conclude, the following FSI research areas seem worth exploring:

- Develop interface treatments that satisfy the largest number of conservation conditions for arbitrary nonmatching meshes, and examine related tradeoffs.
- Study which attributes are *inherited* in the hierarchical "cascading" flowcharted of Fig. 4.
- Investigate links between interface treatments and stability+accuracy of partitioned time-stepping procedures.
- Find out which interface treatment(s) best support reliable and effective energy error measures for adaptive timestep control.
- Find effective LLM frame configuration rules for interacting three-dimensional nonmatching meshes, thus generalizing the ZMR of [30] to 3D.
- Study the possibility of "localizing" global DFMT treatments by matrix filtering to generate more computationally efficient local DFMT schemes.
- Investigate the suitability of the various interface treatments to support Reduced Order Models for one or more problem components.
- Generalize dual and primal-dual interface treatments to handle multiscale effects.

Acknowledgements The first author (CAF) expresses his thanks to the organizers of the International Workshop on Computational Engineering 2009 for their invitation to present a keynote lecture and write the present contribution. The contribution of the second author (KCP) was partly supported by the WCU Program of the Korea Science and Engineering Foundation funded by the Ministry of Education, Science and Technology, Republic of Korea, through Grant R31-2008-000-10045-0. The work of the third author (MRR) was part of his doctoral dissertation while at the Aerospace Engineering Sciences Department, University of Colorado at Boulder. That research was supported by the US National Science Foundation under Grant *High-Fidelity Simulations for Heterogeneous Civil and Mechanical Systems*, CMS-0219422.

References

1. Anonymous. COSMOS Virtual Data Center for Strong Motion, Consortium of Organizations for Strong Motion Observation Systems (COSMOS), Pacific Earthquake Engineering Research Center. Available from http://db.cosmos-eq.org, University of California, Berkeley.
2. S. N. Atluri. On "hybrid" finite-element models in solid mechanics, in: *Advances in Computer Methods for Partial Differential Equations*. Ed. by R. Vichnevetsky, AICA, Rutgers University, 346–356, 1975.
3. F. Baaijens. A fictitious domain/mortar element method for fluid-structure interaction. *Int. J. Numer. Meths. Fluids*. **35**, 743–761, 2001.
4. F. Belgacem. The mortar finite element method with Lagrange multipliers. *Numer. Math.*. **84**, 173–197, 1999.
5. C. Bernardi, Y. Maday, and A. T. Patera. A new nonconforming approach to domain decomposition: the mortar element method. Technical report, Université Pierre at Marie Curie, Paris, France, 1990.
6. D. Braess, W. Dahmen and C. Wieners. A multigrid algorithm for the mortar finite element method. *SIAM Journal on Numerical Analysis*. **37**, 48–69, 2000.
7. F. Casadei, E. Gabellini, G. Fotia, F. Maggio and A. Quarteroni. A mortar spectral/finite element method for complex 2D and 3D elastodynamic problems. *Comp. Meth. Appl. Mech. Engrg.* **191**, 5119–5148, 2002.
8. C. Farhat and F.-X. Roux. Implicit parallel processing in structural mechanics, *Comput. Mech. Advances*, **2**, 1–124, 1994.
9. C. Farhat, M. Lesoinne and N. Maman. Mixed explicit/implicit time integration of coupled aeroelastic problems: three-field formulation, geometric conservation and distributed solution. *Int. J. Numer. Meth. Engrg.*, **21**, 807–835, 1995.
10. C. Farhat, S. Piperno and B. Larrouturu. Partitioned procedures for the transient solution of coupled aeroelastic problems; Part I: model problem, theory and two-dimensional application. *Comp. Meths. Appl. Mech. Engrg.*, **124**, 79–112, 1995.
11. C. Farhat, M. Lesoinne and P. LeTallec. Load and motion transfer algorithms for fluid/structure interaction problems with nonmatching discrete interfaces: Momentum and energy conservation, optimal discretization and application to aeroelasticity. *Comp. Meths. Appl. Mech. Engrg.*, **157**, 95–114, 1998.
12. V. Faucher and A. Combescure. A time and space mortar method for coupling linear modal subdomains and non-linear subdomains in explicit structural dynamics. *Comp. Meths. Appl. Mech. Engrg.*, textbf192, 509–533, 2003.
13. C. A. Felippa and K. C. Park. Staggered transient analysis procedures for coupled dynamic systems: formulation, *Comp. Meths. Appl. Mech. Engrg.*, **24**, 61–112, 1980.
14. C. A. Felippa and J. A. DeRuntz. Finite element analysis of shock-induced hull cavitation, *Comp. Meths. Appl. Mech. Engrg.*, **44**, 297–337, 1984.
15. C. A. Felippa and T. L. Geers. Partitioned analysis of coupled mechanical systems, *Engrg. Comput.*, **5**, 123–133, 1988.
16. C. A. Felippa, K. C. Park and C. Farhat. Partitioned analysis of coupled mechanical systems, *Comp. Meths. Appl. Mech. Engrg.*, **190**, 3247–3270, 2001.
17. C. A. Felippa and K. C. Park. Model-based partitioned analysis of coupled problems, chapter 4 in *Computational Aspects of Structural Dynamics and Vibrations*, ed. by G Sandberg and R. Ohayon, CISM Courses and Lectures, vol. 505, Springer-Verlag, Berlin, 2008, 171–216.
18. T. L. Geers. Residual potential and approximate methods for three-dimensional fluid-structure interaction, *J. Acoust. Soc. Am.*, **49**, 1505–1510, 1971.
19. T. L. Geers. Doubly asymptotic approximations for transient motions of general structures, *J. Acoust. Soc. Am.*, **64**, 1500–1508, 1978.
20. T. L. Geers. Boundary element methods for transient response analysis, in: Chapter 4 of *Computational Methods for Transient Analysis*, ed. by T. Belytschko and T. J. R. Hughes, North-Holland, Amsterdam, 221–244, 1983.

21. B. Herry, L. Di Valentin and A. Combescure. An approach to the connection between subdomains with nonmatching meshes for transient mechanical analysis. *Int. J. Numer. Meth. Engrg.*, **55**, 973–1003, 2002.
22. L. A. Jakobsen. A finite element approach to analysis and sensitivity analysis of time dependent fluid-structure interaction systems. Ph.D. Dissertation, Aalborg University, Denmark, 2002.
23. M. C. Junger and D. Feit, *Sound, Structures and Their Interaction*, MIT Press, Cambridge, Massachussets, 1972.
24. K. C. Park, C. A. Felippa and J. A. DeRuntz. Stabilization of staggered solution procedures for fluid-structure interaction analysis, in: *Computational Methods for Fluid-Structure Interaction Problems*, ed. by T. Belytschko and T. L. Geers, AMD Vol. 26, American Society of Mechanical Engineers, New York, 95–124, 1977.
25. K. C. Park. Partitioned transient analysis procedures for coupled-field problems: stability analysis, *J. Appl. Mech.*, **47**, 370–376, 1980.
26. K. C. Park and C. A. Felippa. Partitioned analysis of coupled systems, Chapter 3 in *Computational Methods for Transient Analysis*, T. Belytschko and T. J. R. Hughes, eds., North-Holland, Amsterdam–New York, 157–219, 1983.
27. K. C. Park and C. A. Felippa. A variational principle for the formulation of partitioned structural systems, *Int. J. Numer. Meth. Engrg.*, **47**, 395–418, 2000.
28. K. C. Park, C. A. Felippa and R. Ohayon. Partitioned formulation of internal fluid-structure interaction problems via localized Lagrange multipliers, *Comp. Meths. Appl. Mech. Engrg.*, **190**, 2989–3007, 2001.
29. K. C. Park, C. A. Felippa and G. Rebel, A simple algorithm for localized construction of non-matching structural interfaces, *Int. J. Numer. Meth. Engrg.*, **53**, 1261–1285, 2002.
30. K. C. Park, C. A. Felippa and G. Rebel. Interfacing nonmatching finite element discretizations: the Zero Moment Rule. In *Trends in Computational Mechanics*, ed. by W. A. Wall et. al., CIMNE, Barcelona, Spain, 355–367, 2001.
31. T. H. H. Pian and P. Tong. Basis of finite element methods for solid continua, *Int. J. Numer. Meth. Engrg.*, **1**, 3–29, 1969
32. S. Piperno and C. Farhat. Partitioned procedures for the transient solution of coupled aeroelastic problems - Part II: energy transfer analysis and three-dimensional applications. *Comp. Meth. Appl. Mech. Engrg.*, **190**, 3147–3170, 2001.
33. W. Prager. Variational principles for linear elastostatics for discontinous displacements, strains and stresses, in *Recent Progress in Applied Mechanics*, The Folke-Odgvist Volume, ed. by B. Broger, J. Hult and F. Niordson, Almqusit and Wiksell, Stockholm, 463–474, 1967.
34. C. R. Rao and S. K. Mitra. *Generalized Inverse of Matrices and its Applications*. Wiley, New York, 1971.
35. G. Rebel, K. C. Park, C. A. Felippa. A contact formulation based on localized Lagrange multipliers: formulation and applications to two-dimensional problems, *Int. J. Numer. Meth. Engrg.*, **54**, 263–297, 2002.
36. M. R. Ross. *Coupling and simulation of acoustic fluid-structure interaction systems using localized Lagrange multipliers*. PhD thesis, University of Colorado at Boulder, 2006. Available from http://caswww.colorado.edu/courses/FSI.d
37. M. R. Ross, C. A. Felippa, K. C. Park and M. A. Sprague. Treatment of acoustic fluid-structure interaction by localized Lagrange multipliers: Formulation, *Comp. Meths. Appl. Mech. Engrg.*, **197**, 2008, 3057–3079.
38. M. R. Ross, M. A. Sprague, C. A. Felippa and K. C. Park. Treatment of acoustic fluid-structure interaction by localized Lagrange multipliers and comparison to alternative interface coupling methods, *Comp. Meths. Appl. Mech. Engrg.*, **198**, 2009, 986–1005.
39. M. A. Sprague and T. L. Geers. A spectral-element method for modeling cavitation in transient fluid-structure interaction. *Int. J. Numer. Meth. Engrg.*. **60**, 2467–2499, 2004.
40. B. Wohlmuth. Hierarchical a posteriori error estimators for finite element methods. *SIAM Journal of Numerical Analysis*. **36**, 1636–1658, 1999.

Computer Modeling and Analysis of the Orion Spacecraft Parachutes

K. Takizawa, C. Moorman, S. Wright, and T.E. Tezduyar

Abstract We focus on fluid-structure interaction (FSI) modeling of the ringsail parachutes to be used with the Orion spacecraft. The geometric porosity of the ringsail parachutes with ring gaps and sail slits is one of the major computational challenges involved in FSI modeling. We address the computational challenges with the latest techniques developed by the Team for Advanced Flow Simulation and Modeling (T★AFSM) in conjunction with the Stabilized Space–Time Fluid–Structure Interaction (SSTFSI) technique. We investigate the performance of the three possible design configurations of the parachute canopy, carry out parametric studies on using an over-inflation control line (OICL) intended for enhancing the parachute performance, discuss rotational periodicity techniques for improving the geometric-porosity modeling and for computing good starting conditions for parachute clusters, and report results from preliminary FSI computations for parachute clusters. We also present a stability and accuracy analysis for the Deforming-Spatial-Domain/Stabilized Space–Time (DSD/SST) formulation, which is the core numerical technology of the SSTFSI technique.

1 Introduction

Various types of fluid–structure interaction (FSI) problems have been addressed and numerous FSI solution techniques have been developed in recent decades (see, for example, [2–11, 15, 17–33, 35, 38, 40, 42–45, 50, 54–57, 59–64, 66–78]). The Team for Advanced Flow Simulation and Modeling (T★AFSM) has addressed many of the challenges involved in FSI modeling of parachutes (see [20,35–37,39,41,45,57–59, 61,66]), with parallel, 3D computations going as far back as 2000. The core technology used in all this FSI modeling is the the Deforming-Spatial-Domain/Stabilized Space–Time (DSD/SST) formulation [46, 47, 51, 52], which was introduced as a

K. Takizawa, C. Moorman, S. Wright, and T.E. Tezduyar
Mechanical Engineering, Rice University, 6100 Main Street, Houston, TX 77005, USA
e-mail: tezduyar@rice.edu

H.-J. Bungartz et al. (eds.), *Fluid Structure Interaction II*, Lecture Notes
in Computational Science and Engineering 73, DOI 10.1007/978-3-642-14206-2_3,
© Springer-Verlag Berlin Heidelberg 2010

general-purpose interface-tracking (moving-mesh) technique for flows with moving boundaries and interfaces, including FSI. In early years of parachute modeling by the T⋆AFSM, a block-iterative FSI coupling technique [48] (see [54, 57] for the terminology) was used (see, for example, [20, 35, 36, 45]). The quasi-direct FSI coupling technique was introduced in [56, 57] and became part of the core technology used in the subsequent parachute FSI simulations of the T⋆AFSM (see, for example, [49, 57, 61, 62]). The stabilized space–time FSI (SSTFSI) technique was introduced in [54]. It is based on the new-generation DSD/SST formulations, which were also introduced in [54], increasing the scope and performance of the space–time FSI techniques developed earlier. The SSTFSI technique is now the core technology used in the parachute FSI computations of the T⋆AFSM (see, for example, [54, 58, 59]). A number of special FSI techniques were introduced in [54, 59] in conjunction with the SSTFSI technique and the DSD/SST formulation, including the the Homogenized Modeling of Geometric Porosity (HMGP). With the HMGP, we bypass the intractable complexities of the geometric porosity by approximating it with an "equivalent", locally-varying fabric porosity.

The HMGP and some of the other special FSI techniques were motivated by the task the T⋆AFSM has undertaken: computer modeling and analysis of the ring-sail parachutes to be used with the Orion spacecraft. In this paper we focus on the performance analysis of those parachutes, comparison of different canopy design configurations, parametric studies on using an over-inflation control line (OICL) considered for enhancing the parachute performance, and modeling of parachute clusters. We also describe the FSI technique we have introduced for more effective parachute modeling and analysis, such as techniques for building a consistent starting condition for the FSI computations and rotational-periodicity techniques for improving the HMGP and for computing good starting conditions for parachute clusters. We include a stability and accuracy analysis for the DSD/SST formulation.

2 Starting Condition

A consistent starting condition is essential for making accurate comparisons in many applications using FSI modeling. Starting conditions are especially important when investigating the transient response. A number of techniques for building FSI starting conditions are reported in [58, 66]. These techniques mostly focus on starting the FSI computations gently. The purpose of further influencing the starting condition with the methods introduced here is primarily related to making the starting conditions consistent and matching the physical conditions observed during NASA drop tests. To build an appropriate starting point for comparing performance of various parachute designs, we first analyzed the parachute drop test data and compared the test results to our earlier computations [66]. Based on this, we concluded that a fully inflated parachute with all sails behaves as follows: the parachute exhibits a periodic breathing motion caused by vortex shedding, and this dynamic nature results in a fluctuating descent speed. Furthermore, during drop tests for a given

Table 1 Computational result reported in [66]. V and T denote velocities and periods. Subscripts "D" and "RH" denote descent and relative horizontal. Subscripts "B" and "S" denote breathing and swinging. NA: Not applicable.

	V_D (ft/s)	V_{RH} (ft/s)	T_B (s)	T_S (s)
Computation	22.0	4 to 12	7.0	NA

range of altitudes, there is a corresponding wind direction, meaning the parachute sees a variable wind. In steady wind, however, the parachute settles to a nearly constant gliding direction after a few seconds. Moreover, when this is the case, the speed of the parachute relative to the surrounding air is the dominant consideration for the flow condition, and any lateral force the parachute sees from the wind is only a function of the relative speed. The computation reported in [66] does not include any side wind, but the parachute exhibits a comparable relative velocity due to the parachute gliding. Table 1 provides the earlier computations reported in [66], which are in close agreement with the drop test data.

Based on the argument above, we infer that the main influence of a side wind is inducing a payload swing. To induce a payload swing in our computations however, instead of adding a variable wind to our model, we introduce an instantaneous "payload swing" effect by giving the payload a one-time horizontal-velocity hike.

We now provide more details on the sequence for arriving at a starting point for FSI computations. After 100 s of "symmetric FSI", which is described in [66], the parachute reaches a settled periodic ("breathing") stage, which we deem to be consistent between the various parachute designs computed. The payload and the parachute have no horizontal speed at the end of the "symmetric FSI" step, which does not match what is observed in the tests. To imitate the naturally-occurring swinging motion, we instantaneously hike the horizontal speed of the payload to 20 ft/s. Simultaneously, we begin the de-symmetrization (see [66]) using a Cosine form which lasts for one breathing period (7 s). Although the vortex shedding pattern behind the parachute is not exactly the same in each computation, the momentum added to the payload with this velocity hike is consistent. We believe imitating the payload momentum in this fashion represents this aspect of the starting condition with a reasonable closeness to the actual conditions.

3 Various Canopy Configurations

The steady descent of a ringsail parachute is dynamic. Through this investigation we seek to reduce oscillations in the descent speed of the payload and reduce the horizontal gliding of a single parachute. The oscillations in descent speed have a direct impact on the maximum descent speed constraint, and the horizontal gliding of the parachute has cluster performance implications. Adjusting the geometric porosity by reconfiguring the canopy can impact both stability and drag performance of the parachute. The baseline and two alternate canopy configurations are investigated. The two alternate configurations are "missing" the 5th and 11th sail, respectively,

Table 2 Number of nodes and elements for each of the three parachute configurations. Here nn and ne are number of nodes and elements, respectively. The structural mechanics mesh consists of four-node quadrilateral membrane elements, two-node cable elements and one-node payload element. The structure interface mesh consists of four-node quadrilateral elements. The fluid volume mesh consists of four-node tetrahedral elements, while the fluid interface mesh consists of three-node triangular elements.

		PA	PM5	PM11
Structure				
	nn	30,722	28,642	28,082
Membrane	ne	26,000	24,080	23,600
Cable	ne	12,521	11,401	12,121
Payload	ne	1	1	1
Interface	nn	29,200	27,120	26,560
	ne	26,000	24,080	23,600
Fluid				
Volume	nn	178,270	192,412	180,917
	ne	1,101,643	1,192,488	1,119,142
Interface	nn	2,140	2,060	2,060
	ne	4,180	3,860	3,860

when numbered starting from the top including the rings. We will refer to the three parachutes as "PA" (all sails are in place), "PM5" (missing the 5th sail) and "PM11" (missing the 11th sail).

3.1 Computational Conditions

The computational conditions are the same as those described in [66]. The number of nodes and elements are given in Table 2. The payload weight is approximately 5,570 lbs. The total weights, including the parachute for each of the three configurations: PA, PM5 and PM11, are approximately 5,725 lbs, 5,720 lbs and 5,715 lbs, respectively. Fig. 1 shows the parachute shape and structure mesh for each configuration. The homogenized porosity distribution used here is based on the computation reported earlier by the T★AFSM, which is given in Table 1 in [66]. The porosity at the edges facing a missing sail is calculated in the same way as the porosity at the edges of the vent and skirt. The porosity coefficient for the edge nodes is set to the fabric porosity, linearly progressing to the homogenized value for the adjacent patch. Figure 2 shows the porosity distribution for each of the three cases.

3.2 Results

An important dynamical feature in parachute flight is the breathing in steady descent. The flow separates near the skirt of the parachute and a ring vortex forms.

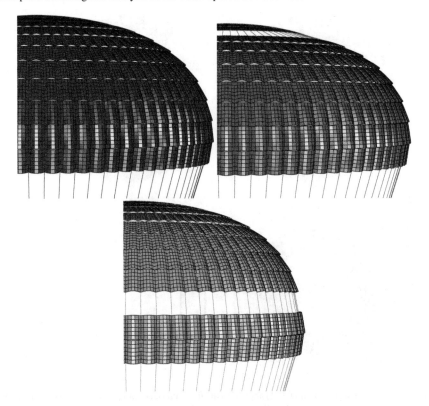

Fig. 1 Parachute shape and structure mesh for PA, PM5 and PM11, shown from left to right and top to bottom.

The vortex creates a large pressure differential near the skirt of the parachute at maximum inflation. The vortex then moves upward causing the maximum pressure differential to move from the skirt to the crown. Figure 3 shows the parachute and flow field for each configuration when the tilt angle, as measured between the vertical axis and a line connecting the payload and vent, is at a maximum. Table 3 provides the results for the three parachute configurations.

Figure 4 shows the payload descent speed and skirt diameter of the three parachutes. The descent speed is directly dependent on the diameter because drag production is dictated by the projected area of the canopy. It can be seen from the plots that the PA and PM5 have similar maximum diameters. Though the diameters are close, PM5 has a slightly higher average descent speed due to the decreased projected area with a missing sail. The similar maximum diameter is due to the fact that the pressurization of the lower sails for PA and PM5 are unaffected by the modification. The PM11 parachute drag performance is hindered largely by a lack of projected area due to the loss of pressurization of the bottom three sails. Figure 5

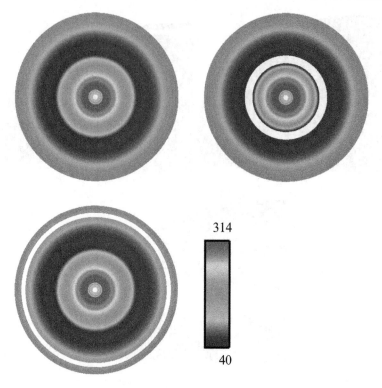

314

40

Fig. 2 Porosity distribution (in CFM) applied to the fluid interface for the three different parachute geometries, "PA", "PM5" and "PM11", from left to right and top to bottom.

shows the horizontal-velocity magnitudes for the three configurations. In this case we define improved static stability as having lower gliding speed. PA and PM5 have larger gliding speeds while the glide speed of PM11 is very low. The static stability of PM11 is much improved over that of PM5 and PA. This may have implications in parachute cluster applications. The horizontal-velocity magnitude for the PM11 payload exhibits sharp kinks near 0 ft/s, indicating that the horizontal velocity is reversing direction. This reversal happens when, in a swinging cycle, the tilt angle is maximum. On the other hand, PA and PM5 payloads are moving on an elliptical trajectory with respect to the vent, as clearly seen in Fig. 6, and therefore do not exhibit sharp kinks. Overall, the study shows that the more stable configurations exhibit a loss of drag. However, it shows that geometric porosity can influence the descent characteristics. Therefore a parachute's performance might be tuned by readjusting the geometric porosity.

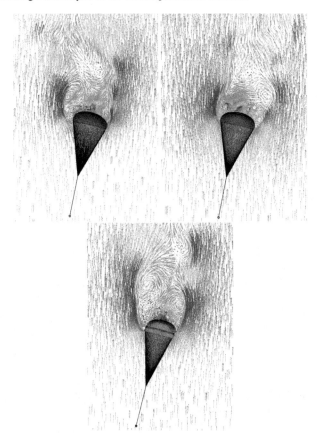

Fig. 3 Parachute and flow field for PA. PM5 and PM11 when the tilt angle is at a maximum.

Table 3 Computational results for the three parachute configurations.

	V_D (ft/s)	V_{RH} (ft/s)	T_B (s)	T_S (s)
PA	21.4	4 to 13	6.7	16.4
PM5	24.0	4 to 13	5.8	16.6
PM11	29.0	0 to 4	NA	17.0

4 Over-Inflation Control Line (OICL)

From drop tests and computations, we observe that the change in projected area due to canopy breathing causes large fluctuations in descent speed. We test several over inflation control lines (OICL) attached at the canopy skirt, which limit inflation. Figure 7 shows the configuration of the OICL.

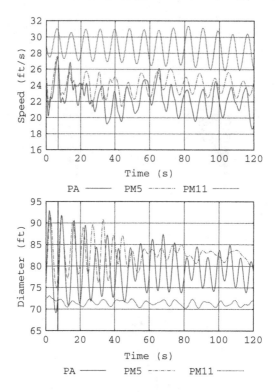

Fig. 4 Payload descent speed and parachute skirt diameter for the three configurations. The thin vertical line at 7 s marks the end of the de-symmetrization.

4.1 Computational Conditions

The computational conditions are the same as those described in [66]. The number of nodes and elements are given in Table 2. The PA canopy configuration is used for this study. We present results for 76, 80 and 84 ft OICL cases. In the case of an OICL smaller than the starting skirt diameter, for example 76 ft, we first compute symmetric FSI with a zero-stiffness OICL. Then we turn on the stiffness when the diameter is at a minimum. In this way we avoid the sudden changes of the internal force balance for the structure.

4.2 Results

Figure 8 shows the parachute and flow field for PA and OICL[1], each at maximum diameter. The descent speed and diameter are shown in Fig. 9. The average inflated

---

[1] The acronym OICL is used also to indicate "PA with OICL"

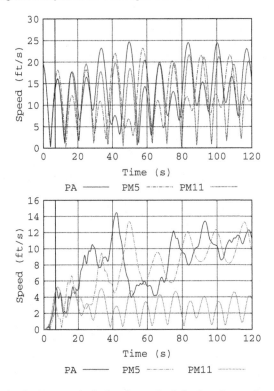

Fig. 5 Horizontal-velocity magnitude for the payload (top) and vent (bottom) for the three parachute configurations. The thin vertical line at 7 s marks the end of the de-symmetrization.

diameter of the unrestricted canopy is 79 ft. We demonstrate that an OICL smaller than the average inflated diameter causes breathing to cease completely. Meanwhile, breathing continues in a damped fashion if the OICL is larger than the average diameter. The descent speed illustrated in Fig. 9 is the superposition of several parachute-dynamics factors. The first one is force oscillations produced by the varying area of the canopy, which leads to acceleration of the entire system. The second one is the geometric coupling between the payload and parachute. As the parachute expands in diameter, the payload moves toward the canopy and vice versa. The third, and the dominant, one is the payload swing.

Figure 10 shows the horizontal-velocity magnitudes for the three OICL cases. The magnitude of the canopy horizontal velocity suggests that the OICL has little impact on the static stability of the parachute. Figure 11 shows the axial force and projected area. It can be seen that the OICL smooths the oscillations in projected area and hence the axial force. The OICL accomplishes its goal of damping the breathing oscillations that cause descent speed spikes, which are undesirable in terms of the maximum descent speed constraint.

We propose that the descent speed of the payload be examined using a different approach. The goal of this effort was to reduce the oscillations in descent speed due

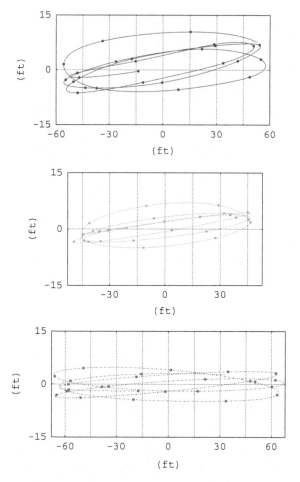

Fig. 6 Payload trajectories (relative to the vent) for PA, PM5 and PM11. Lines are drawn from 60 to 120 s. Dots are placed every 2.3 s. We note that the y scale is stretched to twice that of the x scale.

to breathing. This has been accomplished, but the dominant swinging factor over-whelms the descent speed data. We suggest attempting to separate out the dynamical factors that contribute to the overall payload descent speed to provide a more clear metric for comparison of the parachutes. One has to be careful in trying to choose an optimum value for the OIC, because a shorter OICL reduces average drag as well as descent speed peaks. Reducing the average drag too much would result in a descent speed with little or no oscillation but a maximum descent speed that exceeds the oscillatory peaks of the less constrained cases. Future analysis of the descent data will provide more insight into the optimum length of the OICL.

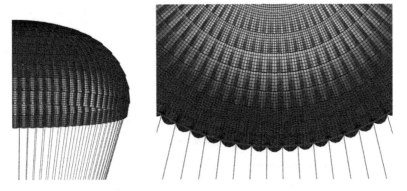

Fig. 7 OICL (red) is attached to the suspension lines. Views from outside (left) and inside (right).

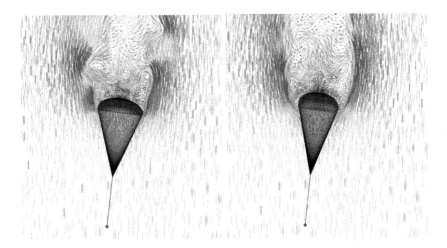

Fig. 8 Parachute and flow field for PA and OICL (76 ft).

5 Rotational Periodicity

In general, periodic domains are useful for reducing computational cost. We benefit from that by applying rotational periodicity to a highly refined parachute model.
In the case of parachute clusters, we also benefit from another aspect of periodic
domains, namely, their ability to stabilize the computation prior to full modeling
by enforcing solution periodicity. We already use similar constraints to stabilize
pre-FSI parachute computations. For example, we are using symmetric FSI to generate a good starting condition. A symmetric FSI computation is more likely to
remain stable because of the symmetric deformation of the parachute, even if we
apply a local artificial perturbation, such as the instantaneous horizontal velocity
hike described in Sect. 2. By imposing a similar stabilizing constraint, periodic
domains are useful for pre-FSI computations.

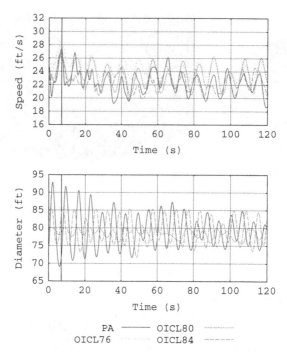

Fig. 9 Payload descent speed and parachute skirt diameter for PA and the three OICL cases. The thin vertical line at 7 s marks the end of the de-symmetrization.

5.1 Periodic n-gore Model

Here our goal is to carry out a fluid mechanics computation with the actual geometric porosity. We use a periodic domain to reduce the computational cost of such high refinement. We realize that the parachute wake and deformation are completely asymmetric. Moreover, the parachutes glides horizontally. Therefore, a periodic assumption is not suitable in terms of global quantities, such as descent speed and drag coefficient. However, it is meaningful when we focus on local phenomena such as flow through the parachute gaps and slits. It is also helpful to improve or verify the HMGP (see [58, 59, 65]). We can afford much higher resolution near the surface because of the limited domain. Figure 12 shows a comparison of the mesh resolution between a HMGP mesh and a periodic four-gore model. In previous computations, slip conditions were applied on the boundaries intersecting the canopy. Here we employ rotational periodicity on those boundaries. One boundary is a master boundary and the other is a slave boundary. The nodes on the slave boundary do not contain any unknowns. Because of this, we changed the mesh generation strategy.

First we extract one gore of the canopy. Figure 13 shows an extracted one-gore surface mesh. Using this surface mesh, we generate a volume mesh with a pie-slice-shaped domain, which is only a 4.5° angle. Then, we repeat the one-gore model n

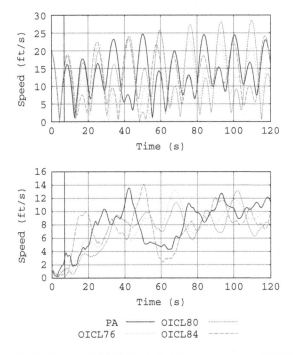

Fig. 10 Horizontal-velocity magnitude for the payload (top) and canopy (bottom) for PA and the three OICL cases. The thin vertical line at 7 s marks the end of the de-symmetrization.

times and merge them as an n-gore model as shown in Fig. 14. A one-gore model cannot be used for computations because the volume mesh may have an element consisting of both master and slave nodes. By merging n one-gore models, the mesh will have at least n elements between the master and slave boundaries.

Figure 15 shows the flow around the rings, which is unsteady. Figure 16 shows the flow around the sails. We observe that some of the jets exit the slits in a downward direction. This means that the momentum exchange between the surrounding air and the canopy is not in the same direction as that represented by the smoothed-surface normal-vector in HMGP. Thus, HMGP should be improved by incorporating the directional nature of geometric porosity, as shown in the periodic n-gore results.

5.2 Periodic FSI

A cluster of three main parachutes will be used to recover the Orion spacecraft. It is necessary to compute this cluster in FSI since there is no direct correlation between single main and cluster performance. Again, starting conditions for FSI computations are very important, especially for highly dynamic problems such as a parachute cluster. We generate a single main parachute mesh at an appropriate coning angle

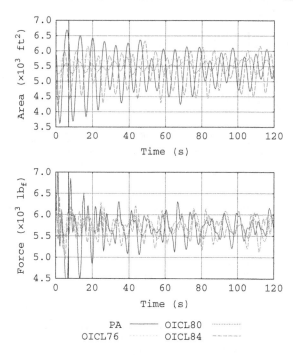

Fig. 11 Projected area (top) and axial force (bottom) for PA and the three OICL cases. The thin vertical line at 7 s marks the end of the de-symmetrization.

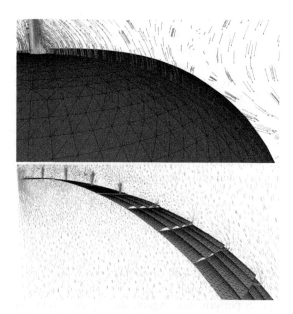

Fig. 12 Mesh resolution comparison. Meshes for the HMGP (top) and four-gore model (bottom).

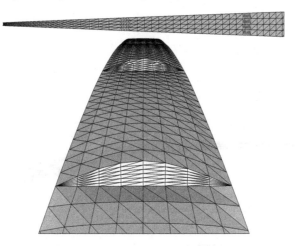

Fig. 13 One-gore surface mesh around the rings (top) and sails (bottom). The surface mesh is shaded and the mesh across the ring gaps and sail slits is unshaded.

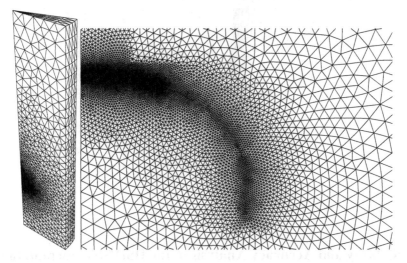

Fig. 14 Four-gore domain (left) and the enlarged view of the periodic boundary around the canopy (right).

in a one-third domain as shown in Fig. 17. We employ rotational periodicity on the fluid dynamics and limit the payload motion to the vertical direction. Figure 18 shows the flow patterns and parachute deformation for this periodic cluster FSI. We merge three one-third periodic domains to form a full-domain solution, which is a good starting condition for full FSI. Figure 19 shows the result from a preliminary full FSI computation.

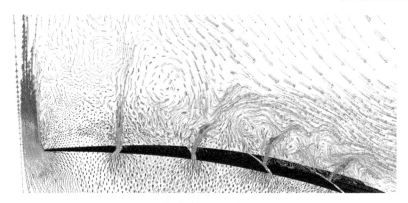

Fig. 15 Periodic four-gore model. Flow around the rings.

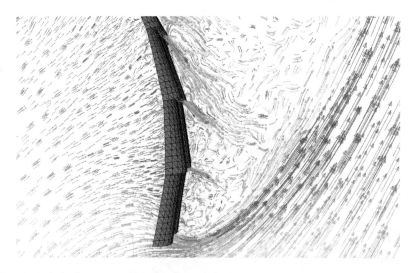

Fig. 16 Periodic four-gore model. Flow around the sails.

6 Stability and Accuracy Analysis of the DSD/SST Formulation

6.1 Space–Time Shape Functions

A space–time shape function can be written as a product of its spatial and temporal parts:

$$N_a^\alpha = T^\alpha(\theta) \, N_a(\boldsymbol{\xi}), \quad a = 1, 2, \ldots, n_{en}, \quad \alpha = 1, 2, \ldots, n_{ent}, \tag{1}$$

where $\boldsymbol{\xi}$ and θ are the spatial and temporal element coordinates, and n_{en} and n_{ent} are the numbers of spatial and temporal nodes. More details can be found in [57].

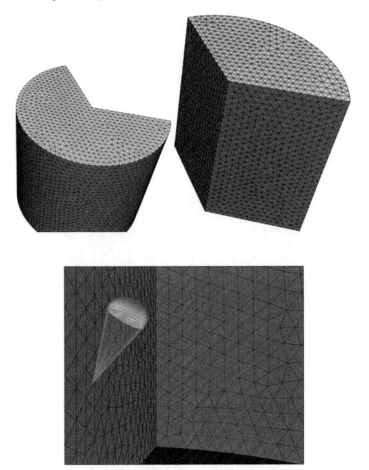

Fig. 17 One-third domain (top-right) and the rest of the whole domain (top-left), which is not used in the computation. The single main parachute at specified coning angle in the one-third domain (bottom).

Both the spatial and temporal shape functions can be either classical finite element or NURBS shape functions. NURBS shape functions can be found in [3, 14].

6.2 Temporal Shape Functions

For $n_{ent} = 2$, both the finite element and B-spline shape functions are the linear kind. Figure 20 shows the temporal shape functions for $n_{ent} = 3$. The quadratic shape functions can be written as

$$T^1(\theta) = -\frac{1}{2}\theta(1-\theta), \quad T^2(\theta) = 1 - \theta^2, \quad T^3(\theta) = \frac{1}{2}\theta(1+\theta), \quad (2)$$

Fig. 18 Periodic FSI computation of cluster using a one-third domain. Pressure on the cutting plane and velocity vectors colored by magnitude.

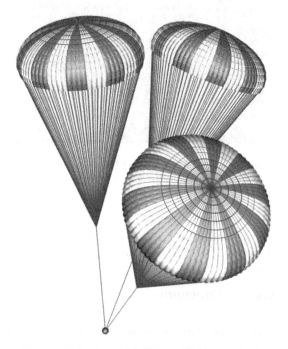

Fig. 19 A cluster of three main parachutes. Preliminary FSI computation.

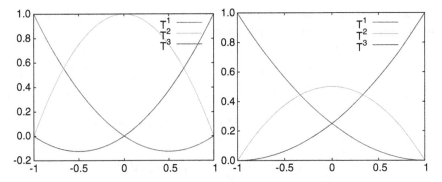

Fig. 20 Quadratic temporal shape functions. Finite element functions (left) and B-splines (right).

and the quadratic B-spline shape functions can be written as

$$T^1(\theta) = \frac{(1-\theta)^2}{4}, \quad T^2(\theta) = \frac{1-\theta^2}{2}, \quad T^3(\theta) = \frac{(1+\theta)^2}{4}. \qquad (3)$$

Since the quadratic shape functions can be represented by the quadratic B-splines, we focus on the B-splines.

6.3 Advection Equation

Our stability and accuracy analysis is based on the advection equation:

$$\frac{\partial \varphi}{\partial t} + \mathbf{u} \cdot \nabla \varphi = 0, \qquad (4)$$

where φ is a scalar and $\mathbf{u}$ is a constant advection velocity. The DSD/SST formulation of Eq. (4) is written as follows:

$$\int_{Q_n} w^h \left(\frac{\partial \varphi^h}{\partial t} + \mathbf{u}^h \cdot \nabla \varphi^h \right) dQ + \int_{\Omega_n} (w^h)_n^+ \left((\varphi^h)_n^+ - (\varphi^h)_n^- \right) d\Omega$$

$$+ \sum_{e=1}^{(n_{el})_n} \int_{Q_n^e} \tau_{\text{SUPG}} \left(\frac{\partial w^h}{\partial t} + \mathbf{u}^h \cdot \nabla w^h \right) \left(\frac{\partial \varphi^h}{\partial t} + \mathbf{u}^h \cdot \nabla \varphi^h \right) dQ = 0, \qquad (5)$$

where w^h is the test function, Q_n is the slice of the space–time domain between the time levels t_n and t_{n+1}, the notation $(\cdot)_n^-$ and $(\cdot)_n^+$ denotes the values as t_n is approached from below and above respectively, $(n_{el})_n$ is the number of elements at time level t_n, and the stabilization parameter

$$\tau_{\text{SUPG}} = \tau_{\text{SUGN12}} = \left(\sum_{\alpha=1}^{n_{ent}} \sum_{a=1}^{n_{en}} \left| \left(\frac{\partial T^{\alpha}}{\partial t} \bigg|_{\boldsymbol{\xi}} N_a + T^{\alpha} \left(\mathbf{u} - \mathbf{v} \right) \cdot \boldsymbol{\nabla} N_a \right) \right| \right)^{-1}. \tag{6}$$

Here $\mathbf{v} \equiv \frac{d\mathbf{x}}{dt}$ is the mesh velocity. The SUPG test function can also be written as

$$\frac{\partial \mathbf{w}^h}{\partial t} + \mathbf{u} \cdot \boldsymbol{\nabla} \mathbf{w}^h = \frac{\partial \mathbf{w}^h}{\partial t} \bigg|_{\boldsymbol{\xi}} + (\mathbf{u} - \mathbf{v}) \cdot \boldsymbol{\nabla} \mathbf{w}^h. \tag{7}$$

The option where the $\frac{\partial \mathbf{w}^h}{\partial t}\big|_{\boldsymbol{\xi}}$ term is excluded was called "WTSE" in Remark 2 of [54], and the option where the $\frac{\partial \mathbf{w}^h}{\partial t}\big|_{\boldsymbol{\xi}}$ term is active "WTSA".

Remark 1 *The τ definition given by Eq. (6) is coming from [47], which is the space–time version of the original definition given in [53]. These definitions sense, in addition to the element geometry, the order of the interpolation functions. Some τ definitions do that and some do not. The definitions given in Sect. 3.3.1 and 3.3.2 of [34], for example, are among those that do not.*

Remark 2 *Remark 1 is applicable also when the interpolation functions are NURBS functions. This includes classical p-refinement and also k-refinement, except when used in conjunction with periodic B-splines.*

6.3.1 Fourier Analysis for One-dimensional Uniform Mesh

A stability and accuracy analysis for the case of exact time-integration and no stabilization was given in [1] for B-splines. In the analysis we present here, we use different temporal and spatial shape function combinations at different CFL values, where CFL $= \frac{c\Delta t}{h}$. Here, h is the element length, Δt is the time step size, and $c = u - v$ is the relative velocity. The element length is equivalent to the distance between the control points for periodic B-splines. Therefore the element length is directly related to the number of unknowns. We test three temporal shape functions: "SV", reported in [54], linear functions and quadratic B-splines, with three spatial shape functions: linear functions and C^1 and C^2 B-splines. Figures 21–28 show, at different CFL values, the phase error, $|1 - \tilde{\omega}/\omega|$, and the algorithmic damping ratio, $\tilde{\xi}/\tilde{\omega}$, as function of the dimensionless wave number kh. More details on this type of analysis can be found in [12, 13, 16]. We note that this type of analysis can only measure phase error within $\pm\pi$.

Remark 3 *Although not shown here, except for the SV option, the space–time formulation involves non-zero algorithmic damping even for the Galerkin method.*

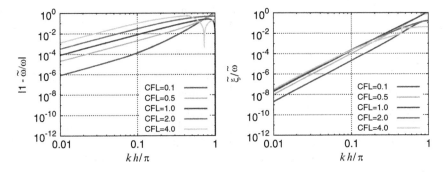

Fig. 21 Phase error (left) and algorithmic damping ratio (right). Linear functions in space and SV.

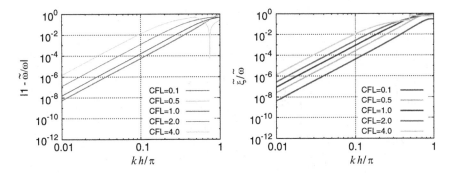

Fig. 22 Phase error (left) and algorithmic damping ratio (right). Linear functions in space and time.

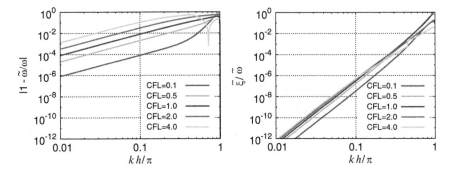

Fig. 23 Phase error (left) and algorithmic damping ratio (right). C^1 B-splines in space and SV.

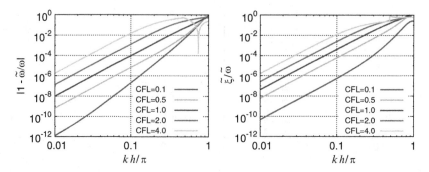

Fig. 24 Phase error (left) and algorithmic damping ratio (right). C^1 B-splines in space and linear functions in time.

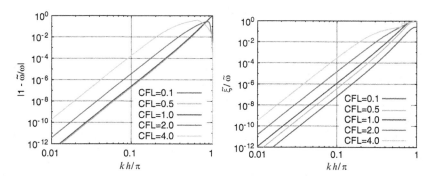

Fig. 25 Phase error (left) and algorithmic damping ratio (right). C^1 B-splines in space and quadratic B-splines in time.

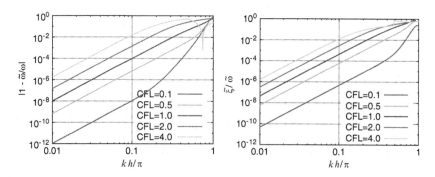

Fig. 26 Phase error (left) and algorithmic damping ratio (right). C^2 B-splines in space and linear functions in time.

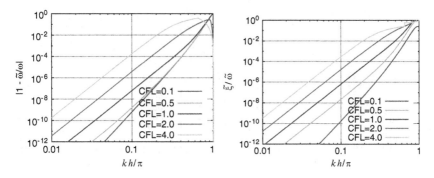

Fig. 27 Phase error (left) and algorithmic damping ratio (right). C^2 B-splines in space and quadratic B-splines in time.

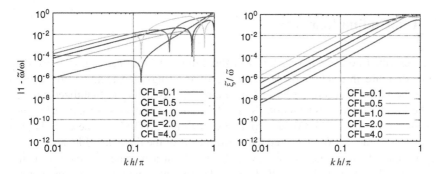

Fig. 28 Phase error (left) and algorithmic damping ratio (right). Linear functions in space and time with the WTSE option.

Remark 4 *Linear functions and C^1 B-splines in space are potentially fourth-order and sixth-order accurate, respectively. But we quickly lose the higher-order accuracy at higher CFL numbers, unless we also use higher-order interpolation in time. For example, SV with C^1 B-spline in space is basically second-order accurate when $CFL > 0.1$, which is what we already have when we use SV with just linear functions in space.*

Remark 5 *Linear functions in space and time with the WTSE option is second-order accurate, as can be seen in Fig. 28. This is consistent with what was reported in [12] using a stabilization equivalent to the WTSE option. The WTSE option is consistently lower-order accurate compared to the WTSA option, except for with linear functions in space, in which case the accuracies are comparable but the phase error directions are different in some regions, as can be seen in Fig. 29.*

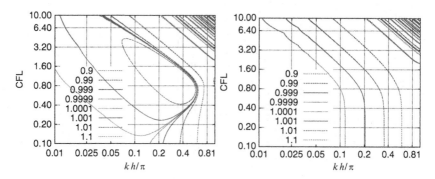

Fig. 29   Frequency ratio $\tilde{\omega}/\omega$ as function of the dimensionless wave number and CFL. Values larger than one indicate that the numerical solution propagates faster than the exact one. Linear functions in space and SV with the WTSE (left) and WTSA (right) options.

7   Concluding Remarks

We described how we have addressed the computational challenges involved in fluid-structure interaction (FSI) modeling of the ringsail parachutes to be used with the Orion spacecraft. Our computations are based on the Stabilized Space–Time Fluid–Structure Interaction (SSTFSI) technique and some recent techniques developed by the T⋆AFSM in conjunction with the SSTFSI. We described some of those techniques. We analyzed the performance of the three possible design configurations of the parachute canopy and carried out a parametric study on using an over-inflation control line (OICL) intended for enhancing the parachute performance. We showed that we have a good computational capability for evaluating parachute design and performance. We compared several aspects of our parachute FSI computations and NASA drop test data, and they are in reasonably good agreement. We described how we can use rotational periodicity techniques for improving the geometric-porosity modeling and for computing a good starting condition for FSI computation of parachute clusters. We presented results from preliminary FSI computations for a cluster of three parachutes. The core numerical technology used in the SSTFSI technique is the Deforming-Spatial-Domain/Stabilized Space–Time (DSD/SST) formulation, which was developed by the T⋆AFSM earlier for flow problems with moving boundaries and interfaces, including FSI problems. We presented a stability and accuracy analysis for various options of the DSD/SST formulation, explaining the interplay between using higher-order interpolation in space and higher-order integration in time.

Acknowledgements   This work was supported in part by NASA Johnson Space Center under grants NNJ06HG84G and NNX09AM89G. It was also supported in part by the Rice Computational Research Cluster funded by NSF under Grant CNS-0821727. We thank Jason Christopher for his contributions at various stages of the parachute computations. We also thank Joel Martin for carrying out some of the PM5 computations.

References

1. Y. Bazilevs, V. M. Calo, J. A. Cottrel, T. J. R. Hughes, A. Reali, and G. Scovazzi. Variational multiscale residual-based turbulence modeling for large eddy simulation of incompressible flows. *Computer Methods in Applied Mechanics and Engineering*, 197:173–201, 2007.
2. Y. Bazilevs, V. M. Calo, T. J. R. Hughes, and Y. Zhang. Isogeometric fluid–structure interaction: theory, algorithms, and computations. *Computational Mechanics*, 43:3–37, 2008.
3. Y. Bazilevs, V. M. Calo, Y. Zhang, and T. J. R. Hughes. Isogeometric fluid–structure interaction analysis with applications to arterial blood flow. *Computational Mechanics*, 38:310–322, 2006.
4. Y. Bazilevs, J. R. Gohean, T. J. R. Hughes, R. D. Moser, and Y. Zhang. Patient-specific isogeometric fluid–structure interaction analysis of thoracic aortic blood flow due to implantation of the Jarvik 2000 left ventricular assist device. *Comp. Meth. in Appl. Mech. and Engng.*, 198:3534–3550, 2009.
5. Y. Bazilevs, Ming-Chen Hsu, D. Benson, S. Sankaran, and A. Marsden. Computational fluid–structure interaction: Methods and application to a total cavopulmonary connection. *Computational Mechanics*, 45:77–89, 2009.
6. Y. Bazilevs, Ming-Chen Hsu, Y. Zhang, W. Wang, X. Liang, T. Kvamsdal, R. Brekken, and J. Isaksen. A fully-coupled fluid–structure interaction simulation of cerebral aneurysms. *Computational Mechanics*, 2009. Published online. DOI:10.1007/s00466-009-0421-4.
7. Kai-Uwe Bletzinger, R. Wüchner, and A. Kupzok. Algorithmic treatment of shells and free form-membranes in FSI. In H.-J. Bungartz and M. Schäfer, editors, *Fluid–Structure Interaction*, volume 53 of *Lecture Notes in Computational Science and Engineering*, pages 336–355. Springer, 2006.
8. M. Brenk, H.-J. Bungartz, M. Mehl, and T. Neckel. Fluid–structure interaction on Cartesian grids: Flow simulation and coupling environment. In H.-J. Bungartz and M. Schäfer, editors, *Fluid–Structure Interaction*, volume 53 of *Lecture Notes in Computational Science and Engineering*, pages 233–269. Springer, 2006.
9. W. Dettmer and D. Peric. A computational framework for fluid-structure interaction: Finite element formulation and applications. *Comp. Meth. in Appl. Mech. and Engng.*, 195:5754–5779, 2006.
10. W. G. Dettmer and D. Peric. On the coupling between fluid flow and mesh motion in the modelling of fluid–structure interaction. *Computational Mechanics*, 43:81–90, 2008.
11. Jean-Frederic Gerbeau, M. Vidrascu, and P. Frey. Fluid–structure interaction in blood flow on geometries based on medical images. *Computers and Structures*, 83:155–165, 2005.
12. G. Hauke and M. H. Doweidar. Fourier analysis of semi-discrete and space–time stabilized methods for the advective-diffusive-reactive equation: I. SUPG. *Computer Methods in Applied Mechanics and Engineering*, 194:45–81, 2005.
13. T. J. R Hughes. *The Finite Element Method. Linear Static and Dynamic Finite Element Analysis*. Prentice-Hall, Englewood Cliffs, New Jersey, 1987.
14. T. J. R. Hughes, J. A. Cottrell, and Y. Bazilevs. Isogeometric analysis: CAD, finite elements, NURBS, exact geometry, and mesh refinement. *Comp. Meth. in Appl. Mech. and Engng.*, 194:4135–4195, 2005.
15. T. J. R Hughes, W. K. Liu, and T. K. Zimmermann. Lagrangian–Eulerian finite element formulation for incompressible viscous flows. *Comp. Meth. in Appl. Mech. and Engng.*, 29:329–349, 1981.
16. T. J. R Hughes and T. E. Tezduyar. Finite element methods for first-order hyperbolic systems with particular emphasis on the compressible Euler equations. *Comp. Meth. in Appl. Mech. and Engng.*, 45:217–284, 1984.
17. J. G. Isaksen, Y. Bazilevs, T. Kvamsdal, Y. Zhang, J. H. Kaspersen, K. Waterloo, B. Romner, and T. Ingebrigtsen. Determination of wall tension in cerebral artery aneurysms by numerical simulation. *Stroke*, 39:3172–3178, 2008.
18. A. A. Johnson and T. E. Tezduyar. Parallel computation of incompressible flows with complex geometries. *International Journal for Numerical Methods in Fluids*, 24:1321–1340, 1997.

19. A. A. Johnson and T. E. Tezduyar. Advanced mesh generation and update methods for 3D flow simulations. *Computational Mechanics*, 23:130–143, 1999.
20. V. Kalro and T. E. Tezduyar. A parallel 3D computational method for fluid–structure interactions in parachute systems. *Comp. Meth. in Appl. Mech. and Engng.*, 190:321–332, 2000.
21. R. A. Khurram and A. Masud. A multiscale/stabilized formulation of the incompressible Navier–Stokes equations for moving boundary flows and fluid–structure interaction. *Computational Mechanics*, 38:403–416, 2006.
22. U. Küttler, C. Förster, and W. A. Wall. A solution for the incompressibility dilemma in partitioned fluid–structure interaction with pure Dirichlet fluid domains. *Computational Mechanics*, 38:417–429, 2006.
23. U. Küttler and W. A. Wall. Fixed-point fluid–structure interaction solvers with dynamic relaxation. *Computational Mechanics*, 43:61–72, 2008.
24. R. Löhner, J. R. Cebral, C. Yang, J. D. Baum, E. L. Mestreau, and O. Soto. Extending the range of applicability of the loose coupling approach for FSI simulations. In H.-J. Bungartz and M. Schäfer, editors, *Fluid–Structure Interaction*, volume 53 of *Lecture Notes in Computational Science and Engineering*, pages 82–100. Springer, 2006.
25. M. Manguoglu, A. H. Sameh, F. Saied, T. E. Tezduyar, and S. Sathe. Preconditioning techniques for nonsymmetric linear systems in computation of incompressible flows. *Journal of Applied Mechanics*, 76:021204, 2009.
26. M. Manguoglu, A. H. Sameh, T. E. Tezduyar, and S. Sathe. A nested iterative scheme for computation of incompressible flows in long domains. *Computational Mechanics*, 43:73–80, 2008.
27. M. Manguoglu, K. Takizawa, A. H. Sameh, and T. E. Tezduyar. Solution of linear systems in arterial fluid mechanics computations with boundary layer mesh refinement. *Computational Mechanics*, published online, DOI: 10.1007/s00466-009-0426-z, October 2009.
28. A. Masud, M. Bhanabhagvanwala, and R. A. Khurram. An adaptive mesh rezoning scheme for moving boundary flows and fluid–structure interaction. *Computers & Fluids*, 36:77–91, 2007.
29. C. Michler, E. H. van Brummelen, and R. de Borst. An interface Newton–Krylov solver for fluid–structure interaction. *International Journal for Numerical Methods in Fluids*, 47:1189–1195, 2005.
30. S. Mittal and T. E. Tezduyar. Massively parallel finite element computation of incompressible flows involving fluid-body interactions. *Comp. Meth. in Appl. Mech. and Engng.*, 112:253–282, 1994.
31. S. Mittal and T. E. Tezduyar. Parallel finite element simulation of 3D incompressible flows – Fluid-structure interactions. *International Journal for Numerical Methods in Fluids*, 21:933–953, 1995.
32. R. Ohayon. Reduced symmetric models for modal analysis of internal structural-acoustic and hydroelastic-sloshing systems. *Comp. Meth. in Appl. Mech. and Engng.*, 190:3009–3019, 2001.
33. T. Sawada and T. Hisada. Fuid–structure interaction analysis of the two dimensional flag-in-wind problem by an interface tracking ALE finite element method. *Computers & Fluids*, 36:136–146, 2007.
34. F. Shakib, T. J. R. Hughes, and Z. Johan. A new finite element formulation for computational fluid dynamics: X. The compressible euler and navier-stokes equations. *Comput. Methods Appl. Mech. and Engrg.*, 89:141–219, 1991.
35. K. Stein, R. Benney, V. Kalro, T. E. Tezduyar, J. Leonard, and M. Accorsi. Parachute fluid–structure interactions: 3-D Computation. *Comp. Meth. in Appl. Mech. and Engng.*, 190:373–386, 2000.
36. K. Stein, R. Benney, T. Tezduyar, and J. Potvin. Fluid–structure interactions of a cross parachute: Numerical simulation. *Comp. Meth. in Appl. Mech. and Engng.*, 191:673–687, 2001.
37. K. Stein, T. Tezduyar, and R. Benney. Computational methods for modeling parachute systems. *Computing in Science and Engineering*, 5:39–46, 2003.
38. K. Stein, T. Tezduyar, and R. Benney. Mesh moving techniques for fluid–structure interactions with large displacements. *Journal of Applied Mechanics*, 70:58–63, 2003.

39. K. Stein, T. Tezduyar, V. Kumar, S. Sathe, R. Benney, E. Thornburg, C. Kyle, and T. Nonoshita. Aerodynamic interactions between parachute canopies. *Journal of Applied Mechanics*, 70:50–57, 2003.

40. K. Stein, T. E. Tezduyar, and R. Benney. Automatic mesh update with the solid-extension mesh moving technique. *Comp. Meth. in Appl. Mech. and Engng.*, 193:2019–2032, 2004.

41. K. R. Stein, R. J. Benney, T. E. Tezduyar, J. W. Leonard, and M. L. Accorsi. Fluid–structure interactions of a round parachute: Modeling and simulation techniques. *Journal of Aircraft*, 38:800–808, 2001.

42. K. Takizawa, J. Christopher, T. E. Tezduyar, and S. Sathe. Space–time finite element computation of arterial fluid–structure interactions with patient-specific data. *Communications in Numerical Methods in Engineering*, published online, DOI: 10.1002/cnm.1241, March 2009.

43. K. Takizawa, C. Moorman, S. Wright, J. Christopher, and T. E. Tezduyar. Wall shear stress calculations in space–time finite element computation of arterial fluid–structure interactions. *Computational Mechanics*, published online, DOI: 10.1007/s00466-009-0425-0, October 2009.

44. T. Tezduyar, S. Aliabadi, M. Behr, A. Johnson, and S. Mittal. Parallel finite-element computation of 3D flows. *Computer*, 26(10):27–36, 1993.

45. T. Tezduyar and Y. Osawa. Fluid–structure interactions of a parachute crossing the far wake of an aircraft. *Comp. Meth. in Appl. Mech. and Engng.*, 191:717–726, 2001.

46. T. E. Tezduyar. Stabilized finite element formulations for incompressible flow computations. *Advances in Applied Mechanics*, 28:1–44, 1992.

47. T. E. Tezduyar. Computation of moving boundaries and interfaces and stabilization parameters. *International Journal for Numerical Methods in Fluids*, 43:555–575, 2003.

48. T. E. Tezduyar. Finite element methods for fluid dynamics with moving boundaries and interfaces. In E. Stein, R. De Borst, and T. J. R. Hughes, editors, *Encyclopedia of Computational Mechanics*, Volume 3: Fluids, chapter 17. John Wiley & Sons, 2004.

49. T. E. Tezduyar. Interface-tracking and interface-capturing techniques for finite element computation of moving boundaries and interfaces. *Comp. Meth. in Appl. Mech. and Engng.*, 195:2983–3000, 2006.

50. T. E. Tezduyar, S. K. Aliabadi, M. Behr, and S. Mittal. Massively parallel finite element simulation of compressible and incompressible flows. *Comp. Meth. in Appl. Mech. and Engng.*, 119:157–177, 1994.

51. T. E. Tezduyar, M. Behr, and J. Liou. A new strategy for finite element computations involving moving boundaries and interfaces – the deforming-spatial-domain/space–time procedure: I. The concept and the preliminary numerical tests. *Comp. Meth. in Appl. Mech. and Engng.*, 94(3):339–351, 1992.

52. T. E. Tezduyar, M. Behr, S. Mittal, and J. Liou. A new strategy for finite element computations involving moving boundaries and interfaces – the deforming-spatial-domain/space–time procedure: II. Computation of free-surface flows, two-liquid flows, and flows with drifting cylinders. *Comp. Meth. in Appl. Mech. and Engng.*, 94(3):353–371, 1992.

53. T. E. Tezduyar and Y. J. Park. Discontinuity capturing finite element formulations for nonlinear convection-diffusion-reaction equations. *Comp. Meth. in Appl. Mech. and Engng.*, 59:307–325, 1986.

54. T. E. Tezduyar and S. Sathe. Modeling of fluid–structure interactions with the space–time finite elements: Solution techniques. *International Journal for Numerical Methods in Fluids*, 54:855–900, 2007.

55. T. E. Tezduyar, S. Sathe, T. Cragin, B. Nanna, B. S. Conklin, J. Pausewang, and M. Schwaab. Modeling of fluid–structure interactions with the space–time finite elements: Arterial fluid mechanics. *International Journal for Numerical Methods in Fluids*, 54:901–922, 2007.

56. T. E. Tezduyar, S. Sathe, R. Keedy, and K. Stein. Space–time techniques for finite element computation of flows with moving boundaries and interfaces. In S. Gallegos, I. Herrera, S. Botello, F. Zarate, and G. Ayala, editors, *Proceedings of the III International Congress on Numerical Methods in Engineering and Applied Science*. CD-ROM, Monterrey, Mexico, 2004.

57. T. E. Tezduyar, S. Sathe, R. Keedy, and K. Stein. Space–time finite element techniques for computation of fluid–structure interactions. *Comp. Meth. in Appl. Mech. and Engng.*, 195:2002–2027, 2006.
58. T. E. Tezduyar, S. Sathe, J. Pausewang, M. Schwaab, J. Christopher, and J. Crabtree. Fluid–structure interaction modeling of ringsail parachutes. *Computational Mechanics*, 43:133–142, 2008.
59. T. E. Tezduyar, S. Sathe, J. Pausewang, M. Schwaab, J. Christopher, and J. Crabtree. Interface projection techniques for fluid–structure interaction modeling with moving-mesh methods. *Computational Mechanics*, 43:39–49, 2008.
60. T. E. Tezduyar, S. Sathe, M. Schwaab, and B. S. Conklin. Arterial fluid mechanics modeling with the stabilized space–time fluid–structure interaction technique. *International Journal for Numerical Methods in Fluids*, 57:601–629, 2008.
61. T. E. Tezduyar, S. Sathe, and K. Stein. Solution techniques for the fully-discretized equations in computation of fluid–structure interactions with the space–time formulations. *Comp. Meth. in Appl. Mech. and Engng.*, 195:5743–5753, 2006.
62. T. E. Tezduyar, S. Sathe, K. Stein, and L. Aureli. Modeling of fluid–structure interactions with the space–time techniques. In H.-J. Bungartz and M. Schäfer, editors, *Fluid–Structure Interaction*, volume 53 of *Lecture Notes in Computational Science and Engineering*, pages 50–81. Springer, 2006.
63. T. E. Tezduyar, M. Schwaab, and S. Sathe. Sequentially-Coupled Arterial Fluid–Structure Interaction (SCAFSI) technique. *Comp. Meth. in Appl. Mech. and Engng.*, 198:3524–3533, 2009.
64. T. E. Tezduyar, K. Takizawa, and J. Christopher. Multiscale Sequentially-Coupled Arterial Fluid–Structure Interaction (SCAFSI) technique. In S. Hartmann, A. Meister, M. Schaefer, and S. Turek, editors, *International Workshop on Fluid–Structure Interaction — Theory, Numerics and Applications*. Kassel University Press, 2009.
65. T. E. Tezduyar, K. Takizawa, J. Christopher, C. Moorman, and S. Wright. Interface projection techniques for complex FSI problems. In T. Kvamsdal, B. Pettersen, P. Bergan, E. Onate, and J. Garcia, editors, *Marine 2009*, Barcelona, Spain, 2009. CIMNE.
66. T. E. Tezduyar, K. Takizawa, C. Moorman, S. Wright, and J. Christopher. Space–time finite element computation of complex fluid–structure interactions. *International Journal for Numerical Methods in Fluids*, published online, DOI: 10.1002/d.2221, 2009.
67. T. E. Tezduyar, K. Takizawa, C. Moorman, S. Wright, and J. Christopher. Multiscale sequentially-coupled arterial FSI technique. *Computational Mechanics*, published online, DOI: 10.1007/s00466-009-0423-2, October 2009.
68. R. Torii, M. Oshima, T. Kobayashi, K. Takagi, and T. E. Tezduyar. Influence of wall elasticity on image-based blood flow simulation. *Japan Society of Mechanical Engineers Journal Series A*, 70:1224–1231, 2004. in Japanese.
69. R. Torii, M. Oshima, T. Kobayashi, K. Takagi, and T. E. Tezduyar. Computer modeling of cardiovascular fluid–structure interactions with the Deforming-Spatial-Domain/Stabilized Space–Time formulation. *Comp. Meth. in Appl. Mech. and Engng.*, 195:1885–1895, 2006.
70. R. Torii, M. Oshima, T. Kobayashi, K. Takagi, and T. E. Tezduyar. Fluid–structure interaction modeling of aneurysmal conditions with high and normal blood pressures. *Computational Mechanics*, 38:482–490, 2006.
71. R. Torii, M. Oshima, T. Kobayashi, K. Takagi, and T. E. Tezduyar. Influence of wall elasticity in patient-specific hemodynamic simulations. *Computers & Fluids*, 36:160–168, 2007.
72. R. Torii, M. Oshima, T. Kobayashi, K. Takagi, and T. E. Tezduyar. Numerical investigation of the effect of hypertensive blood pressure on cerebral aneurysm — Dependence of the effect on the aneurysm shape. *International Journal for Numerical Methods in Fluids*, 54:995–1009, 2007.
73. R. Torii, M. Oshima, T. Kobayashi, K. Takagi, and T. E. Tezduyar. Fluid–structure interaction modeling of a patient-specific cerebral aneurysm: Influence of structural modeling. *Computational Mechanics*, 43:151–159, 2008.
74. R. Torii, M. Oshima, T. Kobayashi, K. Takagi, and T. E. Tezduyar. Fluid–structure interaction modeling of blood flow and cerebral aneurysm: Significance of artery and aneurysm shapes. *Comp. Meth. in Appl. Mech. and Engng.*, 198:3613–3621, 2009.

75. R. Torii, M. Oshima, T. Kobayashi, K. Takagi, and T. E. Tezduyar. Influence of wall thickness on fluid–structure interaction computations of cerebral aneurysms. *Communications in Numerical Methods in Engineering*, published online, DOI: 10.1002/cnm.1289, July 2009.
76. R. Torii, M. Oshima, T. Kobayashi, K. Takagi, and T. E. Tezduyar. Role of 0D peripheral vasculature model in fluid–structure interaction modeling of aneurysms. *Computational Mechanics*, published online, DOI: 10.1007/s00466-009-0439-7, November 2009.
77. E. H. van Brummelen and R. de Borst. On the nonnormality of subiteration for a fluid-structure interaction problem. *SIAM Journal on Scientific Computing*, 27:599–621, 2005.
78. W. A. Wall, S. Genkinger, and E. Ramm. A strong coupling partitioned approach for fluid–structure interaction with free surfaces. *Computers & Fluids*, 36:169–183, 2007.

Stability Issues in Partitioned FSI Calculations

J. Vierendeels, J. Degroote, S. Annerel, and R. Haelterman

Abstract In this chapter a short review will be given on stability issues for fluid-structure interaction (FSI) problems we encountered and studied in the last decade. Based on this, the ideas behind two implicit coupling algorithms, developed in the department, will be explained. The first algorithm is the Interface Quasi-Newton coupling method and the second is the Interface Artificial Compressibility coupling method. Most of the applications that are shown are in the biomechanical field. These are representative for more general strongly coupled problems with incompressible fluids and flexible structures.

1 Stability issues when computing bileaflet heart valve motion

When the motion of the heart valve is simulated with two separate solvers, one has to calculate a new position of the heart valve leaflets for each time step and the flow field has to be updated each time step taking into account the motion of the

J. Vierendeels
Ghent University, Department of Flow, Heat and Combustion Mechanics,
Sint-Pietersnieuwstraat 41, 9000 Gent, Belgium
e-mail: Jan.Vierendeels@UGent.be

J. Degroote
Ghent University, Department of Flow, Heat and Combustion Mechanics,
Sint-Pietersnieuwstraat 41, 9000 Gent, Belgium
e-mail: Joris.Degroote@UGent.be

S. Annerel
Ghent University, Department of Flow, Heat and Combustion Mechanics,
Sint-Pietersnieuwstraat 41, 9000 Gent, Belgium
e-mail: Sebastiaan.Annerel@UGent.be

R. Haelterman
Royal Military Academy, Department of Mathematics, Renaissancelaan 30,
1000 Brussels, Belgium
e-mail: Robby.Haelterman@rma.ac.be

H.-J. Bungartz et al. (eds.), *Fluid Structure Interaction II*, Lecture Notes
in Computational Science and Engineering 73, DOI 10.1007/978-3-642-14206-2_4,
© Springer-Verlag Berlin Heidelberg 2010

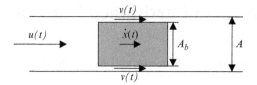

Fig. 1 Motion of a rigid body in a tube, generic one-dimensional test case.

leaflets. Let us assume that a flow solver can calculate the flow field at the new time level when a prescribed motion of the leaflet is provided and that the moment on the leaflet, computed by the flow solver, can be used to integrate the motion of the leaflet in time. The motion of a leaflet with a frictionless hinge can be described by

$$M = I\ddot{\theta},$$
(1)

where M is the moment computed from the flow solver, I is the moment of inertia of the leaflet and $\ddot{\theta}$ is the angular acceleration of the leaflet. The motion of a heart valve with the motion of the surrounding blood is a strongly coupled problem and implicit coupling is necessary to obtain a stable solution [10, 11, 22]. However, with a partitioned method, where subiterations are necessary between the flow solver and the structural solver, it was observed that the stability of these subiterations strongly depends on the density of the fluid, the density of the valve material, the moments of inertia of the leaflets and the gap size between the leaflets and the housing, whereas the time step size had no influence. This can be explained with a stability analysis on a generic one-dimensional test case: the motion of a rigid cylindrical body in a cylindrical tube, shown in Fig. 1.

Between the cylinder and the tube there is a small gap which connects the fluid at the front and the back of the cylinder. There exists an analogy between the moving valve leaflet problem and the 1D test case because during certain phases of the leaflet motion also a small gap is present that connects the fluid on both sides of the leaflets. Also, both problems are described with one degree of freedom. In Fig. 1, $u(t)$ is the velocity of the oncoming fluid, $v(t)$ is the velocity in the gaps, $\dot{x}$ is the velocity of the rigid body, A and A_b are the cross sectional area of the tube and the front area of the rigid body, respectively.

For this generic test case, the conservation of mass is given by

$$Au(t) = A_b\dot{x}(t) + (A - A_b)v(t)$$
(2)

or

$$v(t) = \frac{1}{a}\left(u(t) - (1-a)\dot{x}(t)\right),$$
(3)

with $a = A_g/A$ and $A_g = A - A_b$, the gap size.

Conservation of momentum in the gap is expressed by

$$\frac{\partial v}{\partial t} + \frac{1}{\rho_f}\frac{\partial p}{\partial x} = 0 \tag{4}$$

for inviscid flow. ρ_f denotes the density of the fluid, p is the pressure. In the analysis the influence of the viscous terms is neglected. The driving force for the rigid body motion is given by the pressure difference between the left and right wall of the cylinder. This pressure difference is

$$p_l - p_r = \frac{\partial v}{\partial t}L\rho_f, \tag{5}$$

with L the length of the gap. The force on the solid body is then given by

$$F = A_b(p_l - p_r) = (1 - a)A(p_l - p_r). \tag{6}$$

The equation of motion for the solid body becomes

$$m\ddot{x} = F \tag{7}$$

or

$$\ddot{x} = \frac{\partial u}{\partial t}\frac{\rho_f}{\rho_s}\frac{1}{a} - \ddot{x}\frac{1-a}{a}\frac{\rho_f}{\rho_s}, \tag{8}$$

with ρ_s the density of the solid. Furthermore, K is denoted by

$$K = \frac{1-a}{a}\frac{\rho_f}{\rho_s}. \tag{9}$$

The stability of different coupling strategies for this problem was studied in [22] and the main conclusions are given below. The study was done using a backward Euler time integration scheme for the flow solver. For the structural solver, a class of time integration schemes given by

$$\begin{aligned}\dot{x}^{n+1} &= \dot{x}^n + (1 - \beta)\ddot{x}^n\Delta t + \beta\ddot{x}^{n+1}\Delta t \\ x^{n+1} &= x^n + \dot{x}^n\Delta t + \gamma\ddot{x}^n\Delta t^2 + \alpha\ddot{x}^{n+1}\Delta t^2\end{aligned} \tag{10}$$

was considered. For $\gamma = 1/2 - \alpha$, this corresponds to the class of Newmark schemes. For $\gamma = 0$ and $\alpha = \beta = 1$, this corresponds to the backward Euler scheme. Indeed, with the latter choice, the scheme can be rewritten as

$$\begin{aligned}\dot{x}^{n+1} &= \dot{x}^n + \ddot{x}^{n+1}\Delta t, \\ x^{n+1} &= x^n + \dot{x}^{n+1}\Delta t.\end{aligned} \tag{11}$$

1.1 Explicit coupling

When explicit coupling is used ($\alpha = \beta = 0$), it was shown that the stability condition for γ can only be fulfilled if K is smaller than 3, which means that for large K an explicit scheme is unstable and cannot be used. K becomes large for this case when the gap size is small or when the density of the fluid is large relative to the density of the solid. For this model, the size of the time step has no influence on the stability of this partitioned scheme.

1.2 Implicit coupling

When implicit coupling is used, it was shown that even for large values of K stable schemes could be obtained. However, only when the backward Euler scheme for the structure was used, corresponding to the time integration scheme used in the flow solver, the spurious modes are damped critically, which means that spurious modes will disappear in the next time step. However, when other time integration schemes are used, the spurious modes are not damped critically and it can take a lot of time steps before a spurious mode disappears, depending of the amount of numerical damping in the time integration scheme. This was verified for several time integration schemes of the class mentioned above, including the second-order Newmark scheme, the fully implicit Newmark scheme and a damped Newmark scheme. The spurious oscillations were due to the choice of a different time integration scheme for the flow and the structural solver and not because of the partitioned approach since the coupling iterations within the time steps were completely converged. This behaviour was also numerically verified with the heart valve simulation [22]. Below, the influence of the choice of the coupling algorithm on the convergence of the coupling iterations is described.

1.2.1 Gauss-Seidel coupling iterations

When Gauss-Seidel iterations are used to converge the partitioned problem within one time step (explicit coupling iterations), it was shown that the underrelaxation factor that has to be used for optimal convergence of these subiterations scales proportionally with $1/K$ for large values of K. An optimal value can be derived for this generic test case, but when the gap size is not constant as a function of time, e.g. with leaflets moving around a hinge, the optimal value is changing with time.

1.2.2 Newton and quasi-Newton coupling iterations

When the underrelaxation factor is computed, which can e.g. be done by computing a numerical derivative of the change in moment with respect to a change in angular acceleration, the coupling iterations become implicit. When this numerical derivative is computed in each coupling iteration, the Newton-Raphson method is obtained. Otherwise, also a secant method can be used. If the derivative is estimated from the previous time steps or if it is only computed once, a quasi-Newton method is obtained. This idea will be extended further when dealing with more than one degree of freedom at the interface between the flow problem and the structural problem.

1.2.3 Example of divergence with Gauss-Seidel coupling iterations

It is shown in the following example why Gauss-Seidel coupling iterations can become unstable. Consider the computation of the motion of the leaflets of a bileaflet mechanical heart valve starting from a closed position and the fluid at rest. Consider the computation of a time step when the valve is opening but when the gaps are still small. A velocity profile changing with time is applied at the inlet. In Fig. 2, the geometry and the plane in which pressure and velocity fields will be analyzed are shown. In Fig. 3, pressure and velocity contours are shown in this plane on the old time level and the new time level, obtained with a convergent coupling scheme.

Figure 4 shows the velocity and the pressure field during Gauss-Seidel coupling iterations without underrelaxation. When the flow solver is called the first time for

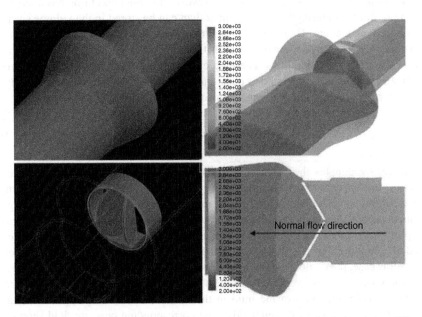

Fig. 2 Bileaflet mechanical heart valve, 3D geometry and cut plane in which the results of Fig. 3 and Fig. 4 are shown.

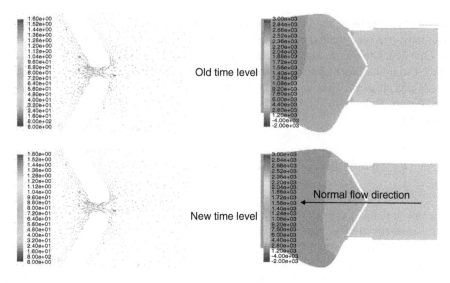

Fig. 3 Velocity and pressure field in the cut plane on the old time level and on the new time level (converged result).

the computation of this time step, the leaflets are put in a position extrapolated from the positions of the previous time steps. However, the inlet condition has changed and more fluid is entering the fluid domain than can be stored in the volume swept by the moving leaflets. Therefore, the incompressible fluid is squeezed through the gaps between the leaflets, such that high flow velocities are present in the gaps, giving rise to high accelerations. These accelerations can be related to high pressure gradients, since in an accelerated or decelerated flow the pressure gradient is mainly caused by the acceleration or deceleration and not by the viscous terms, i.e.

$$\rho_f \frac{D\mathbf{v}}{Dt} \approx -\nabla p, \tag{12}$$

where D/Dt denotes the material or substantial derivative. Thus, the computed pressure in the region near the inlet will be much higher than the pressure computed behind the valve which can be seen in the pictures. This will result in a large moment around the hinge, resulting in a large angular acceleration. Thus, in the next coupling iteration, when this predicted value for the angular acceleration has been used to compute a new position of the leaflets, the position of the leaflets can be too far. In this case, the computed flow field shows a backflow through the gaps because then the volume that is swept by the leaflets is larger than the amount of fluid entering the domain, thus part of the fluid in the volume that is swept by the leaflets has to flow back towards the inlet side. The corresponding pressure field shows a large gradient in the opposite direction because in the gap between the leaflets, the flow is accelerated now in the opposite direction. In the next coupling iteration, the leaflets will now be moved in the opposite direction with a certain error with respect

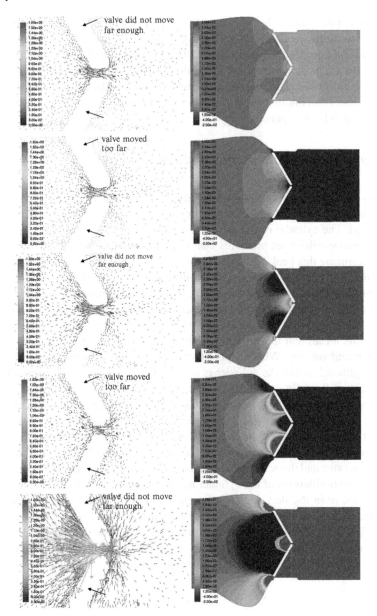

Fig. 4 Velocity and pressure field in the cut plane during Gauss-Seidel coupling iterations without underrelaxation. Divergence is detected.

to the correct position. When the amplitude of this error is increasing during the coupling iterations, the coupling method is unstable and a solution for the next time step cannot be found, although it exists. Underrelaxation can be used to make these coupling iterations convergent, but the underrelaxation factor cannot be estimated in advance, thus one has to use a dynamic underrelaxation factor (e.g. Aitken relaxation [15,16]) or an implicit coupling method within the coupling iterations in order to obtain a good convergence.

1.3 Spurious oscillations due to initial conditions

Spurious oscillations can also be introduced in a calculation due to non-physical initial conditions. Consider as an example the impact of a cylinder on a water surface (slamming). The cylinder is hollow and rather thin and can deform during impact, giving rise to smaller peak pressures than when the cylinder is rigid [8]. When one wants to estimate this peak pressure, one has to take into account the fluid-structure interaction. The interaction takes place during impact, thus it would be interesting if the calculation could be started when the cylinder is just above the surface in order to reduce the computational time. However, it has then already an initial velocity and also the air around the cylinder is not at rest, but when starting a calculation, typically the air around the cylinder is initialized to be at rest and the cylinder is given an initial velocity. What happens then is that during the first time step the air in the neighbourhood of the cylinder is accelerated to a velocity corresponding to the velocity of the falling cylinder. If the time step is taken small, the corresponding acceleration is rather high and also the pressures acting on the cylinder are high, although the cylinder is only in contact with air. Of course this situation is not physical, but the result is that the cylinder starts to oscillate because of this pressure pulse. In the next time steps these high pressures are absent, because the air is moving after the first time step and the accelerations in the air are rather small until impact occurs. However, the oscillations of the cylinder remain, because there is little damping in the structure or in the flow (due to viscosity) and when looking at the force acting on the cylinder as a function of time, these spurious oscillations remain present (Fig. 5). When keeping the cylinder as a rigid body during the first time steps and by accelerating it until it has the correct speed, the spurious oscillations are eliminated. In these simulations no damping in the material is modelled.

2 Stability issues when computing flexible structures

Consider the following 2D problem: a plate is present in a channel as shown in Fig. 6. The flow and the plate are at rest and should remain at rest when performing fluid-structure interaction calculations, since the velocity is kept zero at the inlet and the pressure at the outlet is kept constant. We calculate from this initial condition the next time step, but we perturb the position of the plate before starting the coupling

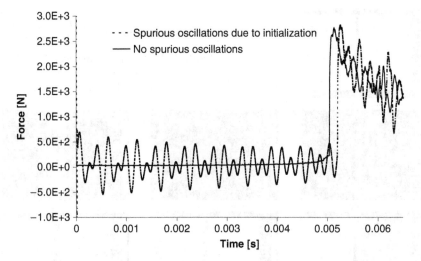

Fig. 5 Force acting on a hollow cylinder during slamming simulations. The spurious oscillations present during the whole simulations are initiated due to the initial conditions.

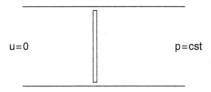

Fig. 6 Plate in a channel. The flow is initially at rest and the velocity is kept zero at the inlet. The pressure is initialized with the outlet pressure and the outlet pressure is kept constant. Everything should stay at rest during FSI calculations with these initial and boundary conditions.

iterations. What should happen is that the perturbation should damp out during the coupling iterations.

Let's consider a perturbation with a small wave number as seen in the left panel of Fig. 7. When the flow solver is called with this perturbed position, it computes a flow through the gaps, of course, the fluid initially present in the volume swept by the plate must be transfered to the other side of the plate within this time step, again giving rise to a large acceleration in the gap and a large pressure difference across the plate. With this pressure load, the structural solver is called, and depending on the properties of fluid and solid and geometry the plate can move in the other direction but with an increased amplitude of the perturbation, leading to divergence in subsequent coupling iterations. However, when a perturbation with a high wave number is applied as seen in the middle and the right panel of Fig. 7, it can be observed that the computed flow field results in more local flow behaviour and local pressure differences with much smaller amplitudes. Thus, with the same fluid and structural parameters, the coupling iterations can be convergent. What can be concluded from this numerical experiment is that errors on the position of the interface with small

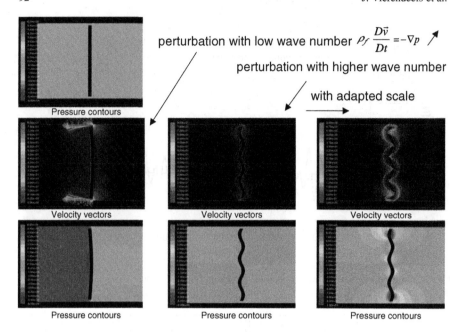

Fig. 7 Plate in a channel. The velocity and pressure field is shown for the first coupling iteration. Before we start the coupling iteration, the position of the plate is perturbed with a small wave number perturbation (left panel) and a high wave number perturbation (middle and right panels). The right panel shows the same velocity and pressure fields as the middle panel but on a different scale.

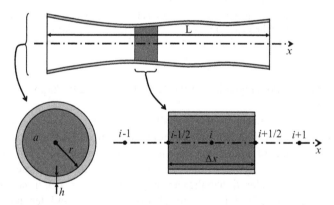

Fig. 8 Flexible tube used for the 1D Fourier analysis.

wave number are likely to be unstable in Gauss-Seidel coupling iterations, while errors with high wave number are more likely to be stable.

The same conclusion can be drawn from a stability analysis on the flow through an elastic tube (Fig. 8). Such an analysis was performed in [2, 6, 18, 19]. The analysis was based on a Fourier decomposition of the error modes with respect to the

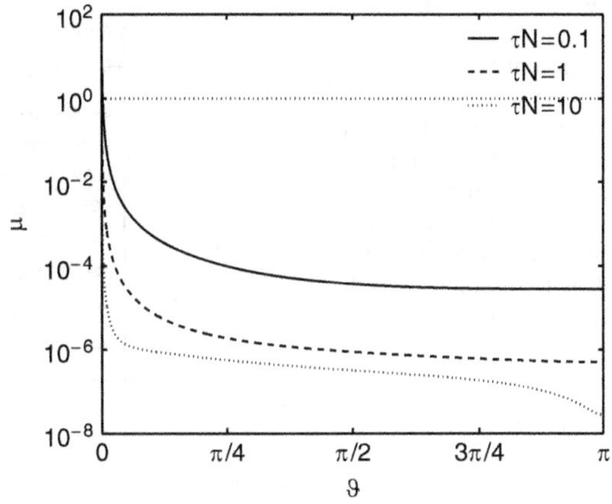

Fig. 9 Amplification factor as a function of the wave number computed with a 1D Fourier analysis for the flexible tube problem. A high dimensionless stiffness is used. Mass in the structure is neglected. τ denotes the dimensionless time step and N is the number of grid cells.

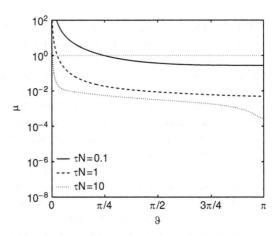

Fig. 10 Amplification factor as a function of the wave number computed with a 1D Fourier analysis for the flexible tube problem. A low dimensionless stiffness is used. Mass in the structure is neglected. τ denotes the dimensionless time step and N is the number of grid cells.

coupled solution. The behaviour of the amplification factor μ of these error modes is studied for a coupling algorithm with Gauss-Seidel iterations without underrelaxation. Linearisation was done around a state with constant pressure p_0, radius r_0 and velocity v_0. The results are obtained as a function of a dimensionless stiffness κ and a dimensionless time step $\tau = u_0 \Delta t / L$ with L the length of the tube, u_0 the velocity inside the tube and Δt the time step size. θ represents the dimensionless wave number and ranges between 0 and π, with $\theta = \pi$ the highest wave number that can be represented on the computational grid. From the results in Fig. 9 and Fig. 10, it can be indeed observed that error modes with small wave number are less stable than error modes with higher wave number. The influence of the stiffness and the time step is also shown. When the stiffness is increased, the amplification factor becomes smaller and when the time step decreases, the amplification factor increases. The same happens when the length L of the tube increases. A similar behaviour was also observed in [1, 12]. In Fig. 9 and Fig. 10, the influence of the mass in the structure is neglected. When taking into account the mass of the structure, it can be seen that for a decreasing time step, the amplification factor stops increasing (Fig. 11).

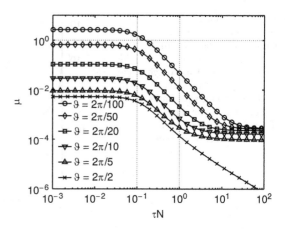

Fig. 11 Amplification factor as a function of the dimensionless time step τ for different wave numbers computed with a 1D Fourier analysis for the flexible tube problem. A small dimensionless stiffness is used. Mass in the structure is taken into account. N is the number of grid cells.

This behaviour can be explained as follows. For very small time steps the displacement is mainly controlled by the acceleration term in the structural equations and this term starts to balance the load term, resulting in smaller displacements for smaller time steps. Thus, for very small time steps, an error in the load will result in an error on the displacement, scaled with the square of the time step. On the other hand, in the flow solver, an error in the displacement will introduce an error on the load, amplified inversely with the square of the time step size. Thus, for small time steps there is no amplification related to the time step size on the displacement error after one coupling iteration. For larger time steps the displacement computed by the structural solver is rather controlled by a balance between the stiffness and the load term. However, in the flow solver the displacement stays in balance with the load with a factor inversely with the square of the time step. This is because the viscosity term is typically very small in comparison with the acceleration term. Thus, for larger time steps an error in the displacement will be amplified after a coupling iteration with an amplification factor depending on the time step size. This analysis is verified with numerical simulations. In Fig. 12 the required number of coupling iterations is shown as a function of the time step size for this flexible tube problem. This demonstrates that Gauss-Seidel iterations become unstable for smaller time steps.

Therefore, a coupling method which treats at least the error modes with small wave number implicitly has a chance to be a convergent method. Since implicitness has to be obtained by computing derivatives numerically (when black box solvers are used), such semi-implicit methods (such as quasi-Newton methods which make use of increasing-rank Jacobians) can be interesting to use, compared to full implicit methods such as Newton-Raphson methods (e.g. Newton-GMRES). The Interface Quasi Newton technique with approximation of the Inverse of the Jacobian form a Least-Squares model (IQN-ILS) [7] is such a method. For the approximation of the Jacobian, data of previous time steps can be reused. Figure 12 shows that for the IQN-ILS method the number of coupling iterations rises if the time step size is decreased because the number of unstable modes rises which have to be treated implicitly. For very small time steps the number of coupling iterations does not

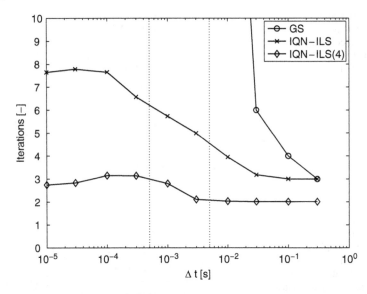

Fig. 12 Number of coupling iterations as a function of the time step size for the flexible tube problem. A small dimensionless stiffness is used. Mass in the structure is taken into account. GS: Gauss-Seidel, IQN-ILS(x): Interface Quasi-Newton with approximation of the Inverse of the Jacobian from a Least-Squares model, x: number of time steps of which data is reused when calculating the inverse of the Jacobian.

vary anymore, which confirms the analysis above. With reuse of data of previous time steps (IQN-ILS(x), x denotes the number of previous time steps of which data is reused), the number of coupling iterations becomes rather independent of the number of unstable modes in this test case. In this chapter the number of coupling iterations corresponds to the number of flow and structural solver calls, so it is directly related to the runtime of the computation, since the amount of work outside the solvers is negligible.

3 Coupling methods

A short overview of some coupling methods is given below together with a discussion on the performance of these methods, related to the stability issues mentioned above. The solution of the following fixed point problem is sought:

$$X = S(F(X)), \tag{13}$$

with X the position of the nodes at the interface between the fluid and the structure, F the flow solver which computes the load on the interface and S the structural solver which computes the new position of the interface. In Fig. 13 the Gauss-Seidel

Fig. 13 Fixed point
iterations with the
Gauss-Seidel coupling
method without
underrelaxation.

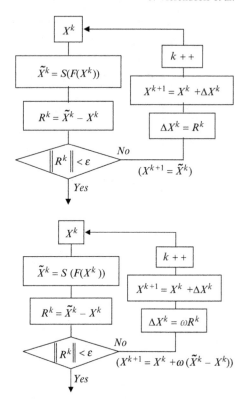

Fig. 14 Fixed point
iterations with the
Gauss-Seidel coupling
method with underrelaxation.
The underrelaxation factor
can be dynamically computed
with Aitken's method.

method is shown.[1] Here, $X^k = X^{n+1,k}$ where $n + 1$ denotes the new time level and
k denotes the coupling iteration level when looking for the solution on time level
$n + 1$. The notation $n + 1$ will be omitted. R denotes the residual of the coupling iteration. In the Gauss-Seidel method an underrelaxation parameter ω can be used.[2] The
Gauss-Seidel coupling method with Aitken underrelaxation [15, 16] is an example
where dynamic underrelaxation is used (Fig. 14).

Two classes of methods developed in our research group are the Interface Artificial Compressibility (IAC) methods [9, 18, 19, 21] and the Interface Quasi-Newton
(IQN) methods [3, 7] which were originally developed as a coupling method based
on Reduced Order Models for the behaviour at the interface of the flow and the
structural solver [23, 24]. In the IAC method (Fig. 15) the flow solver is altered during the coupling iterations in such a way that the altered flow solver converges to
the original flow solver together with the convergence of the coupling iterations.

Of course it is important to alter the flow solver in such a way that convergence of
the coupled problem is enhanced. This can be done by introducing information from

---

[1] This corresponds to the Richardson's method for a fixed point problem, but in partitioned FSI
calculations the iterations are also called Gauss-Seidel iterations.

[2] This corresponds to the modified Richardson's method for fixed point problems.

Fig. 15 Fixed point
iterations with the Interface
Artificial Compressibility
(IAC) method. The flow
solver is altered in such a way
that convergence of the
coupling iterations is
enhanced. The altered flow
solver also converges towards
the real flow solver during the
convergence of the coupling
iterations [9].

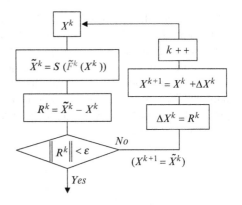

the structural solver in the flow solver. The idea on which the IAC method is based
can be best explained when looking at the problem of the flexible tube. Consider an
equilibrium flow in a straight flexible tube. From a certain moment on the inflow
condition is changed, such that a wave starts travelling from the inlet to the outlet.
Consider the first time step where e.g. a rise of the velocity at the inlet occurs. In
this time step the wave will travel over a certain distance and locally near the inlet
the velocity and the pressure will rise. If the time step is small enough, the wave
will not be noticed yet at the outlet. However, when computing this solution with a
partitioned approach with Gauss-Seidel iterations, the flow solver does not know yet
that the tube wall has to move and due to the incompressibility constraint the extra
amount of fluid entering the tube will immediately flow out. Therefore, the fluid in
the whole tube has to accelerate which results in a pressure gradient in the whole
tube. This results in a solution of the first coupling iteration where a big pressure
difference will be present between the inlet and the outlet, the pressure difference
needed to accelerate all the fluid between inlet and outlet. When the structural solver
is called, the large pressure can cause the tube wall to move much too far, resulting
in unstable coupling iterations. When the flow solver is modified in such a way that
the incoming fluid does not have to leave the tube immediately, but can be stored
locally, the coupling method can be stabilized. When a source term is added to the
continuity equation of the cells adjacent to the fluid-structure interface

$$\rho \frac{V^{n+1,k} - V^n}{\Delta t} + \Sigma_f \rho(\mathbf{v}_f - \mathbf{v}_{grid,f})\mathbf{n}_f S_f = -\rho \frac{\partial V}{\partial p} \frac{p^{n+1,k+1} - p^{n+1,k}}{\Delta t},$$

(14)

the computed pressure field $p^{n+1,k+1}$ will be altered in such a way that it takes into
account the effect of storing fluid near the boundary. In this equation, V denotes
the control volume, $\mathbf{v}$ and $\mathbf{v}_{grid,f}$ denote the flow velocity and the grid velocity
at the face f with area S_f of the control volume. n and $n + 1$ denote the old and
the new time level, k denotes the coupling iterations level (not the iteration level in
the flow solver). In the flow solver, $V^{n+1,k}$ is known (computed from the structure
solver) and a solution is sought for $p^{n+1,k+1}$ and $\mathbf{v}^{n+1,k+1}$. The structural solver

then computes the new position of the interface, resulting in an updated position of the control volume, now at coupling iteration level $k + 1$: $V^{n+1,k+1}$. When due to the source term, this altered pressure field is much closer to the coupled solution of the time step, and when this pressure field is subsequently applied to the structural solver, the computed tube wall displacement will also be close to the coupled solution. The question is: how to alter the flow solver in order to obtain such a pressure field? The answer can be found by analyzing the source term in eq. (14), which can be approximated as

$$-\rho\frac{\partial V}{\partial p}\frac{p^{n+1,k+1} - p^{n+1,k}}{\Delta t} \approx -\rho\frac{V^{n+1,k+1} - V^{n+1,k}}{\Delta t}, \tag{15}$$

such that the continuity equation that is solved for is approximated by

$$\rho\frac{V^{n+1,k+1} - V^n}{\Delta t} + \Sigma_f \rho(\mathbf{v}_f - \mathbf{v}_{grid,f})\mathbf{n}_f S_f \approx 0. \tag{16}$$

This equation takes into account that the interface will move and, consequently, that the control volume will change due to a change in the pressure field. When the coupling iterations converge, the source term becomes zero, the approximation becomes an equality and the incompressible solution is found.

Thus, the source term is a vanishing term which approximates the amount of mass that will be stored near the boundary due to the motion of the boundary in one coupling iteration, although the boundary motion has not been computed yet when the new pressure and velocity fields are computed. Therefore, one has to estimate what the displacement of the boundary will be. When the applied load changes and when the pressure is the main reason for the boundary displacement (which is typically the case), one has to know how a local pressure change will cause a local volume change. This information is derived from the structural solver by perturbing the current pressure distribution with a small value and by calculating the corresponding local volume change. From this information $\partial V/\partial p$ is computed for each boundary cell (Fig. 16).

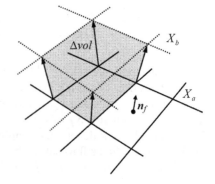

Fig. 16 Computation of $\partial V/\partial p$ by perturbing the current pressure distribution, resulting in a motion of the boundary from X_a to X_b, computed by the structure solver. The computed volume change (Δvol) in each cell, together with $\partial V/\partial p$ that are used in the continuity equation of the flow solver.

Fig. 17 Fixed point
iterations with the Interface
Quasi-Newton method with
approximation of the Inverse
of the Jacobian from a
Least-Squares model
(IQN-ILS). [3, 7]

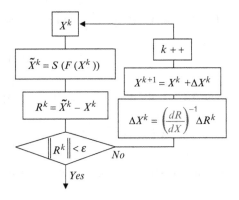

For a one-dimensional problem it was analyzed in [9, 18, 19] that this method stabilized the Gauss-Seidel iterations and in such a way that convergence is rapidly obtained. Also in 3D calculations [9], this was observed.

The Newton-Raphson technique can also be applied to this fixed point problem as shown in Fig. 17. The problem here is the calculation of the inverse of the Jacobian of the residual with respect to the position of the interface nodes, especially when black box solvers are used. In the Newton GMRES method, ΔX^k is computed with GMRES from a linearized system, but in each iteration within the GMRES solver, a call to the linearized flow and structural solver is needed. In the interface GMRES method [20], ΔX^k is computed with GMRES applied on the original black box flow and structural solver. Thus, GMRES is applied on a non-linear problem and also within each iteration of the GMRES solver a call to the flow and structural solver is needed.

When the update ΔX^k is computed with an approximation for the inverse of the Jacobian, the method is called a quasi-Newton method. We developed a method [3, 7] in which the inverse of the Jacobian is directly computed with a least squares approach in such a way that, if the method is applied to a linear problem, the full Jacobian is obtained after $N + 1$ iterations with N the number of degrees of freedom at the interface (size of ΔX) [13, 14]. It was also shown that the method is a generalized secant method. The method can be summarized as follows [7]: calculate

$$\Delta X^k = \left(\frac{dR}{dX}\right)^{-1} \Delta R^k, \quad \Delta R^k = R^{k+1} - R^k = 0 - R^k = -R^k \quad (17)$$

by approximating the inverse of the Jacobian as

$$\left(\frac{dR}{dX}\right)^{-1} = \frac{dX}{dR} = \frac{d(\tilde{X} - R)}{dR} = \frac{d\tilde{X}}{dR} - I. \quad (18)$$

With

$$V^k = \left[\Delta R^0 \dots \Delta R^{k-1}\right], W^k = \left[\Delta \tilde{X}^0 \dots \Delta \tilde{X}^{k-1}\right], \alpha^k = [\alpha_0 \dots \alpha_{k-1}], \quad (19)$$

solve $\Delta R^k = V^k \alpha^k$ for α^k with the least squares method:

$$\alpha^k = \left(V^{k^T} V^k \right)^{-1} V^{k^T} \Delta R^k. \tag{20}$$

Then $\Delta \tilde{X}^k \approx W^k \alpha^k$, which results in an expression for ΔX^k as a function of ΔR^k:

$$\Delta X^k = \Delta \tilde{X}^k - \Delta R^k = W^k \left(V^{k^T} V^k \right)^{-1} V^{k^T} \Delta R^k - \Delta R^k, \tag{21}$$

such that an approximation of the inverse of the Jacobian $(dR/dX)^{-1}$ is given by

$$\left(\frac{dR}{dX} \right)^{-1} \approx W^k \left(V^{k^T} V^k \right)^{-1} V^{k^T} - I. \tag{22}$$

However, this matrix with size corresponding to the number of degrees of freedom on the interface is never computed nor stored, since only a multiplication of this matrix with a vector is needed. In practice a QR decomposition of V is used when calculating ΔX [3, 7]. Remark that Eq. (21) can also be written as

$$\Delta X^k = R^k + \left(\frac{d\tilde{X}}{dR} \right) \Delta R^k, \tag{23}$$

which shows that the Gauss-Seidel iteration is used when the second term is zero.

One can also decompose the residual in small wave number errors and high wave number errors. From previous analysis it is expected that the errors with small wave number will be amplified in the Gauss-Seidel iteration, but once they are incorporated in the Jacobian, they will be treated implicitly in subsequent coupling iterations. The error modes that are not incorporated yet, will not give a contribution to the second term and thus these modes are treated with the Gauss-Seidel approach. Since typically only a few error modes are unstable, one does not need a full Jacobian in order to converge the coupling iterations. When the Jacobian is being built, it is also easy to include information from the previous time steps under the assumption that the Jacobian is not changing much with time. When small time steps are taken, the coupling problem can become tougher as shown above, but with small time steps, it takes more time steps for the Jacobian to change, thus with small time steps also more time steps can be used to compute the Jacobian. The effect of the reuse of data of previous time steps was already shown in Fig. 12 for the flexible tube problem. Other examples of the performance of the method can be found in [3–5, 8].

4 Conclusion

Stability analysis performed on coupling methods for partitioned coupled problems revealed that one must be cautious when using different time integration schemes for the flow and the structural solver. It revealed also that typically only a few error modes of the interface problem are unstable when using Gauss-Seidel iterations. These can be stabilized by using (dynamic) underrelaxation. Based on the insights of stability analysis, one can understand why the Interface Artificial Compressibility method can perform well and why the Interface Quasi-Newton method can converge well within a limited amount of coupling iterations, especially with reuse of data of previous time steps.

Acknowledgements J. Degroote gratefully acknowledges a PhD fellowship of the Research Foundation-Flanders (FWO).

References

1. Causin, P., Gerbeau, J.F., Nobile, F.: Added-mass effect in the design of partitioned algorithms for fluid-structure problems. Comput. Meth. Appl. Mech. Eng. **194** 4506–4527 (2005)
2. Degroote, J., Bruggeman, P., Haelterman, R., Vierendeels, J.: Stability of a coupling technique for partitioned solvers in FSI applications. Comput. Struct. **86** 2224–2234 (2008)
3. Degroote, J., Bathe, K.J., Vierendeels, J.: Performance of a new partitioned procedure versus a monolythic procedure in fluid-structure interaction. Comput. Struct. **87** 793–801 (2009)
4. Degroote, J., Bruggeman, P., Vierendeels, J.. A coupling algorithm for partitioned solvers applied to bubble and droplet dynamics. Comput. Fluids **38** 613–624 (2009)
5. Degroote, J., Bruggeman, P., Haelterman, R., Vierendeels, J.: Bubble simulations with an interface tracking technique based on a partitioned fluid-structure interaction algorithm. J. Comput. Appl. Math. (2010) doi: 10.1016/j.cam.2009.08.096
6. Degroote, J., Annerel, S., Vierendeels, J.: Stability analysis of Gauss-Seidel iterations in a partitioned simulation of fluid-structure interaction. Comput. Struct. (88) 263–271 (2010)
7. Degroote, J., Haelterman, R., Bruggeman, P., Vierendeels, J.: Performance of partitioned procedures in fluid-structure interaction. Comput. Struct. (88) 446–457 (2010)
8. Degroote, J., Souto-Iglesias, A., Van Paepegem, W., Annerel, S., Bruggeman, P., Vierendeels J.: Partitioned simulation of the interaction between an elastic structure and free surface flow. Comput. Meth. Appl. Mech. Eng. (2010) In press.
9. Degroote, J., Swillens, A., Bruggeman, P., Haelterman, R., Segers, P., Vierendeels, J.: Simulation of fluid-structure interaction with the interface artificial compressibility method. Commun. Numer. Methods Eng. **26** 276–289 (2010)
10. Dumont, K., Vierendeels, J.A.M., Segers, P., Van Nooten, G.J., Verdonck, P.R.: Predicting ATS Open PivotTM heart valve performance with computational fluid dynamics. J. Heart Valve Dis. **14**, 393–399 (2005)
11. Dumont, K., Vierendeels, J., Kaminsky, R., Van Nooten, G., Verdonck, P., Bluestein, D.: Comparison of the hemodynamic and thrombogenic performance of two bileaflet mechanical heart valves using a CFD/FSI model. J. Biomech. Eng.-Trans. ASME **129** 558–565 (2007)
12. Förster, C., Wall, W.A., Ramm, E.: Artificial added mass instabilities in sequential staggered coupling of nonlinear structures and incompressible viscous flows. Comput. Meth. Appl. Mech. Eng. **196** 1278–1293 (2007)

13. Haelterman, R., Degroote, J., Van Heule, D., Vierendeels, J.: The Quasi-Newton Least Squares Method: A New and Fast Secant Method Analyzed for Linear Systems. SIAM J. Numer. Anal. **47** 2347–2368 (2009)
14. Haelterman, R., Degroote, J., Van Heule, D., Vierendeels, J.: On the similarities between the quasi-Newton inverse least squares method and GMRes. SIAM J. Numer. Anal. **47** 4660–4679 (2010)
15. Küttler U., Wall W.A.: Fixed-point fluid-structure interaction solvers with dynamic relaxation. Comput. Mech. **43** 61–72 (2008)
16. Mok, D.P., Wall, W.A., Ramm, E.: Accelerated iterative substructuring schemes for instationary fluid-structure interaction. In: Bathe, K.-J. (ed.) Computational Fluid and Solid Mechanics, pp. 1325–1328. Elsevier (2001)
17. Riemslagh, J., Vierendeels, J., Dick, E.: Coupling of a Navier-Stokes solver and an elastic boundary solver for unsteady problems. In: Proc. of the Fourth ECCOMAS Computational Fluid Dynamics Conference, Athens, September 1998, Computational Fluid Dynamics '98, pp. 1040-1045, Chichester, John Wiley & Sons. (2002) ISBN: 0-471-98579-1
18. Riemslagh, K., Vierendeels, J., Dick, E.: A simple but efficient coupling procedure for flexible wall fluid-structure interaction. In: Proc. of the Fluids 2000 AIAA Computational Fluid Dynamics Conference, Denver, CO, 2000. Paper AIAA 2000-2336, 8 pp. (2000) CD-ROM ISBN: 1-56347-433-6
19. Riemslagh, K., Vierendeels, J., Dick, E.: An efficient coupling procedure for flexible wall fluid-structure interaction. In: Proc. of the Fifth ECCOMAS Computational Fluid Dynamics Conference, Barcelona, September 2000, Computational Fluid Dynamics '00, 2000. 13 pp. (2000) CD-ROM ISBN: 84-89925-70-4
20. van Brummelen, E.H., Michler, C., de Borst, R.: Interface-GMRES(R) acceleration of subiteration for fluid-structure-interaction problems. Report DACS-05-001, January 2005. Available from: http://www.em.lr.tudelft.nl/downloads/DACS-05-001.pdf.
21. Vierendeels, J.A., Riemslagh, K., Dick, E., Verdonck, P.R.: Computer simulation of intraventricular flow and pressure gradients during diastole. J. Biomech. Eng.-Trans. ASME **122** 667–674 (2000)
22. Vierendeels, J., Dumont, K., Dick, E., Verdonck, P.: Analysis and stabilization of fluid-structure interaction algorithm for rigid-body motion. AIAA J. **43** 2549–2557 (2005)
23. Vierendeels, J., Lanoye, L., Degroote, J., Verdonck, P.: Implicit coupling of partitioned fluid-structure interaction problems with reduced order models. Comput. Struct. **85** 970–976 (2007)
24. Vierendeels, J., Dumont, K., Verdonck, P.: A partitioned strongly coupled fluid-structure interaction method to model heart valve dynamics. J. Comput. Appl. Math. **215** 602–609 (2008)

Hydroelastic Analysis and Response of Pontoon-Type Very Large Floating Structures

C.M. Wang and Z.Y. Tay

Abstract Pontoon-type very large floating structures (VLFS) are giant plates resting on the sea surface. As these structures have a large surface area and a relatively small depth, they behave elastically under wave action. This type of fluid-structure interaction has being termed hydroelasticity. Hydroelastic analysis is thus necessary to be carried out for VLFS designs in order to assess the dynamic motion and stresses due to wave action. This paper presents the mathematical formulation for the hydroelastic analysis of VLFS. Hydroelastic responses and mitigation methods in reducing the structural response are discussed using some example problems.

1 Introduction

The rising interest in very large floating structures (VLFS) over the past two decades was triggered by the corresponding increase in urbanization and human population in land-scarce countries such as Japan, Singapore and The Netherlands. The conventional solution of creating additional land through land reclamation is not cost effective when the water depth is large. Moreover, land reclamation works generally have a negative environmental impact on coastlines and marine eco-system. VLFS on the other hand has advantages over the traditional land reclamation solution with respect to the following aspects: they are cost effective when the water depth is large and the seabed is soft; environmentally friendly as they do not damage the marine eco-system, or silt-up deep harbours or disrupt the ocean currents; they are easy and fast to construct and therefore the investment may be monetized more rapidly; they

C.M. Wang

Professor, Centre for Offshore Research and Engineering, Department of Civil Engineering, National University of Singapore, Kent Ridge, Singapore, 119260
e-mail: cvecwm@nus.edu.sg

Z.Y. Tay

Research Fellow, Centre for Offshore Research and Engineering, Department of Civil Engineering, National University of Singapore, Kent Ridge, Singapore, 119260
e-mail: cvetzy@nus.edu.sg

H.-J. Bungartz et al. (eds.), *Fluid Structure Interaction II*, Lecture Notes
in Computational Science and Engineering 73, DOI 10.1007/978-3-642-14206-2_5,
© Springer-Verlag Berlin Heidelberg 2010

can be easily removed or expanded; and the structure on VLFS is protected from seismic shocks since VLFSs are inherently base isolated.

VLFS may be categorised into the semisubmersible-type and the pontoon-type. The semisubmersible-type VLFS such as the mobile offshore base (MOB) had been proposed by the US Navy as early as in the pre-cold war era to support military operations where conventional land bases were not available. Such semisubmersible-type VLFS with a raised platform above sea level using column tubes are suitable for deployment in high seas with large waves. In contrast, the pontoon-type VLFS are intended for deployment in calm water such as in a cove, lagoon or harbour. Examples of these pontoon-type VLFS are the Mega Float (a runway test model in the Tokyo Bay, see Fig. 1a), emergency rescue bases (moored in Tokyo Bay, Ise Bay and Osaka Bay), floating storage facilities (in Shirashima and Kamigoto Islands, see Fig. 1b), floating bridges (in Seattle, USA, see Fig. 1c) and floating ferry piers (in Ujina Port, Hiroshima, see Fig. 1d).

Singapore has also recently constructed the world's largest floating performance stage at the Marina Bay (Fig. 2a) and feasibility studies are underway to build a mega floating fuel storage facility (FFSF) (Fig. 2b) to cater for the increasing demand for oil storage capacity. Such FFSF may double up as bunker cum mooring

(a) Mega-Float, Tokyo Bay.

(b) Floating oil storage base at Kamigoto Island (Source:http://www.mhi.co.jp/en/products/ detail/oil_storage_terminal_kamigoto.html).

(c) Lacey V. Murrow Bridge and the Third Washington Bridge at Seattle (Source:http:// en.wikipedia.org/Lacey_V._Murrow_Memorial_ Bridge).

(d) Floating ferry pier at Ujina Port, Hiroshima.

Fig. 1 Applications of pontoon-type VLFSs.

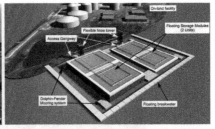

(a) Floating performance stage @ Marina Bay, Singapore.

(b) Proposed floating fuel storage facility (Photo courtesy of JCPL).

(c) Proposed floating cruise terminal.

(d) Proposed mega floating crab restaurant.

Fig. 2 VLFSs in Singapore.

system for ships, thereby relieving traffic congestions in the Singapore harbour and decreasing the turnaround time for ships. The first author has proposed the use of VLFS as a floating-type cruise terminal (Fig. 2c) and a mega floating crab restaurant (Fig. 2d) as opposed to the conventional onshore design in order to create iconic structures for Singapore to attract tourists.

VLFS technology has also made possible future large human habitation on the ocean surface. The Lilypad Floating Ecopolis (Fig. 3), proposed by the Belgium architect Vincent Callebaut, is an example of a visionary proposition to house the city population on a huge floating lily-shaped island. More concepts of floating cities are given in a recent paper by Pernice [1].

As VLFSs have a large surface area and a relatively small depth, they behave elastically under wave action. The fluid-structure interaction has been termed as hydroelasticity. Hydroelastic analysis is thus necessary to be carried out for VLFS in order to assess the dynamic motion and stresses due to wave action design [2]. This paper focuses on the hydroelastic analysis of pontoon-type VLFSs based on the frequency domain approach. Frequency domain analysis has been applied extensively to problems of floating structure dynamics and is particularly useful for the determination of long term responses [3].

Fig. 3 Lilypad Floating Ecopolis. Source: www.vincent.callebaut.org.

2 Literature Review on Hydroelastic Analysis of Floating Structures

The conventional method of solving the motion of floating bodies is to convert the Laplace equation for the velocity potential into a boundary value problem. The boundary conditions are the Neumann condition at the seabed and the wetted surface of the floating body, the linearised free surface condition and the radiation conditions at infinity. The earliest solution to this boundary value problem was given by John [4, 5] in which he used the Green's function within a boundary integral formulation to solve for the wave scattering from floating bodies. A detailed description of the linear wave theory was published by Wehausen and Laiton [6] in their remarkable review article 'Surface Waves'. This review article contains benchmark solutions for wave-structure interactions problems [3,7–11]. However, earlier works on the wave-structure interactions problems only consider the floating structure as a rigid body. With the increasing interest in VLFS as one of the future solutions for creating land for land-scarce countries, hydroelastic analysis on floating structures emerged as a new research area in the early 1990s. To name a few, among the pioneers working on the hydroelastic theory of VLFS are Ertekin *et al.* [12], Yago and Endo [13], Utsunomiya *et al.* [14], Kashiwagi [15] and Ohmatsu [16]. The development of hydroelastic theory should also be attributed to Meylan and Squires [17] and Meylan [18, 19] who studied ice-floe problems which are similar to VLFS problems [20].

The Mega-Float project in Japan from 1995 to 2001 triggered extensive studies on the hydroelastic response of the VLFS. As the pontoon-type VLFS has a small draft compare to its length, the common approach is to model the entire floating structure by a single plate based on the Kirchhoff plate theory (see for example, refs. [14,15,21]) while the water wave is modelled by using the linear wave theory.

Usually the frequency domain approach is used instead of the time domain approach when determining the hydroelastic response amplitude operator of the plate model because of its simplicity and ability to capture the pertinent response parameters in a steady state condition.

More sophisticated hydroelastic analyses have also been performed for a complete 3D floating structure, in order to obtain the deflections and stresses for the secondary structural components in the VLFS [22–25] as well as the local strength of the structural members [26, 27]. The hydroelastic analyses of the VLFS in the the presence of breakwaters [14, 16] and even attached anti-motion devices in the form of submerged plates [28–31] have been carried out. Researchers found that breakwaters and anti-motion devices have a profound effect in reducing the structural motion due to attenuated wave forces. In order to solve for aircraft landing or wave impact problems on VLFS, Kim and Webster [32], Watanabe *et al.* [33] and Kashiwagi [34, 35] investigated the transient response using the time-domain approach.

Mega floating fuel storage modules, as shown in Figs. 1b and 2b, have larger draft to length ratios as opposed to the mat-like VLFS. This necessitates the modelling of the floating modules as a thick plate according to the Mindlin plate theory [36–39]. The use of the Mindlin plate theory not only leads to more accurate prediction of the deflections, but it also provides a better prediction for the stress-resultants. Studies on hydroelastic interactions of two large box-like floating storage modules such as the one shown in Fig. 2b have also been carried out recently by Tay *et al.* [39]. The adjacent module affects each other due to diffracted waves, radiated waves and waves been squeezed into a channel formed by the floating modules being placed side-by-side. Khabakhpasheva and Korobkin [40] propounded an auxiliary floating structure that is connected to the main floating structure in order to reduce the deflection of the main floating structure. Wang *et al.* [41] then proposed a novel way in reducing the hydroelastic response of VLFS through the design of semi-rigid mechanical joints that connect the auxiliary structure to the main floating structure. The design involves suitable rotational stiffness and location of such mechanical joints.

3 Problem Definition

Fig. 4 shows the components of a typical pontoon-type very large floating structure, which consist of the floating structures, access-bridge, breakwater and mooring facility. Breakwaters are constructed around the floating structure to attenuate the wave forces that impact the floating structure. The VLFS is usually moored by the dolphin-frame guide mooring system that restrains the horizontal movement of the floating structure but allows the structure to freely move in the vertical direction according to the tidal variation and varying payload.

The problem at hand is to determine the hydroelastic response of a VLFS under wave action. Also we aim to show the reduction of the hydroelastic responses of floating structures with the attachment of submerged plates and the appropriate design of semi-rigid connections.

Fig. 4 Components of a
pontoon-type VLFS.

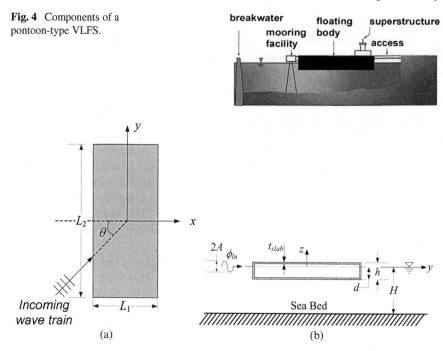

Fig. 5 Figure depicting the pontoon-type very large floating structure in (a) plan view and (b) side view.

4 Mathematical Formulations

4.1 Plate-Water Model

Consider a pontoon-type very large floating structure with a rectangular plan area of $L_1 L_2$ as shown in Fig. 5a. The box-like floating structure has a constant height h, top, bottom and side wall thickness t_{slab}, and draft d as shown in Fig. 5b. The seabed is assumed to be level and the water depth is H. The structure is subjected to an incoming wave of wave period T and a wave height $2A$ that impacts the structure at a wave angle θ with respect to the x-axis.

For hydroelastic analysis, the box-like floating structure is commonly modelled as an equivalent solid plate (see Fig. 6a), by keeping the lengths and the height to be the same as the actual structure (Fig. 5) but its Young's modulus E and Poisson's ratio v of the equivalent solid plate are tweaked to match the natural frequencies and vibration modes of the actual structure. The equivalent plate is assumed to be perfectly flat with free edges. The plate material is commonly assumed to be isotropic or orthotropic and obeys Hooke's law. For the hydroelastic analysis, most researchers modelled the plate using the classical thin plate theory, but more recently the use of Mindlin plate theory is getting popular due to its ability to provide a better

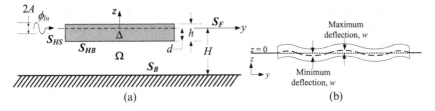

Fig. 6 Schematic diagram depicting (a) cross section of equivalent solid plate (b) deflection of plate with respect to water surface.

prediction to the stress resultants and its allowance for the effects of transverse shear deformation and rotary inertia. Unlike the classical thin plate theory where the equations of motion are described solely by the deflection w variable, the motion of the Mindlin plate is represented by the vertical deflection $W(x, y; t)$, the rotation about the x-axis $\Psi_y(x, y; t)$ and the rotation about the y-axis $\Psi_x(x, y; t)$. In the analysis, the plate is assumed to be restrained in the x-y plane by the station keeping system and the plate can only deformed in the vertical direction (i.e. z-direction). The equations for plate motion will be given in Sect. (4.3.1).

The water is assumed to be a perfect fluid with no viscosity and incompressible and the fluid motion to be irrotational. Based on these assumptions, the fluid motion may be represented by a velocity potential Φ. It is also assumed that the fluid motion is small so that the linear potential theory can be applied for formulating the fluid-motion problem. We also consider the steady-state harmonic motions of the fluid and the structure, with the circular frequency ω. The equations for fluid motion will be given in Sect. (4.3.2).

The water-plate interaction is realized by assuming the deformed surface of the plate to be coincident with the water surface in contact with the plate, i.e. there are no gaps between the plate surface and the water surface.

4.2 Symbols and Non-Dimensional Variables

Fig. 6a shows the cross section of the equivalent solid plate. The symbol Δ denotes the floating plate domain while the water domain is denoted by Ω. The symbols S_F, S_B, S_{HB} and S_{HS} represent the free surface, the seabed, the wetted bottom surface of the floating body and the wetted side surface of the floating body, respectively. The free and undisturbed water surface is at $z = 0$. The deflection w is the vertical elevation of the plate measured from the free and undisturbed water surface as shown in Fig. 6b.

In the subsequent mathematical formulation, we shall use the following non-dimensionalised spatial variables: $\bar{x} = x/L$, $\bar{y} = y/L$, $\bar{z} = z/L$, $\bar{w} = W/L$, $\bar{h} = h/L$, $\bar{d} = d/L$ and the time variables: $\bar{t} = t\sqrt{g/L}$, $\bar{\varphi} = \Phi/(L\sqrt{gL})$ and $\bar{\omega} = \omega\sqrt{L/g}$, where $L = L_2$ of the floating structure as shown in Fig. 5a, g is the gravitational acceleration and ω the wave frequency.

We consider the steady state motions of a harmonically excited system at a circular frequency $\bar{\omega}$. Thus, we can separate the time variable from the spatial variables as follows:

$$\frac{W(x,y,t)}{L} = Re[\bar{w}(\bar{x},\bar{y})e^{-i\bar{\omega}\bar{t}}],$$ (1)

$$\Psi_x(x,y,t) = Re[\bar{\psi}_x(\bar{x},\bar{y})e^{-i\bar{\omega}\bar{t}}],$$ (2)

$$\Psi_y(x,y,t) = Re[\bar{\psi}_y(\bar{x},\bar{y})e^{-i\bar{\omega}\bar{t}}],$$ (3)

$$\frac{\Phi(x,y,z,t)}{\sqrt{gL}} = Re[\bar{\varphi}(\bar{x},\bar{y},\bar{z})e^{-i\bar{\omega}\bar{t}}],$$ (4)

where $\bar{w}(\bar{x},\bar{y})$, $\bar{\psi}_x(\bar{x},\bar{y})$, $\bar{\psi}_y(\bar{x},\bar{y})$ denote the dimensionless Mindlin plate deflection, rotation about the y-axis and rotation about the x-axis, respectively, and $\bar{\varphi}(\bar{x},\bar{y},\bar{z})$ is the dimensionless water velocity potential. Re denotes the real part of the complex number.

4.3 Mathematical Formulation for Plate and Water Motions

4.3.1 Eq. of Motion and Boundary Conditions for Plate

The equations of motion of the Mindlin plate are given as [42]

$$\kappa^2 G\bar{h}\left[\left(\frac{\partial^2 \bar{w}}{\partial \bar{x}^2} + \frac{\partial^2 \bar{w}}{\partial \bar{y}^2}\right) + \left(\frac{\partial \bar{\psi}_x}{\partial \bar{x}} + \frac{\partial \bar{\psi}_y}{\partial \bar{y}}\right)\right] + \gamma\bar{h}\bar{\omega}^2\bar{w} = \bar{w} - i\bar{\omega}\bar{\varphi},$$ (5)

$$\frac{D(1-v)}{2}\left(\frac{\partial^2 \bar{\psi}_x}{\partial \bar{x}^2} + \frac{\partial^2 \bar{\psi}_x}{\partial \bar{y}^2}\right) + \frac{D(1+v)}{2}\left(\frac{\partial^2 \bar{\psi}_x}{\partial \bar{x}^2} + \frac{\partial^2 \bar{\psi}_y}{\partial \bar{x}\partial \bar{y}}\right) - \kappa^2 G\bar{h}\left(\bar{\psi}_x + \frac{\partial \bar{w}}{\partial \bar{x}}\right)$$

$$= -\frac{\gamma\bar{h}^3}{12}\bar{\omega}^2\bar{\psi}_x,$$ (6)

$$\frac{D(1-v)}{2}\left(\frac{\partial^2 \bar{\psi}_y}{\partial \bar{x}^2} + \frac{\partial^2 \bar{\psi}_y}{\partial \bar{y}^2}\right) + \frac{D(1+v)}{2}\left(\frac{\partial^2 \bar{\psi}_y}{\partial \bar{y}^2} + \frac{\partial^2 \bar{\psi}_x}{\partial \bar{y}\partial \bar{x}}\right) - \kappa^2 G\bar{h}\left(\bar{\psi}_y + \frac{\partial \bar{w}}{\partial \bar{y}}\right)$$

$$= -\frac{\gamma\bar{h}^3}{12}\bar{\omega}^2\bar{\psi}_y,$$ (7)

where $D = E\bar{h}^3/[12(1-v^2)]$ is the flexural rigidity, E the Young's modulus, $G = E/[2(1+v)]$ the shear modulus, v the Poisson ratio, γ the specific gravity of the plate material and κ^2 the shear correction factor (normally taken as 5/6).

Since all the edges of the plate are free, the bending moment, twisting moment and transverse shear force must vanish at the edges, i.e.

$$\text{Bending moment } M_{nn} = D \left[\frac{\partial \bar{\psi}_n}{\partial n} + \nu \frac{\partial \bar{\psi}_s}{\partial s} \right] = 0, \tag{8}$$

$$\text{Twisting moment } M_{ns} = D \left(\frac{1-\nu}{2} \right) \left[\frac{\partial \bar{\psi}_n}{\partial s} + \frac{\partial \bar{\psi}_s}{\partial n} \right] = 0, \tag{9}$$

$$\text{Shear force } Q_n = \kappa^2 G \bar{h} \left[\frac{\partial \bar{w}}{\partial n} + \bar{\psi}_n \right] = 0, \tag{10}$$

where s and n denote the tangential and normal directions to the cross-section of the plate.

For convenience, the overbar embellishment for the variables in the foregoing equations will be dropped from hereon.

4.3.2   Eq. of Motion for Water

The linear wave potential theory is used to model the fluid-motion problem by using the velocity potential φ. The single frequency velocity potential $\varphi(x, y, z)$ of the water must satisfy the Laplace's equation [43],

$$\nabla^2 \varphi = 0 \text{ in } \Omega, \tag{11}$$

where $\nabla^2 (\bullet) = \partial^2 (\bullet) / \partial x + \partial^2 (\bullet) / \partial y + \partial^2 (\bullet) / \partial z$ and the boundary conditions are

$$\frac{\partial \varphi}{\partial z} = -i \omega w \text{ on } S_{HB}, \tag{12}$$

$$\frac{\partial \varphi}{\partial n} = 0 \text{ on } S_{HS}, \tag{13}$$

$$\frac{\partial \varphi}{\partial z} = \omega^2 \varphi \text{ on } S_F, \tag{14}$$

$$\frac{\partial \varphi}{\partial z} = 0 \text{ on } S_B, \tag{15}$$

where n is the unit normal vector to the surface S. The wave velocity potential must also satisfy the Sommerfeld radiation condition as $|\mathbf{x}| \rightarrow \infty$ [44],

$$\lim_{|\mathbf{x}| \rightarrow \infty} \sqrt{|\mathbf{x}|} \left(\frac{\partial}{\partial |\mathbf{x}|} - i k \right) (\varphi - \varphi_{In}) = 0 \text{ on } S_{\infty}, \tag{16}$$

where $\mathbf{x} = (x, y)$ and S_{∞} the artificial fluid boundary at infinity. The wave number k satisfies the dispersion relationship

$$k \tanh(kH) = \omega^2, \tag{17}$$

and φ_{In} is the incident velocity potential given by

$$\varphi_{In} = \frac{A}{\omega} \frac{\cosh(k(z+H))}{\cosh kH} e^{ik(x\cos\theta + y\sin\theta)}, \qquad (18)$$

where A is the wave amplitude.

4.4 Modal Expansion Method for Solving Coupled Water-Plate Problem

The plate equations (5) to (7) indicates that the response of the plate w is coupled with the fluid motions (or velocity potential φ). On the other hand, the fluid motion can only be obtained when the plate deflection w is specified in the boundary condition (12). In order to decouple this interaction problem into a hydrodynamic problem in terms of the velocity potential and a fluid-plate vibration problem in terms of the generalized displacement, we adopt the modal expansion method as proposed by Newman [7]. In this method, the plate deflection is expanded by a series of the products of the modal functions $c_l^w(x, y)$ and their corresponding complex amplitudes ζ_l^w,

$$w(x, y) = \sum_{l=1}^{N} \zeta_l^w c_l^w(x, y), \qquad (19)$$

where N denotes the total number of modes.

As the problem is linear, the total velocity potential can be represented by a linear superposition of the diffracted part φ_D ($= \varphi_{In} + \varphi_S$, where φ_S is the scattered potential) and the radiated part φ_R. By using the modal expansion method, the total velocity potential φ may be expressed as

$$\varphi(x, y, z) = \varphi_D(x, y, z) + \varphi_R(x, y, z) = \varphi_D(x, y, z) + \sum_{l=1}^{N} \zeta_l^{\varphi_l} \varphi_l(x, y, z), \quad (20)$$

where $\varphi_{l=1,2,...,N}$ is the radiated potential corresponding to the unit-amplitude motion of the l-th modal function of the freely vibrating plate. Note that the complex amplitudes $\zeta_l^{\varphi_l}$ in Eq. (20) have been assumed to be equal to ζ_l^w in Eq. (19).

By substituting Eqs. (19) and (20) into the Laplace equation (11) and the fluid boundary condition [Eqs. (12) to (16)], we arrive at the following governing equation and boundary conditions for each of the unit-amplitude radiated potential (i.e. for $l = 1, 2, ..., N$) and the diffracted potential (i.e. for $l = D$).

$$\nabla^2 \varphi = 0 \text{ in } \Omega, \qquad (21)$$

$$\frac{\partial \varphi_l}{\partial z} = \begin{cases} -i\omega c_l^w(x, y) & \text{for } l = 1, 2, ..., N \\ 0 & \text{for } l = D \end{cases} \quad \text{on } S_{HB}, \tag{22}$$

$$\frac{\partial \varphi_l}{\partial n} = 0 \text{ on } S_{HS}, \tag{23}$$

$$\frac{\partial \varphi_l}{\partial z} = \omega^2 \varphi_l \text{ on } S_F, \tag{24}$$

$$\frac{\partial \varphi_l}{\partial z} = 0 \text{ on } S_B, \tag{25}$$

and the Sommerfeld radiation condition

$$\lim_{|\mathbf{x}| \to \infty} \sqrt{|\mathbf{x}|} \left(\frac{\partial}{\partial |\mathbf{x}|} - ik \right) (\varphi - \varphi_{In}) = 0 \text{ on } S_\infty. \tag{26}$$

Now, the boundary value problems for each of the unit-amplitude radiated potential and diffracted potential are given explicitly in an uncoupled form.

5 Method of Solution

The method of solution for solving the hydroelastic problem is shown in the appendices and involves the following steps:

Step 1   Transform the Laplace equation (21) with the boundary conditions given by Eqs. (22) to (26) into a boundary integral equation. This step can be best done by using the boundary element method (BEM) because this method can easily handle the Sommerfeld radiation condition using the free surface Green function. For details on the transformation of the Laplace equation into a boundary integral equation using the free surface Green function, please refer to Appendix 1.

Step 2   The governing equation for the plate [(5) to (7)] is solved by using the finite element method. The plate is discretised into a finite number of 8-node Mindlin plate elements. Substitute the velocity potential φ_l obtained from Step 1 into the discretised equation of motion (assembled in the global form). This step is shown in Appendix 2.

Step 3   Expand the plate displacement field $\hat{\mathbf{w}}$ as a series of products of vibration modes $\hat{c}_l$ and complex amplitudes ζ_l. The vibration modes (eigenvectors) could be obtained by performing a free vibration analysis of the plate. Solve the assembled equation of motion for the plate in Step 2 for the complex amplitudes ζ_l. This step is shown in Appendix 2.

Step 4   Multiply the complex amplitudes ζ_l computed in Step 3 with the vibration modes $\hat{c}_l$ to obtain the plate deflection (hydroelastic response). The stress resultants may be readily obtained from taking appropriate derivatives of the deflection and rotations.

6 Results and Discussions

Some example problems are treated to demonstrate the modelling and the hydroelastic analysis of pontoon-type VLFS with and without the presence of breakwaters. Moreover, the effects of submerged-type anti-motion device and semi-rigid connection joints in reducing the hydroelastic response of the VLFS will be discussed.

6.1 Numerical Modelling of VLFS

The detailed model of an actual VLFS (with bulkheads and stiffeners) consists of a huge number of degrees of freedoms, thereby making the hydroelastic analysis on such a model very time-consuming. In order to reduce computational time, the detailed structural model is usually replaced by an equivalent solid plate with the same dimensions (i.e. length, width and height being kept the same as the actual structure), but its Young's modulus and Poisson's ratio are tweaked in order to match the vibration modes and natural frequencies of the actual structure.

We shall use the example problem of the Marina Bay floating performance stage [38] (see Fig. 2a), to describe the procedure in obtaining the equivalent solid plate that modeled the actual floating structure. The floating performance stage consists of 15 pontoons of 40m × 16.6m × 1.2m laid out in a 5 by 3 configuration to produce a large platform measuring 120m × 83m × 1.2m. The thickness of the top surface slab is 12mm and all the side and bottom slabs are 8mm. The pontoon has a system of stiffeners to stiffen the structure as shown in Fig. 7.

Wang *et al.* [38] created a 3-D finite element model of the floating stage in the software ABAQUS as shown in Fig. 8a. The stiffeners, top and bottom slabs, side slabs as well as interior webs are discretized with a combination of plate and beam elements. The side and corner (diamond shape) connectors are used to hold the pontoons together and these connectors are also modeled in detail.

Figure 8b shows the simplified model of the floating performance stage, consisting of 15 equivalent solid isotropic plates that model the pontoons, 20 corner connectors and plate strips along their boundaries that simulate the connector regimes. The properties of the components in the simplified plate model are given in Table 1. These properties are obtained by keeping the length, width and height dimensions of the simplified model to be the same but the Young's modulus and Poisson's ratio adjusted to match the first three vibration frequencies f_i and modes of the 3-D finite element model as shown in Fig. 9.

By establishing the validity of the simplified plate model, the hydroelastic analysis of the floating performance stage may be carried out with a considerably less computational effort. Studies have shown that hydroelastic responses obtained from such simplified plate models are in good agreement with experimental test results (see for example, refs. [13–15]).

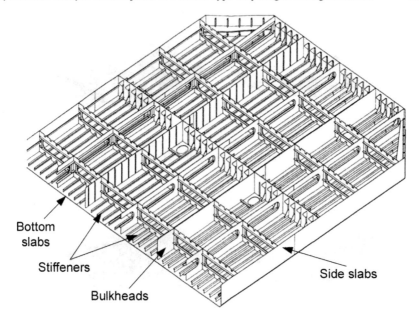

Fig. 7   System of stiffeners in the floating performance platform.

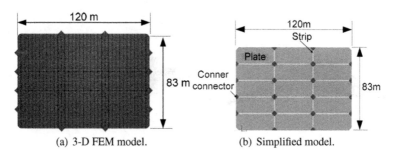

(a) 3-D FEM model.   (b) Simplified model.

Fig. 8   Fig. depicting plan view of (a) 3-D FEM model (b) simplified model.

Table 1   Young's modulus, Poisson's ratio and equivalent mass density for the equivalent plate model components.

Material	Pontoon	Strip	Connector
Young's modulus (GPa)	10	0.05	0.3
Poisson's ratio	0.1	0.1	0.1
Mass density (kg m^{-3})	233.22	233.22	233.22

6.2   Hydroelastic Response of VLFS

The hydroelastic response of the floating structure depends on the operating sea states and water depths. By using the same example problem of the Marina Bay floating performance stage, we shall now discuss the hydroelastic response of the

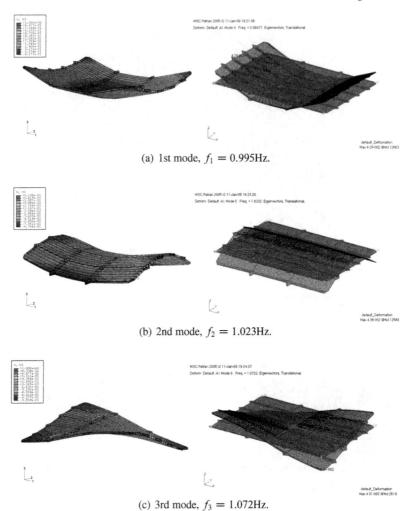

(a) 1st mode, $f_1 = 0.995$Hz.

(b) 2nd mode, $f_2 = 1.023$Hz.

(c) 3rd mode, $f_3 = 1.072$Hz.

Fig. 9 First three vibration modes and natural frequencies of (a) 3-D FEM model (b) simplified plate model.

simplified model (Fig. 8b) operating at a sea state with a constant water depth H of 3m and a wave period T of 2.4s. The deflection for a unit wave amplitude w_{max}/A of the simplified model under head sea and beam sea conditions are shown in Fig. 10. The hydroelastic responses due to the wave loads (shown in Fig. 10) are rather small compared to the deflections resulting from the heavy static loads as reported in [38]. This is because of the benign sea state condition on site. Note that the wave length λ/L is only 0.07.

Next, we shall show the effect of wavelength on the hydroelastic response of VLFS treated by Yago and Endo [13]. The VLFS measures 300m × 60m × 2m and floats in very deep waters (i.e. $H = \infty$). Figure 11 shows the deflection

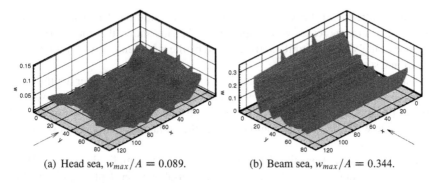

(a) Head sea, $w_{max}/A = 0.089$.   (b) Beam sea, $w_{max}/A = 0.344$.

Fig. 10 Deflection at the bottom plate of floating stage under (a) head sea (b) beam sea. Wave period $T = 2.4$s. Water depth $H = 3$m.

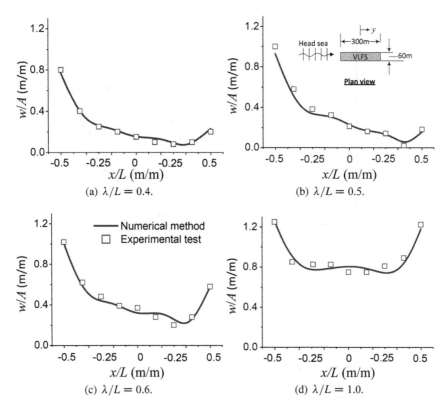

(a) $\lambda/L = 0.4$.   (b) $\lambda/L = 0.5$.

(c) $\lambda/L = 0.6$.   (d) $\lambda/L = 1.0$.

Fig. 11 Vertical deflection along the centreline of the VLFS at (a) $\lambda/L = 0.2$ (b) $\lambda/L = 0.4$ (c) $\lambda/L = 0.6$ (d) $\lambda/L = 1.0$.

along the centreline of the VLFS under a head sea condition. The response of the floating structures increases with respect to λ/L. This shows that the hydroelastic response of the VLFS is significantly affected by the wave lengths λ/L found at the operational sea state.

Utsunomiya and Watanabe [45] also investigated the effect of water depth on the hydroelastic response of VLFS. They considered a 1500m × 150m × 1m pontoon-type VLFS floating on a relatively shallow constant water depth $H = 8$m as well as a variable water depth. The contour plot of the variable water depth is shown in Fig. 12.

A comparison between the deflection (for unit wave amplitude) of the VLFS under the wave angle $\theta = \pi/4$ in constant and variable water depth is presented in Fig. 13. A considerable difference in the response characteristic is observed when the VLFS is located in either a constant or a variable water depth. Thus, in the relatively shallow water case, the effect of a water depth under the VLFS becomes significant and must be accounted for in the hydroelastic analysis.

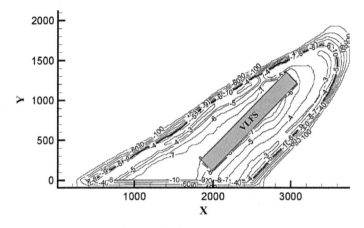

Fig. 12 Contour plot of the variable water depth.

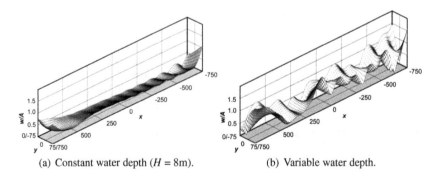

(a) Constant water depth ($H = 8$m).   (b) Variable water depth.

Fig. 13 Deflection of VLFS under wave angle $\theta = \pi/4$ and wavelength $\lambda = 156.8m$ at (a) constant water depth (b) variable water depth.

6.3   Mitigation of Hydroelastic Response

The previous section shows that the hydroelastic response of VLFS could be very large especially when subjected to sea states with large wavelengths or when the VLFS is floating on shallow water with a variable seabed topology. This has prompt researchers to figure out ways to mitigate the response of the structure under wave action. The conventional way of reducing the response of the structure under wave action is by constructing breakwaters close to the VLFS in order to attenuate the wave forces. Utsunomiya et al. [14] carried out hydroelastic analysis on VLFS with and without the presence of bottom-founded type breakwater. Tay et al. [39] showed that the floating-type breakwater is effective in attenuating the wave forces impacting on the two adjacently placed floating fuel storage modules. The effects of breakwaters on the responses of their respective VLFS are shown in Fig. 14.

The use of breakwaters to attenuate the wave forces impacting on VLFS is relatively expensive and requires more time for construction. This prompted engineers to invent anti-motion devices that are attached to the VLFS. For example, Takagi et al. [29] proposed a submerged box-shape device attached to the fore-end of the floating structure (see Fig. 15a). He carried out hydroelastic analysis and showed that the anti-motion performance of the box-shape device is effective in reducing the deformation, shearing force and bending moment of the floating structure (see Figs. 15b-d).

Ohta et al. [28] also carried out hydroelastic analysis as well as experimental tests to investigate the effect of a submerged vertical plate (Fig. 16a) on the hydroelastic deflection of VLFS. The reduction in deflection was found to increase with increasing submerged plate depth and decreasing wave length. Ohta et al. [28] and

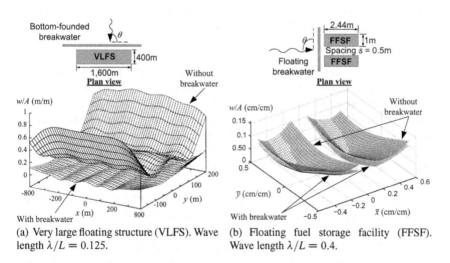

(a) Very large floating structure (VLFS). Wave length $\lambda/L = 0.125$.

(b) Floating fuel storage facility (FFSF). Wave length $\lambda/L = 0.4$.

Fig. 14 Comparison of deflection amplitude with and without breakwater for (a) very large floating structure (b) floating fuel storage facility.

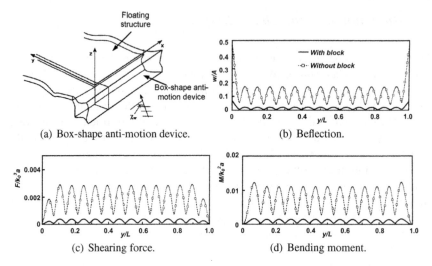

<table>
<tr><td>(a) Box-shape anti-motion device.</td><td>(b) Beflection.</td></tr>
<tr><td>(c) Shearing force.</td><td>(d) Bending moment.</td></tr>
</table>

Fig. 15 (a) box-shape anti-motion device. Effect of box-shape anti-motion device towards reduction in (b) deflection (c) shearing force (d) bending moment. Wave period $T = 6$s. Wave direction: Head sea.

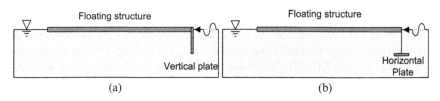

(a)   (b)

Fig. 16 Very large floating structure with attached submerged anti-motion device in the form of (a) vertical plate (b) horizontal plate.

Watanabe *et al.* [30,31] carried out hydroelastic analysis on VLFS with a submerged horizontal plate attached at the fore-end of the structure (Fig. 16b). They found that the horizontal plate is effective in reducing the deflection of the floating structure if the plate is placed at an appropriate depth from the floating structure. However, the depth requirement of these box-shape and submerged plate anti-motion devices will result in an increase in steady drift forces. Hence the size and cost of the mooring dolphin system will also increase correspondingly.

Khabakhpasheva and Korobkin [40] proposed a novel design in reducing the hydroelastic response of the VLFS by connecting the VLFS with an auxiliary module using hinge or semi rigid joints. They found that the auxiliary structure is able to reduce the deflection of the main floating structure significantly. Wang *et al.* [41] then proposed the use of semi-rigid mechanical joints to connect the modules as shown in Fig. 17a. They found that by using an appropriate rotational stiffness ε and location of such a mechanical joint, the hydroelastic response for the interconnected floating beams can be significantly reduced as shown in Fig. 17b.

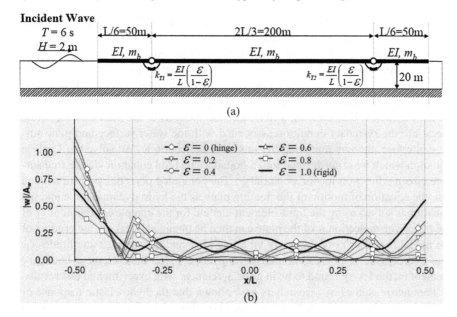

Fig. 17 (a) Schematic diagram of interconnected beams (b) Normalised maximum deflection of interconnected beams with varying rotational stiffness ξ.

Wang *et al.* [41] claimed that the semi-rigid mechanical joint is more effective in reducing the hydroelastic response when compared to the rigid joint or a simple hinge mechanical joint.

7 Summary and Future Studies on Hydroelastic Analysis/Response of VLFS

We have presented the mathematical formulation for the hydroelastic analysis of pontoon-type VLFS in the frequency domain framework. In order to reduce computational time, a simplified plate model having the same dynamic properties (in terms of vibration modes and natural periods) as the actual VLFS is generally used for the hydroelastic analysis.

In the hydroelastic analysis, we have employed the coupled finite-element boundary-element (FE-BE) method to solve the coupled water-plate problem. We assume the water to be a perfect fluid and the fluid motion is irrotational, so that a velocity potential φ exists. The fluid motion is governed by the Laplace equation $\nabla^2 \varphi = 0$ and the potential φ must satisfy the Neumann boundary conditions on the wetted surface of the floating bodies, the free surface and the seabed as well as the Sommerfeld radiation condition at the far field. On the other hand, the plate is assumed to be perfectly flat and may be modelled using the Mindlin plate theory. To obtain the deflection w of the floating plate, one needs to solve the plate equation of motion that unfortunately involves the fluid velocity potential φ. In order to determine the velocity potential φ, one needs to solve the Laplace equation which in turn has a surface boundary condition in contact with the floating plate that contains the unknown deflection w. In order to decouple this water-plate interaction problem, we have adopted the modal expansion method. In this method, we take a series of modal functions obtained from a free vibration analysis of the plate and substitute them into the boundary condition associated with the water surface and plate surface interface, thereby removing the unknown deflection w. We solve the Laplace equation which was transformed into a boundary integral equation via the free surface Green's function for the potential φ. The computed potential φ is substituted into the equation of motion of the floating plate as the hydrodynamic force and the equation is solved using the finite element method for the complex amplitudes ζ_l of the plate. The deflection w of the plate can then be obtained by taking the product of the modal functions c_l of the freely vibrating plate and the complex amplitudes ζ_l. The hydroelastic responses computed from this coupled finite-element boundary element method were found to be in good agreement with experimental test results.

Parametric studies by researchers have shown that the hydroelastic response of VLFS is significantly affected by the flexural rigidity of the floating structure, wave length and variation in water depth due to the undulating sea bed.

Various methods for reducing the hydroelastic response of the VLFS were presented. Breakwaters were found to be effective in attenuating the wave forces but are expensive and time consuming to construct, especially the bottom-founded type. On the other hand, the submerged anti-motion devices that are attached to the floating structure are relatively easy to install and effective in reducing motion. Herein, we have discussed three different types of submerged anti-motion devices (i.e. the box-shape device, submerged vertical and horizontal plate). These anti-motion devices were found to be effective in reducing the hydroelastic responses of floating structures if properly designed. However, these anti-motion devices may induce large steady drift forces on the VLFS due to their relatively large submerged surfaces. A novel method in mitigating the response of the structure is to attach auxiliary horizontal plates to the VLFS using mechanical joints of varying rotational stiffness. Preliminary studies have shown that semi-rigid mechanical joints are more effective in reducing the hydroelastic deflection of the VLFS as compared to purely hinge connection or rigid connection.

Despite considerable research effort being made in developing hydroelastic analysis tools for VLFS, studies are still required. These future studies include

- *Non-linear analysis* The frequency domain analysis is not valid in extreme situations such as in a large storm. Such extreme events have to be considered in the VLFS design for safety and survivability reasons. More work should be done to investigate the non-linear responses such as the transient response of VLFS under large wave impact [46].
- *Arbitrary shaped VLFS* As opposed to the conventional rectangular shaped VLFS, architects prefer aesthetically pleasing shapes of VLFS such as the complicated shape for the floating crab restaurant as shown in Fig. 2d. Research work

is needed to develop efficient hydroelastic analysis method to handle complicated shapes of VLFS.

- *Wave-structure-liquid interaction* For VLFS used as floating fuel storage modules, it is necessary to consider the free surface effect due to the sloshing behavior of fuel in the storage modules. The free surface effect might be crucial towards the response of the floating storage module especially when resonance occurs. Some research studies have been made on the sloshing of liquid in tankers but without considering the coupled interaction of liquid-structure with the wave action. Moreover, the structure is usually assumed to be a rigid body in these studies. Wave-structure-fluid interaction that takes into account the elastic behavior of the structure is a potential line of investigation.

- *Articulated VLFS* While studies on the interconnected floating beams (Fig. 17) only focused on 2-D problems and a study has been done on articulated floating plates [47], future studies could be made on optimiziting the rotational stiffness of the connections and their positions in order to develop more effective connector system for reducing the hydroelastic responses.

- *Hydroelastic analysis using Navier-Stokes equation* Vorticity and viscous effects are usually neglected in the hydroelastic analysis using the potential theory. However, high vorticity of fluid is expected from tsunami waves impacting VLFS or in the case of VLFS fitted with anti-motion devices such as bilge keels and submerged plates. In such situations, the fluid has to be represented by the Navier-Stokes (NS) equation that allows for rotational flow.

- *Slowly varying drift forces* The mooring dolphin systems that hold the floating modules in place are usually designed based on the drift/mean forces. Further research studies are needed to include the slowly varying drift forces in the design of the mooring dolphin system for the VLFS located in multi-directional random seas.

References

1. Pernice R (2009) Japanese urban artificial islands: an overview of projects and schemes for marine cities during 1960-1990s. J Archit Plann, AIJ, 74(641):1847–1855
2. Suzuki H. and Yoshida K. Design flow and strategy for safety of very large floating structure. Proc of Int Wksp on Very large Floating Struct, VLFS'96, Hayana, Japan, 1996, 21–27
3. Chakrabarti SK (1988) Hydrodynamics of Offshore Structures. WIT Press, UK
4. John F (1949) On the motion of floating bodies. Com on Pure and Appl Math, Part I, 2:13–57
5. John F (1950) On the motion of floating bodies. Com on Pure and Appl Math, Part II, 3:45–100
6. Wehausen J and Laitone E (1960) Surface wave. Fluid Dynamics III, Handbuch der Physik, Springer, Berlin
7. Newman JN (1977) Marine Hydrodynamics. The MIT Press, Cambridge
8. Newman JN (1985) Algorithm for the free-surface Green function. J Engrg Math, 19:57–67
9. Newman JN (1994) Wave effects on deformable bodies. Appl Ocean Res, 16:47–59
10. Noblesse F (1979) Potential theory of steady motion of ships. Department of Ocean Engineering Report, MIT
11. Mei CC (1989) The Applied Dynamics of Ocean Surface Waves - Advance Series on Ocean Engineering, Volume 1, World Scientific, Singapore

12. Ertekin RC, Riggs HR, Che XL and Du SX (1993) Efficient methods for hydroelastic analysis of very large floating structures. J Ship Res, SOEST No. 3155, 37(1):58–76
13. Yago K and Endo H (1996) On the hydroelastic response of box-shaped floating structure with shallow draft. J Soc Nav Arch Japan, 180:341–352
14. Utsunomiya T, Watanabe E and Eatock Taylor R. Wave response analysis of a box-like VLFS close to a breakwater. Proc of 17th Int Conf on Offshore Mech and Arctic Engrg, Lisbon, Portugal, July 5-9, 1998, OMAE98-4331
15. Kashiwagi M (1998) A b-spline Galerkin scheme for calculating the hydroelastic response of a very large floating structure in waves. J Mar Sci Technol, 3:37–49
16. Ohmatsu S. Numerical calculation method of hydroelastic response of a pontoon-type VLFS close to a breakwater. In: Ertekin RC, Kim JW (eds). Proc 3rd Int Wksp Very Large Floating Struct, University of Hawaii at Manao, Honolulu, Hawaii, USA, September 22-24, 1999, 2:805–811
17. Meylan MH and Squire VA (1996) Response of a circular ice floe to ocean waves. J Geophys Res, 101(C4):8869–8884
18. Meylan MH (1997) The forced vibration of a thin plate floating on an infinite liquid. J Sound and Vib, 205(5):581–591
19. Meylan MH (2001) A variation equation for the wave forcing of floating thin plates. J Appl Ocean Res, 23(4):195–206
20. Squire VA (2008) Synergies between VLFS hydroelasticity and sea ice research. Int J Offshore and Polar Engrg, 18(3):1–13
21. Hermans AJ (2000) A boundary element method for the interaction of free-surface waves with a very large floating flexible platform. J Fluids Struct, 14:943–956
22. Inoue K (2001) Stress analysis of Mega-Float in waves by two-step method. Proc of 20th Int Conf on Offshore Mech and Arctic Engrg, Rio de Janeiro, Brazil, June 3-8, 2001, OMAE2001/OSU-5209
23. Kada K, Fujita T and Kitabayashi K (2002) Stress analysis for structurally discontinuous parts in a Mega-Float structure. Int J Offshore and Polar Engrg, 12:48–55
24. Seto H, Ochi M, Ohta M and Kawakado M. Hydrodynamic response analysis of real very large floating structures in regular waves in open/sheltered sea. Proc of Int Symp on Ocean Space Utilization Tech, Tokyo, 2003, 85–93
25. Seto H, Ohta M, Ochi M and Kawakado M (2005) Integrated hydrodynamic-structural analysis of very large floating structures VLFS. Mar Struct, 18(2):181–200
26. Sasajima H. Local structural analysis of large floating structures. In: Ertekin RC, Kim JW (eds). Proc 3rd Int Wksp Very Large Floating Struct, University of Hawaii at Manao, Honolulu, Hawaii, USA, September 22-24, 1999, 2:602–606
27. Inoue K, Nagata S and Niizato H (2003) Stress analysis of detailed structures of Mega-Float in irregular waves using entire and local structural models. Proc 4th Int Wksp Very Large Floating Struct, Tokyo, Japan, 219–228
28. Ohta H, Torii T, Hayashi N, Watanabe E, Utsunomiya T, Sekita K and Sunahara S. Effect of attachment of a horizontal/vertical plate on the wave response of a VLFS. In: Ertekin RC, Kim JW (eds). Proc 3rd Int Wksp Very Large Floating Struct, University of Hawaii at Manao, Honolulu, Hawaii, USA, September 22-24, 1999, 265–274
29. Takagi K, Shimada K and Ikebuchi T (2000) An anti-motion device for a very large floating structure. Mar Struct, 13:421–436
30. Watanabe E, Utsunomiya T, Kuramoto M, Ohta H, Torii T and Hayashi N. Wave response analysis of VLFS with an attached submerged plate. Proc 12th Int Offshore Polar Engrg, Kitakyushu, Japan, May 26-31, 2002:319–326
31. Watanabe E, Utsunomiya T, Ohta H and Hayashi N. Wave response analysis of VLFS with an attached submerged plate: verification with 2-D model and some 3-D numerical examples. Proc Int Symp Ocean Space Utilisation technol, National Maritime Research Institute, Tokyo, Japan, January, 2003, 147–154
32. Kim JW and Webster WC (1998) The drag on an airplane taking off from a floating runway. J Mar Sci Tech, 3(2):76–81.

33. Watanabe E, Utsunomiya T and Tanigaki S (1998) A transient response analysis of a very large floating structure by finite element method. Structural Engrg/Earthquake Engrg, JSCE;15(2):155–163

34. Kashiwagi M (2000) A time-domain mode-expansion method for calculating transient elastic responses of a pontoon-type VLFS. J Mar Sci Technol, 5:89–100

35. Kashiwagi M (2004) Transient responses of a VLFS during landing and take-off of an airplane. J Mar Sci Tech, 9(1):14–23.

36. Watanabe E, Utsunomiya T, Wang CM and Le Thi Thu Hang (2000) Benchmark hydroelastic responses of circular VLFS under wave action. Engrg Struct, 28(6):423–30

37. Wang CD and Wang CM. A comparative study on the linear wave response of a very large floating body modeled by a plate based on Kirchhoff and Mindlin plate theories. Proc of ICETECH: Structures in ice/iceberg populated waters, Banff, Canada, July 16-19, 2006, ICETECH06-157-RF

38. Wang CM, Song JH, Utsunomiya T, Koh HS and Lim YB. Hydroelastic analysis of floating performance stage @ Marina Bay, Singapore. Proc of 27th Int Conf on Offshore Mech and Arctic Engrg, Estoril, Portugal, June 15-20, 2008, OMAE2008-57324

39. Tay ZY, Wang CM and Utsunomiya T (2009) Hydroelastic responses and interactions of floating fuel storage modules placed side-by-side with floating breakwater. Mar Struct, 22(3):633–658

40. Khabakhpasheva TI and Korobkin AA (2002) Hydroelastic behavior of compound floating plate in waves. J Engrg Maths, 44:21–24

41. Wang CM, Muhammad R and Choo YS. Reducing hydroelastic response of interconnected floating beams using semi-rigid connections. Proc of 28th Int Conf on Offshore Mech and Arctic Engrg, Honolulu, Hawaii, May 31-June 5, 2009, OMAE2009-79692

42. Mindlin RD (1951) Influence of rotary inertia and shear on flexural motions of isotropic, elastic plates. J Appl Mech, 18:31–38

43. Wang CM, Watanabe E and Utsunomiya T (2008) Very Large Floating Structures. Routledge, UK: Taylor and Francis

44. Sarpkaya T and Isaacson M (1981) Mechanics of Wave Forces on Offshore Structrures. Van Nostrand Reinhold, New York, USA

45. Utsunomiya T and Watanabe E (2006) Fast multipole method for wave diffraction/radiation problems and its application to VLFS. Int J Offshore and Polar Engrg, 16(4):253–260

46. Watanabe E, Utsunomiya T and Wang CM (2003) Hydroelastic analysis of pontoon-type VLFS: a literature survey. Eng Struc, 26:245–256

47. Xia D, Kim JW and Cengiz Ertekin R (2000) On the hydroelastic behavior of two-dimensional articulated plates. Mar Struct, 13:261–278

48. Linton CM (1999) Rapidly convergent representation for Green's functions for Laplace's equation. Proc Roy Soc, London, A(455):1767–1797

49. Petyt M.(1990) Introduction to Finite Element Vibration Analysis. Cambridge University Press, Cambridge, UK

50. Cengiz Ertekin R, Wang SQ, Che SL and Riggs HR (1995) On the application of the Haskind-Hanaoka relations to hydroelastic problems. Mar Struct, 8:617–629

Appendix 1 Solution for Velocity Potential Components: Radiated and Diffracted Potentials

The boundary value problem defined by Eqs. (21) – (26) can be solved by using the boundary element method. To transform the Laplace equation into a boundary integral equation, we use the free-surface Green function that satisfies the boundary conditions given by Eqs. (22) – (26). The boundary integral equation for each velocity potential component can be defined as [43]

$$-2\pi\varphi_l(\mathbf{x}) + \int_S \frac{\partial G(\mathbf{x}, \mathbf{x}')}{\partial n}\varphi_l(\mathbf{x}')\,dS = \int_S G(\mathbf{x}, \mathbf{x}')\frac{\partial \varphi_l(\mathbf{x}')}{\partial n}\,dS, \qquad (27)$$

where $\mathbf{x} = (x, y, z)$ is the source point and $\mathbf{x}' = (x', y', z')$ the field point. $G(\mathbf{x}, \mathbf{x}')$ is a free-surface Green's function for water of finite depth and is given by [48]

$$G(\mathbf{x}, \mathbf{x}') = \sum_{m=0}^{\infty} \frac{2K_0(k_m R)}{\frac{h}{2}\left(1 + \frac{\sin 2k_m H}{2k_m H}\right)} \cos k_m(z + H)\cos k_m(z' + H), \qquad (28)$$

where k_m is a positive root number satisfying $k_m \tan k_m H = -\omega^2$, with $(m \geq 1)$ and $k_0 = ik$. K_0 is the modified Bessel function of the second kind and R represents the horizontal distance between $\mathbf{x}$ and $\mathbf{x}'$.

We refer to Eq. (27) as the integral equation for the water. By further introducing the boundary conditions given by Eq. (22) for $\partial\varphi/\partial n$ into Eq. (27), Wang et al. [43] obtained

$$-2\pi\varphi_l(\mathbf{x}) + \int_{S_{HS} \cup S_{HB}} \frac{\partial G(\mathbf{x}, \mathbf{x}')}{\partial n}\varphi_l(\mathbf{x}')\,dS$$
$$= \begin{cases} -i\omega \int_{S_{HB}} G(\mathbf{x}, \mathbf{x}')c_l^w(\mathbf{x}')\,dS & \text{for } l = 1, 2, ..., N \\ 4\pi\varphi_{In} & \text{for } l = D \end{cases}, \qquad (29)$$

where S_{HS} is the wetted side surface of the floating body and S_{HB} the wetted bottom surface of the floating body. The integral equation (29) is then solved for φ_l by using the constant panel method [44]. The integration in Eq. (29) is performed using the Gauss quadrature method.

With the computed φ_l, the radiated potential φ_R and diffracted potential φ_D (introduced in Eq. 20) can be written in the matrix form as

$$\{\varphi_R\} = \{\varphi_l\}\{\zeta_l\} = -i\omega\{\tilde{\varphi}_l\}\{\hat{c}_l^w\}\{\zeta_l\}$$
$$= -i\omega\left(\frac{1}{2\pi}\left[-[\mathbf{I}] + \frac{1}{2\pi}\left[\frac{\partial\mathbb{G}}{\partial n}\right]\right]^{-1}[\mathbb{G}]\right)\{\hat{c}_l^w\}\{\zeta_l\}, \qquad (30)$$

$$\{\varphi_D\} = 2\left[-[\mathbf{I}] + \frac{1}{2\pi}\left[\frac{\partial\mathbb{G}}{\partial n}\right]\right]^{-1}\{\varphi_{In}\}, \qquad (31)$$

where $[\mathbb{G}]$ is the global matrix for the free surface Green function, $[\mathbf{I}]$ the identity matrix, $\{\hat{c}_l^w\}$ the mode shapes (eigenvectors) obtained by performing a free vibration analysis on the plate and

$$\{\tilde{\varphi}_l\} = \frac{1}{2\pi}\left[-[\mathbf{I}] + \frac{1}{2\pi}\left[\frac{\partial\mathbb{G}}{\partial n}\right]\right]^{-1}[\mathbb{G}]. \qquad (32)$$

Appendix 2 Solving Plate Eq. using Finite Element Method

The governing equations [(5) to (7)] of motion of the Mindlin plate are solved by using the finite element method (FEM). For a rectangular plate, consider using the 8-node rectangular element of area $2a \times 2b$ as shown in Fig. (18a) (see ref. [49] for details). By mapping into the $\xi\eta$-space [Fig. (18b)] where $\xi = x/a$ and $\eta = y/b$, the displacement field vector $\mathbf{w}^e$ in the element domain Δ_e may be discretised as

$$\mathbf{w}^e = \sum_{j=1}^{8} \left[\hat{N}_j^e\right]\{\hat{\mathbf{w}}_j^e\} = \sum_{j=1}^{8} \begin{bmatrix} N_j^e & 0 & 0 \\ 0 & N_j^e & 0 \\ 0 & 0 & N_j^e \end{bmatrix} \{\hat{w}_j^e \ \hat{\psi}_{xj}^e \ \hat{\psi}_{yj}^e\}^T, \tag{33}$$

where $\hat{w}_j^e$ is the transverse deflection, $\hat{\psi}_{xj}^e$ the rotation about the η-axis and $\hat{\psi}_{yj}^e$ the rotation about the ξ-axis at the j-th node of the plate element e [see Fig. (18b)]. The basis functions N_j^e are given by

$$N_j^e(\xi, \eta) = \frac{1}{4}\left(1 + \xi_j\xi\right)\left(1 + \eta_j\eta\right)\left(\xi_j\xi + \eta_j\eta - 1\right) \text{ for nodes } j = 1, 2, 3, 4, \tag{34}$$

$$N_j^e(\xi, \eta) = \frac{1}{2}\left(1 - \xi^2\right)\left(1 + \eta_j\eta\right) \text{ for nodes } j = 5, 6, \tag{35}$$

$$N_j^e(\xi, \eta) = \frac{1}{2}\left(1 + \xi_j\xi\right)\left(1 - \eta^2\right) \text{ for nodes } j = 7, 8. \tag{36}$$

In view of the assumed displacement field, the elemental matrices for flexural stiffness $\left[k_f\right]$, shear stiffness $[k_s]$ and mass $[m]$ are given by

$$\left[k_f\right]_{24\times24} = \int_{\Delta_e} \frac{h^3}{12}\left[B_f^e\right]^T [D]\left[B_f^e\right] dA_e, \tag{37}$$

$$[k_s]_{24\times24} = \int_{\Delta_e} \kappa^2 h \left[B_s^e\right]^T [D_s]\left[B_s^e\right] dA_e, \tag{38}$$

$$[m]_{24\times24} = \int_{\Delta_e} \gamma \left[\hat{N}^e\right]^T \begin{bmatrix} h & 0 & 0 \\ 0 & \frac{h^3}{12} & 0 \\ 0 & 0 & \frac{h^3}{12} \end{bmatrix} \left[\hat{N}^e\right] dA_e, \tag{39}$$

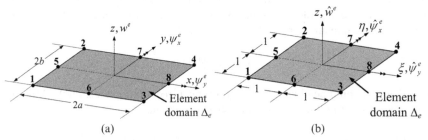

(a)　　　　　　　　　　　　　　　　　　(b)

Fig. 18 8-node rectangular Mindlin plate element (a) in physical space and (b) mapped into $\xi\eta$-space.

where $dA_e = d\xi d\eta$. The flexural rigidity $[D]$, shear rigidity $[D_s]$, flexural strain-displacement matrix $\left[B_f^e\right]$, the shear strain-displacement matrix $\left[B_s^e\right]$ and the basis function matrix $\left[\hat{N}^e\right]$ are given by

$$[D]_{3\times3} = \frac{E}{(1-v^2)} \begin{bmatrix} 1 & v & 0 \\ v & 1 & 0 \\ 0 & 0 & \frac{1-v}{2} \end{bmatrix}, \tag{40}$$

$$[D_s]_{2\times2} = \frac{E}{2(1+v)} \begin{bmatrix} 1 & 0 \\ 0 & 1 \end{bmatrix}, \tag{41}$$

$$\left[B_f^e\right]_{3\times24} = \begin{bmatrix} 0 & 0 & -\frac{1}{a}\frac{dN_1^e}{d\xi} & \cdots & 0 & 0 & -\frac{1}{a}\frac{dN_8^e}{d\xi} \\ 0 & \frac{1}{b}\frac{dN_1^e}{d\eta} & 0 & \cdots & 0 & \frac{1}{b}\frac{dN_8^e}{d\eta} & 0 \\ 0 & \frac{1}{a}\frac{dN_1^e}{d\xi} & -\frac{1}{b}\frac{dN_1^e}{d\eta} & \cdots & 0 & \frac{1}{a}\frac{dN_8^e}{d\xi} & -\frac{1}{b}\frac{dN_8^e}{d\eta} \end{bmatrix}, \tag{42}$$

$$\left[B_s^e\right]_{2\times24} = \begin{bmatrix} \frac{1}{a}\frac{dN_1^e}{d\xi} & 0 & N_1^e & \cdots & \frac{1}{a}\frac{dN_8^e}{d\xi} & 0 & N_8^e \\ \frac{1}{b}\frac{dN_1^e}{d\eta} & N_1^e & 0 & \cdots & \frac{1}{b}\frac{dN_8^e}{d\eta} & N_8^e & 0 \end{bmatrix}, \tag{43}$$

$$\left[\hat{N}^e\right]_{3\times24} = \begin{bmatrix} N_1^e & 0 & 0 & \cdots & N_8^e & 0 & 0 \\ 0 & N_1^e & 0 & \cdots & 0 & N_8^e & 0 \\ 0 & 0 & N_1^e & \cdots & 0 & 0 & N_8^e \end{bmatrix}. \tag{44}$$

The equation of motion for the Mindlin plate element is given by

$$\left([k_f] + [k_s] + [k_{rf}] - \omega^2 [m]\right) \{\hat{w}^e\} = \{f\}, \tag{45}$$

where $\left[k_{rf}\right] = \int_{A_e} \left[\hat{N}^e\right]^T \left[\hat{N}^e\right] dA_e$ is the restoring force, $\{f\} = i\omega \int_{A_e} \{\varphi^e\} dA_e$ the hydrodynamic force and ω the angular velocity of the wave. By assembling Eq. (45) into the global form, one obtains

$$\left([\mathbb{K}_f] + [\mathbb{K}_s] + [\mathbb{K}_{rf}] - \omega^2 [\mathbb{M}] + \omega^2 [\mathbb{M}_a]\right) \{\hat{w}\} = \{\mathbb{F}\}, \tag{46}$$

where $\left[\mathbb{K}_f\right]$, $[\mathbb{K}_s]$, $\left[\mathbb{K}_{rf}\right]$, $[\mathbb{M}]$ and $[\mathbb{M}_a]$ are the global flexural stiffness matrix, the global shear stiffness matrix, the global restoring force matrix, the global mass matrix and the global added mass matrix, respectively. The displacement vector $\{\hat{w}\}$ may be expanded in an appropriate set of modes $\{\hat{c}_l\}$ as

$$\{\hat{w}\} = \{\hat{c}_l\}\{\zeta_l\}. \tag{47}$$

Note that $\{\hat{c}_l\}$ could be obtained by performing a free vibration test on the Mindlin plate. By substituting Eq. (47) into Eq. (46), one obtains

$$\{\hat{c}_l^T\} \left([\mathbb{K}_s] + \left[\mathbb{K}_f\right] + \left[\mathbb{K}_{rf}\right] - \omega^2 [\mathbb{M}] + \omega^2 [\mathbb{M}_a]\right) \{\hat{c}_l\}\{\zeta_l\} = \{\hat{c}_l^T\}\{\mathbb{F}\} = \{\tilde{\mathbb{F}}\}, \tag{48}$$

where $l = 1, 2, ..., N$ and $\{\tilde{\mathbb{F}}\} = \{\hat{c}_l^T\}\{\mathbb{F}\}$ is the exciting force. By using the computed velocity potentials from Eqs. (30) and (31), the global added mass $[\mathbb{M}_a]$ and the exciting force $\{\tilde{\mathbb{F}}\}$ can be calculated from

$$[\mathbb{M}_a] = \int_{S_{HB}} \{\tilde{\varphi}_l\} dS = \sum_{i=1}^{M} u_i \{\tilde{\varphi}_l(\mathbf{x}_i)\}, \tag{49}$$

$$\{\tilde{\mathbb{F}}\} = i\omega\{\hat{c}_l^T\} \int_{S_{HB}} \{\varphi_D\} dS = i\omega\{\hat{c}_l^T\} \sum_{i=1}^{M} u_i \{\varphi_D(\mathbf{x}_i)\}, \tag{50}$$

where $\mathbf{x} \in \Delta_e$; $\mathbf{x}_i$ and u_i are sets of M integration points and their corresponding weights over Δ_e. The generalized added mass is given as a complex matrix, and thus it includes the effect of radiation damping [43].

It should be noted that $\{\hat{c}_l\}$ is a $P \times N$ matrix, $\{\mathbb{F}\}$ is a $P \times 1$ matrix and the bracketed ($\bullet$) term on the left hand side of Eq. (48) is a $P \times P$ matrix, where P is the number of nodes in the plate domain and N the total number of modes. By multiplying the inverse of the bracketed ($\bullet$) term on the left hand side of Eq. (48) with $\{\tilde{\mathbb{F}}\}$, the complex amplitudes $\{\zeta_l\}$ which is a $P \times 1$ matrix could be obtained. These complex amplitudes $\{\zeta_l\}$ are then back-substituted into Eq. (47) for the plate deflection $\{\hat{\mathbf{w}}\}$.

Note that one can also express the exciting force in Eq. (50) in terms of the unit-amplitude radiated potential $\varphi_r = \varphi_{l=1,2,...,N}$ and the incident potential φ_{In}, thereby eliminating the need to solve for the scattered potential φ_S. The alternative expression of the exciting force is derived as follows. The exiting force in Eq. (50) can be written in the integral form as

$$F = i\omega c_l^T \int_{S_{HB}} \varphi_D dS_{HB}. \tag{51}$$

By substituting the bottom hull surface boundary condition for the unit-amplitude radiated potential (i.e. $\partial\varphi_r/\partial n = -i\omega c_l = -i\omega c_l^T$) into Eq. (51), the exciting force F can be re-written as

$$F = -\frac{\partial\varphi_r}{\partial n} \int_{S_{HB}} \varphi_D \, dS_{HB} = -\frac{\partial\varphi_r}{\partial n} \int_{S_{HB}} (\varphi_{In} + \varphi_S) \, dS_{HB}$$

$$= -\int_{S_{HB}} \left(\varphi_{In} \frac{\partial\varphi_r}{\partial n} + \varphi_S \frac{\partial\varphi_r}{\partial n} \right) dS_{HB}. \tag{52}$$

The application of the Green's second identity over the unit-amplitude radiated potential φ_r and the scattered potential φ_S gives

$$\int_{S_{HB}} \left(\varphi_S \frac{\partial\varphi_r}{\partial n} - \varphi_r \frac{\partial\varphi_S}{\partial n} \right) dS_{HB} = 0 \Longrightarrow \int_{S_{HB}} \varphi_S \frac{\partial\varphi_r}{\partial n} \, dS_{HB}$$

$$= \int_{S_{HB}} \varphi_r \frac{\partial\varphi_S}{\partial n} \, dS_{HB}. \tag{53}$$

By substituting Eq. (53) into Eq. (52) and using the bottom hull surface boundary condition for the scattered potential $\partial \varphi_S / \partial n = -\partial \varphi_{In} / \partial n$, one obtains the exciting force in the form of

$$F = \int_{S_{HB}} \left(\varphi_r \frac{\partial \varphi_{In}}{\partial n} - \varphi_{In} \frac{\partial \varphi_r}{\partial n} \right) dS_{HB}. \tag{54}$$

This exciting force expression (54), in terms of the unit-amplitude radiated potential and incident potential, is referred to as the Haskind-Hanaoka relation [50]. The Haskind-Hanaoka relation may be used to check the exciting force obtained from the pressure distribution method.

Efficient Numerical Simulation and Optimization of Fluid-Structure Interaction

M. Schäfer, D.C. Sternel, G. Becker, and P. Pironkov

Abstract The paper concerns the efficient numerical simulation and optimization of fluid-structure interaction (FSI) problems. The basis is an implicit partitioned solution approach involving the finite-volume flow solver FASTEST, the finite-element structural solver FEAP, and the coupling interface MpCCI. Special emphasis is given to the grid moving techniques for which algebraic and elliptic approaches are considered. The possibilities for accelerating the computations by the usage of multigrid methods, adaptive underrelaxation, and displacement prediction are discussed. A concept for integrating the FSI solver into an optimization procedure for FSI problems is presented. Numerical results are given to illustrate the capabilities of the approaches considered.

1 Introduction

Fluid-structure interaction (FSI) problems occur in many applications in industry and science. The numerical efficiency of the underlying coupled solution algorithm is essential for a realistic numerical simulation of such kind of problems. In particular, this is important for a simulation based optimization of FSI problems requiring multiple runs of the coupled solver. In the present paper we consider an implicit partitioned solution approach, which combines the advantages of weakly and strongly coupled schemes. The method is realized based on the finite-volume flow solver FASTEST [1] involving an ALE formulation, the finite-element structural solver FEAP [28], and the coupling interface MpCCI [2]. For each time step the implicit solution procedure consists in the application of different nested iteration processes for linearization, pressure-velocity coupling, and linear system solving, which are linked by an iterative fluid-structure coupling procedure. Several techniques are considered to improve the efficiency and stability of the coupled solution procedure:

M. Schäfer, D.C. Sternel, G. Becker, and P. Pironkov
Institute of Numerical Methods in Mechanical Engineering, Technische Universität Darmstadt,
Dolivostr. 15, 64293 Darmstadt, Germany
e-mail: schaefer@fnb.tu-darmstadt.de, www.fnb.tu-darmstadt.de

H.-J. Bungartz et al. (eds.), *Fluid Structure Interaction II*, Lecture Notes
in Computational Science and Engineering 73, DOI 10.1007/978-3-642-14206-2_6,
© Springer-Verlag Berlin Heidelberg 2010

- advanced grid moving techniques,
- adaptive under-relaxation of the structural displacements,
- prediction of structural displacements,
- multigrid methods.

Special emphasis is given to the grid moving which is a crucial issue when large deformations of the structure are involved. Different strategies based on linear interpolation, transfinite mapping, and Laplace equations are considered and combined in order to ensure a dynamic grid movement procedure, which preserves the quality of the block-structured fluid grid even for large three-dimensional structure deformations.

Based on the FSI solver an integrated approach for the shape optimization of FSI problems is presented. Corresponding approaches have not been deeply investigated yet in the literature [15]. We employ NURBS surfaces for the shape representation and a derivative-free mathematical optimization technique. A key feature of the integrated method is the unified treatment of structural deformation due to FSI and shape variation for fulfilling the optimization objectives.

For all aspects mentioned, results for representative test cases are given which illustrate the functionality and performance of the approaches considered.

2 Governing Equations

For the fluid domain part Ω_f we assume a flow of an incompressible Newtonian fluid. In this case the basic conservation equations governing transport of mass and momentum for a fluid control volume V_f with surface S_f are given by

$$\int_{S_f} \mathbf{v} \cdot \mathbf{n} \, dS_f = 0 \,, \tag{1}$$

$$\frac{\partial}{\partial t} \int_{V_f} \rho_f \mathbf{v} \, dV_f + \int_{S_f} \rho_f (\mathbf{v} - \mathbf{v}^g)(\mathbf{v} \cdot \mathbf{n}) \, dS_f = \int_{V_f} \rho_f \mathbf{f}_f \, dV_f + \int_{S_f} \mathbf{T}_f \cdot \mathbf{n} \, dS_f \,, \tag{2}$$

where $\mathbf{v}$ denotes the velocity vector with respect to Cartesian coordinates $\mathbf{x}$, t denotes the time, ρ_f the fluid density, $\mathbf{n}$ the outward normal vector, and $\mathbf{f}_f$ are external volume forces (e.g., buoyancy forces). $\mathbf{v}^g$ is the velocity with which S_f may move (grid velocity) due to displacements of solid parts. The Cauchy stress tensor $\mathbf{T}_f$ for incompressible Newtonian fluids is defined by

$$\mathbf{T}_f = \mu_f \left(\nabla \mathbf{v} + \nabla \mathbf{v}^T \right) - p\mathbf{I} \tag{3}$$

with the pressure p, the dynamic viscosity μ_f, the vector gradient ∇, and the identity tensor $\mathbf{I}$.

For the structure we denote a material point in the reference configuration as $\mathbf{X}$ whose position in the current configuration is given by

$$\mathbf{x} = \chi(\mathbf{X}, t). \tag{4}$$

The displacements are evaluated by

$$\mathbf{u} = \mathbf{x} - \mathbf{X}. \tag{5}$$

For more details see [18, 30]. The basic balance equation for momentum for the solid domain Ω_s can be written as

$$\nabla \cdot (\mathbf{F}_s \mathbf{S}_s^T) + \rho_s \mathbf{f}_s = \rho_s \ddot{\chi}, \tag{6}$$

where $\ddot{\chi} = \partial^2 \chi(\mathbf{X}, \mathbf{t})/\partial t^2$ denotes the acceleration, $\mathbf{S}_s$ the second Piola-Kirchhoff stress tensor, ρ_s the density of the solid, and $\mathbf{f}_s$ external volume forces acting on the solid (e.g., gravitational forces). $\mathbf{F}_s = \partial \chi/\partial \mathbf{X}$ represents the deformation gradient.

In the present investigation we consider the Saint Venant-Kirchhoff material law

$$\mathbf{S}_s = \lambda_s \operatorname{tr}(\mathbf{E})\, \mathbf{I} + 2\mu_s \mathbf{E} \tag{7}$$

with the Green-Lagrangian strain tensor

$$\mathbf{E} = \frac{1}{2}(\mathbf{F}_s^T \mathbf{F}_s - \mathbf{I}), \tag{8}$$

as kinematic property. λ_s and μ_s are the two Lamé constants.

The problem formulation has to be closed by prescribing suitable boundary and interface conditions. On solid and fluid boundaries standard conditions as for individual solid and fluid problems can be prescribed. For the velocities and the stresses on a fluid-solid interface we have the conditions

$$\mathbf{v} = \dot{\chi} = \mathbf{v}_b \quad \text{and} \quad \mathbf{T}_f \mathbf{n} = \mathbf{T}_s \mathbf{n}, \tag{9}$$

where $\mathbf{v}_b$ is the velocity of the interface and $\mathbf{T}_s = \mathbf{F}_s \mathbf{S}_s \mathbf{F}_s^T / \det \mathbf{F}_s$ is the Cauchy stress tensor of the solid.

3 Basic Numerical Techniques

Solid parts of the problem domain are treated by the finite-element solver FEAP (see [28]). For the fluid parts the block-structured parallel multigrid finite-volume flow solver FASTEST is employed (see [1, 23]). Both solvers involve second-order spatial discretizations and fully implicit second-order time discretizations.

The fluid-structure coupling is realized via an implicit partitioned approach (see also [22]). In Fig. 1 a schematic view of the iteration process, which is performed for

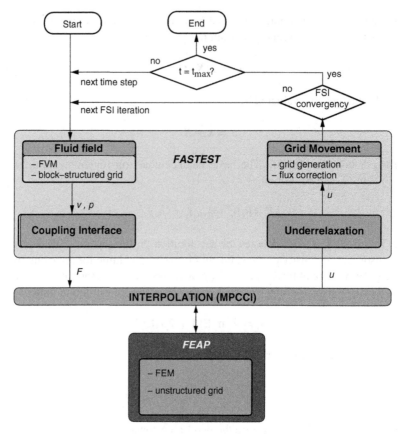

Fig. 1 Flow chart of coupled solution procedure.

each time step, is given. After the initializations the flow field is determined in the actual flow geometry. Based on this the friction and pressure forces on the interacting walls are computed and passed to the structural solver as boundary conditions. The structural solver computes the deformations, which are used to modify the fluid mesh. Then the flow solver is started over again.

For the mesh deformation algebraic and elliptic approaches are employed which are described in detail in Sect. 4. Within the fluid solver a discrete form of the space conservation law

$$\frac{\partial}{\partial t} \int_{V_f} dV_f - \int_{S_f} \mathbf{v}^g \cdot \mathbf{n} \, dS_f = 0 , \tag{10}$$

is used to compute the additional convective fluxes in Eqs. (1) and (2). This is done using the swept volumes δV_c of the control volume faces for which one has the relation (see [9]):

$$\sum_c \frac{\delta V_c^n}{\Delta t_n} = \frac{V_f^n - V_f^{n-1}}{\Delta t_n} = \sum_c (\mathbf{v}^g \cdot \mathbf{n} S_f)_c^n , \tag{11}$$

where the summation index c loops over the faces of the control volume, the index n denotes the time level t_n and Δt_n is the time step size. By this way interface displacements enter the fluid problem part in a manner strictly ensuring mass conservation.

The fluid-structure interaction (FSI) iteration routine is repeated until a convergence criterion is reached, which is defined by the change of the mean displacements:

$$\mathfrak{R}^{FSI} = \frac{1}{N} \sum_{k=1}^{N} \frac{\|\mathbf{u}^{k,m-1} - \mathbf{u}^{k,m}\|_\infty}{\|\mathbf{u}^{k,m}\|_\infty} < \varepsilon , \qquad (12)$$

where m counts the FSI iterations, N denotes the number of interface nodes, and $\|\cdot\|_\infty$ the infinity norm.

The data transfer between the flow and the solid solvers within the partitioned solution procedure is performed via an interface realized by the coupling library MpCCI (see [2]). MpCCI is used for controlling the data communication as well as for carrying out the interpolations of the data within (possibly) non-matching fluid and solid grids. The initialization routine provides MpCCI the geometry information at the fluid-solid interface for both grids. Using these geometry informations the forces at the control volume centers of the fluid grid interface are interpolated onto the strucural grid and passed to the structural solver. The displacements from the structural solver at the nodes are transfered via MpCCI to the control volume vertices of the fluid grid interface. Afterwards the complete fluid grid is adapted (see Sect. 4) and the corresponding coordinates of the control volume centers are computed. An update of the geometry information of the solid grid is not necessary, since the structural finite-element computations always relate to the original solid grid. Note that nearly independent discretizations for the fluid and solid subproblems can be used. For the necessary interpolations MpCCI employs a conservative linear approach (see [2]).

4 Grid movements strategies

Several different strategies for the grid movement based on linear interpolation, transfinite mapping, and Laplace equations are used. Aim is to provide a dynamic grid distortion procedure, which preserves the grid quality even for large three dimensional structural deformations. As illustrated in Fig. 1 every FSI iteration causes a shape altering of the FSI interface which is followed by an update of the spatial discretization of the flow domain. This implies a blockwise movement of the grid which also takes the parallel implementation into account.

The information delivered by MpCCI from FEAP to FASTEST are the displacements of the coupling faces which are the basis of the grid movement. At first the grid of all blocks which contain a moved coupling face (primary blocks) should be updated. In addition, it is essential that the grid of the blocks sharing an edge with the coupling face are updated, too. Furthermore, a conditional face movement can

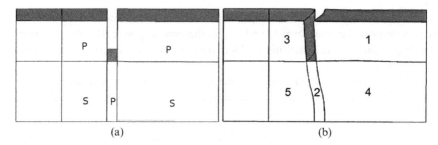

Fig. 2 (a) Primary and secondary blocks (initial block positions). (b) Order of grid adaption (distorted block positions).

be useful. For instance, the opposite face of a primary face can be distorted parallel to the primary face or angles of edges can be preserved. This causes also the grid updates within blocks containing no coupling face or an edge of a coupling face (secondary blocks). In Fig. 2 an example for primary and secondary blocks of a moveable grid is shown. The movement is induced by the structure inside the gap, the coupling faces are the three faces around the gap. Also the order of the treatment of the blocks must be defined (see Fig. 2).

In Fig. 3 the structure of the complete grid distortion procedure is shown. It works blockwise in the defined order and also takes parallelization into account. The procedure starts with the distortion of the edges of the moving block. The second step is the distortion of the block faces. Finally, the new coordinates of the faces are used as boundary conditions for the distortion of the internal grid. We briefly describe the components of the procedure. More details can be found in Yigit [33].

Let us consider a block with nodal coordinates $\mathbf{x}_{i,j,k} = (x_{i,j,k}, y_{i,j,k}, z_{i,j,k})$ with $i = 1, \ldots, N_i$, $j = 1, \ldots, N_j$, and $k = 1, \ldots, N_k$. A simple approach for the distortion of an edge is linear interpolation. For instance, for the edge along i-direction with $j = k = 1$, when the end points $\mathbf{x}_{1,1,1}$ and $\mathbf{x}_{N_i,1,1}$ are already distorted the linear interpolation for the internal points reads

$$\mathbf{x}_{i,1,1} = k_i (\mathbf{x}_{N_i,1,1} - \mathbf{x}_{1,1,1}) + \mathbf{x}_{1,1,1}, \tag{13}$$

where

$$k_i = \frac{1}{L} \sum_{m=2}^{i} \| \mathbf{x}_{m,1,1} - \mathbf{x}_{m-1,1,1} \|, \quad L = \sum_{i=2}^{N_i} \| \mathbf{x}_{i,1,1} - \mathbf{x}_{i-1,1,1} \|. \tag{14}$$

A cubic spline interpolation is employed, which preserves the angle between the edge and the connecting face. We define

$$\mathbf{a}_1 = -3\mathbf{x}_{1,1,1} + 3\mathbf{x}_{N_i,1,1} - 2\mathbf{T}_1 + \mathbf{T}_2, \tag{15}$$

$$\mathbf{a}_2 = 2\mathbf{x}_{1,1,1} - 2\mathbf{x}_{N_i,1,1} + \mathbf{T}_1 - \mathbf{T}_2 \tag{16}$$

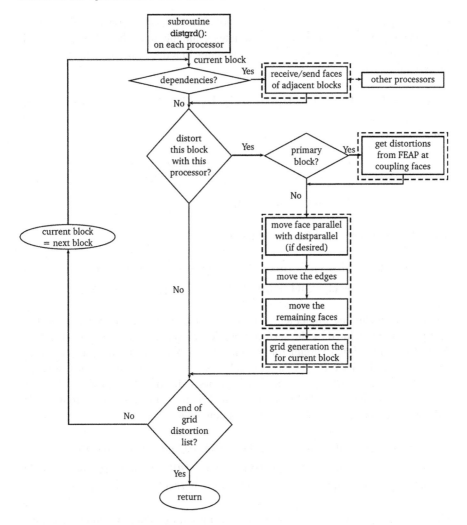

Fig. 3 Flow chart of the grid distortion routine.

and distort the internal points according to

$$\mathbf{x}_{i,1,1} = \mathbf{x}_{1,1,1} + k_i^3 \mathbf{a}_2 + k_i^2 \mathbf{a}_1 + k_i \mathbf{T}_1. \tag{17}$$

$\mathbf{T}_1$ and $\mathbf{T}_2$ are the direction vectors at the end points of the edge which are chosen such that the angles between the edge and connecting face does not change.

For the distortion of both, faces and blocks three different approaches are employed. The methods for the faces are two-dimensional variations of the (three-dimensional) methods used for blocks. The simplest technique is the linear interpolation, where Eq. (13) is used along each of the parametrical directions. Another algebraic approach is the transfinite interpolation (TFI). We consider a mapping between the physical domain with coordinates $\mathbf{x}$ and the unit cube as computational domain with coordinates (ξ, η, ζ). The mapping function according to transfinite interpolation can be written as

$$\mathbf{x}(\boldsymbol{\xi}) = \mathbf{U}_1 + \mathbf{U}_2 + \mathbf{U}_3 - \mathbf{U}_1 \cdot \mathbf{U}_2 - \mathbf{U}_2 \cdot \mathbf{U}_3 - \mathbf{U}_3 \cdot \mathbf{U}_1 + \mathbf{U}_1 \cdot \mathbf{U}_2 \cdot \mathbf{U}_3, \quad (18)$$

with

$$
\begin{aligned}
\mathbf{U}_1 &= (1 - \xi)\mathbf{x}(0, \eta, \zeta) + (1 + \xi)\mathbf{x}(1, \eta, \zeta), \\
\mathbf{U}_2 &= (1 - \eta)\mathbf{x}(\xi, 0, \zeta) + (1 + \eta)\mathbf{x}(\xi, 1, \zeta), \quad (19) \\
\mathbf{U}_3 &= (1 - \zeta)\mathbf{x}(\xi, \eta, 0) + (1 + \zeta)\mathbf{x}(\xi, \eta, 1),
\end{aligned}
$$

where the dot denotes the binary product, e.g.,

$$
\begin{aligned}
\mathbf{U}_1 \cdot \mathbf{U}_2 = {} & (1 - \xi)(1 - \eta)\mathbf{x}(0, 0, \zeta) + (1 + \xi)(1 - \eta)\mathbf{x}(1, 0, \zeta) \quad (20) \\
& + (1 - \xi)(1 + \eta)\mathbf{x}(0, 1, \zeta) + (1 + \xi)(1 + \eta)\mathbf{x}(1, 1, \zeta).
\end{aligned}
$$

The third technique implemented is an elliptic grid generation method on the minimal surface defined by the surface bounded by six bounding faces (four edges in the two-dimensional variant for the faces) with zero mean curvature. Following Spekreijse [26] we define the unit cube as a parameter space with Cartesian coordinates (s, t, u), where s, t, and u are the normalized arclengths along the four corresponding edges with the same direction. The mapping between computational and parameter spaces $(\xi, \eta, \zeta) \rightarrow (s, t, u)$ is defined as

$$
\begin{aligned}
s &= s_1(\xi)(1 - t)(1 - u) + s_2(\xi)t(1 - u) + s_3(\xi)(1 - t)u + s_4(\xi)tu, \\
t &= t_1(\eta)(1 - s)(1 - u) + t_2(\eta)s(1 - u) + t_3(\eta)(1 - s)u + t_4(\eta)su, \quad (21) \\
u &= u_1(\zeta)(1 - s)(1 - t) + u_2(\zeta)s(1 - t) + u_3(\zeta)(1 - s)t + u_4(\zeta)st,
\end{aligned}
$$

where s_i, t_i, and u_i ($i = 1, \ldots, 4$) are the grid point distributions along the edges. According to this bilinear transformation the new coordinates of the interior nodes of the 6 faces are computed and used as a boundary conditions. The edges are computed via a linear interpolation; thus closing the system of equations.

The grid is obtained by solving the following equation:

$$
\begin{aligned}
& a^{11}x_{\xi\xi} + 2a^{12}x_{\xi\eta} + a^{13}x_{\xi\zeta} + a^{22}x_{\eta\eta} + 2a^{23}x_{\eta\zeta} + a^{33}x_{\zeta\zeta} \\
& + (a^{11}P_{11}^1 + 2a^{12}P_{12}^1 + 2a^{13}P_{13}^1 + a^{22}P_{22}^1 + 2a^{23}P_{23}^1 + 2a^{33}P_{33}^1)x_\xi \\
& + (a^{11}P_{11}^2 + 2a^{12}P_{12}^2 + 2a^{13}P_{13}^2 + a^{22}P_{22}^2 + 2a^{23}P_{23}^2 + 2a^{33}P_{33}^2)x_\eta \quad (22) \\
& + (a^{11}P_{11}^3 + 2a^{12}P_{12}^3 + 2a^{13}P_{13}^3 + a^{22}P_{22}^3 + 2a^{23}P_{23}^3 + 2a^{33}P_{33}^3)x_\zeta = 0,
\end{aligned}
$$

with the control functions

$$P_{11} = -T^{-1} \begin{pmatrix} s_{\xi\xi} \\ t_{\xi\xi} \\ u_{\xi\xi} \end{pmatrix}, \quad P_{12} = -T^{-1} \begin{pmatrix} s_{\xi\eta} \\ t_{\xi\eta} \\ u_{\xi\eta} \end{pmatrix}, \quad P_{13} = -T^{-1} \begin{pmatrix} s_{\xi\zeta} \\ t_{\xi\zeta} \\ u_{\xi\zeta} \end{pmatrix},$$

$$P_{22} = -T^{-1} \begin{pmatrix} s_{\eta\eta} \\ t_{\eta\eta} \\ u_{\eta\eta} \end{pmatrix}, \quad P_{23} = -T^{-1} \begin{pmatrix} s_{\eta\zeta} \\ t_{\eta\zeta} \\ u_{\eta\zeta} \end{pmatrix}, \quad P_{33} = -T^{-1} \begin{pmatrix} s_{\zeta\zeta} \\ t_{\zeta\zeta} \\ u_{\zeta\zeta} \end{pmatrix}, \quad (23)$$

where

$$T = \begin{pmatrix} s_\xi & s_\eta & s_\zeta \\ t_\xi & t_\eta & t_\zeta \\ u_\xi & u_\eta & u_\zeta \end{pmatrix}, \quad (24)$$

and a^{ij} are the contravariant metric base vectors of the mapping $(\xi, \eta, \zeta) \to \mathbf{x}$.

The Poisson problem (22) is discretized by a central difference scheme and the resulting discrete system is linearized by a Picard iteration scheme. The resulting linear system is solved by the Gauß-Seidel method in each iteration.

All grid movement techniques described for edges, faces, and blocks can be combined flexibly via an user input file. As an example the corresponding combination of techniques for the configuration of the FSI reference experiment (see [10]) is shown in Fig. 4. We achieve fixed wall distances by definining parallel edge movements, preserved angles near the rear of the structure, transfinite interpolation, elliptically smoothed grids, and fixed grids in not affected blocks.

Due to the employment of these techniques reliable simulations involving turbulent flows and large deformations are possible (that would fail if just simple grid movement were employed). As an example, in Fig. 5 results within the flow domain of a large eddy simulation of the turbulent FSI reference experiment are shown.

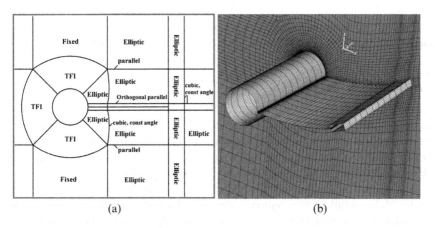

(a)   (b)

Fig. 4 (a) Configuration for turbulent FSI test case. (b) Grid of the turbulent FSI test case.

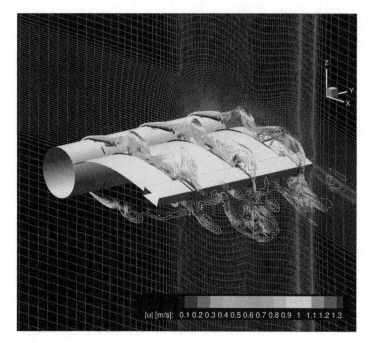

Fig. 5 Snapshot of isolines of the mean velocity for turbulent FSI reference experiment obtained with LES.

Further examples and systematic investigations for the grid movement techniques can be found in [34].

5 Acceleration Techniques

In this section some investigations concerning the possibilities of accelerating partitioned FSI computations are summarized. Further results can be found in [27].

5.1 Multigrid Method

In order to reduce the computational effort multigrid methods can be invoked into the coupled solution procedure. This can be done at the subproblem level by solving the fluid and/or solid subproblem with a multigrid technique. Here, we employ such a kind of partitioned multigrid method for the fluid subproblem (see [32]).

Concerning the multigrid components a standard geometric approach is employed, i.e., a full approximation storage (FAS) scheme for the treatment of non-linearities, a V-cycle strategy for the movement through the grid levels, and bilinear interpolation for prolongation and restriction. Details can be found, for instance, in [11].

Fig. 6 Configuration of
cavity test case.

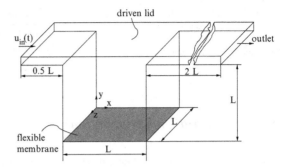

As test case for the partitioned multigrid method a lid driven cavity with a flexible membrane at the bottom is considered (see Fig. 6). The side length of the cube is $L = 1$ m. The membrane thickness is $t_{mem} = 0.1$ m. The membrane is fixed at the edges and the lid moves with a time dependent lid velocity v_{lid} defined by

$$v_{lid}(t) = 0.5 \left(1.5 - \cos\left(2\pi t / T_0\right)\right) \tag{25}$$

with period $T_0 = 5$ s. The inlet and outlet have 10 % of the total height $h_{in} = h_{out} = 0.1$ m and are situated in the upper part on the left and right sides, respectively. The inflow velocity corresponds to the lid velocity. At the outlet a zero gradient condition for the velocity is used.

The material parameters for the structure and the fluid are defined as:

- Youngs modulus $E_s = 50000\,\text{N/m}^2$
- Poisson ratio $v_s = 0.3$
- Structure density $\rho_s = 100\,\text{kg/m}^3$
- Fluid density $\rho_f = 100\,\text{kg/m}^3$
- Fluid dynamic viscosity $\mu_f = 0.01\,\text{kg/(ms)}$
- static pressure $p_{static} = 0.1$ Pa

E_s and v_s are related to the Lamé constants in Eq. (7) by

$$\lambda_s = \frac{E_s v_s}{(1 + v_s)(1 - 2v_s)} \quad \text{and} \quad \mu_s = \frac{E_s}{2(1 + v)}. \tag{26}$$

For the following investigation, three successive refined fluid grids with 2496, 19968, and 159744 control volumes, respectively, are considered. The multigrid computations are performed with three grid levels for each grid set. The structure discretization is fixed for all computations, i.e., linear solid brick elements are utilized with a discretization of $20 \times 2 \times 20$ for the x-, y-, and z-directions. In each test case a simulation time of 1.5 s from a predefined flow field is computed. The computer employed for all computations was a *Intel Pentium4* PC with a clock rate of 2539 MHz. The computing times for the single grid and partitioned multigrid computations are indicated in Table 1. One can observe a typical multigrid performance with increasing acceleration on an increasing problem size.

Table 1 Computing times [s] for single grid and partitioned multigrid methods for different fluid grid sizes.

Control volumes	2496	19968	159744
Single grid	2278	15294	260483
Partitioned multigrid	1202	1943	21445
Acceleration	1.9	7.9	12.1

5.2 Structural Underrelaxation

The partitioned coupling scheme might be rather sensitive with respect to the deformations, in particular, within the first FSI iterations. This may lead to instabilities or even the divergence of the iteration process. In order to counteract this effect an underrelaxation is employed. The actually computed displacements u^{act} are (linearly) weighted with the values u_i^{old} from the preceding iteration to achieve the new displacements u_i^{new}:

$$\mathbf{u}^{\text{new}} = \alpha_{\text{FSI}}\mathbf{u}^{\text{act}} + (1 - \alpha_{\text{FSI}})\mathbf{u}^{\text{old}} \,,$$

where $0 < \alpha_{\text{FSI}} \leq 1$ is the underrelaxation factor.

For the adaptive determination of α_{FSI}^m different methods are known. We utilize an approach based on the Aitken method, which is an extrapolation approach frequently applied in the context of Newton-Raphson iterations. The basis of this approach was proposed by Aitken [3] and later improved by Irons and Tuck [12]. It was identified as very efficient for computations in the field of fluid-structure interaction by Mok [16].

Employing the values from two preceding iterations the Aitken factor γ^m is extrapolated by:

$$\gamma^m = \gamma^{m-1} + (\gamma^{m-1} - 1)\frac{\left(\Delta\mathbf{u}^{m-1} - \Delta\mathbf{u}^m\right)^T \cdot \Delta\mathbf{u}^m}{\left(\Delta\mathbf{u}^{m-1} - \Delta\mathbf{u}^m\right)^2} \tag{27}$$

with $\Delta\mathbf{u}^{m-1} = \mathbf{u}^{m-2} - \tilde{\mathbf{u}}^{m-1}$ and $\Delta\mathbf{u}^m = \mathbf{u}^{m-1} - \tilde{\mathbf{u}}^m$. The actual underrelaxation factor α_{FSI}^m is defined by

$$\alpha_{\text{FSI}} = 1 - \gamma^m \,. \tag{28}$$

As the first Aitken factor in each time step γ^0 the last one from the preceding time step can be taken. For the first time step $\gamma^0 = 0$ can be chosen.

As test case for the performance of the adaptive underrelaxation the lid driven cavity problem (see previous section) is considered. The results for the single grid and partitioned multigrid methods are indicated in Table 2. The average numbers of FSI iterations per time step hardly vary between the single grid and multigrid methods. In both cases the number is nearly halved by using the adaptive Aitken method. The computation using multigrid and adaptive underrelaxation is more than 14 times faster than the single grid computation with fixed underrelaxation.

Table 2 Average number of FSI iterations for single grid and multigrid and acceleration with fixed and adaptive underrelaxation.		Single grid	Multigrid
	$\alpha_{FSI} = 0.8$ fixed	15294	1943
	α_{FSI} adaptive	8550	1057
	Acceleration	1.8	1.8

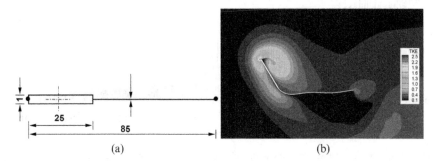

25

85

(a)   (b)

Fig. 7 (a) Configuration for turbulent FSI test case. (b) Turbulent kinetic energy and deformation for turbulent FSI test case.

5.3 Displacement Prediction

A straightforward and simple approach to reduce the number of FSI iterations is the extrapolation of the starting value for the structural displacement at the beginning of a new time step based on values from previous time steps. Employing two time steps, the extrapolation (for equidistant time step size Δt) reads:

$$\mathbf{u}^{n+1} = \mathbf{u}^n + \frac{\Delta t}{2}(\mathbf{u}^n - \mathbf{u}^{n-1}). \tag{29}$$

In particular, the extrapolation is useful for large deformations and/or large time step sizes.

To illustrate possible acceleration effects we consider the FSI of a turbulent flow around a structure plate with one rotational degree of freedom and an elastic membrane (see Fig. 7). The Reynolds number based on the total length of the structure is Re=170 000. 640 hexaeder elements with enhanced strain formulation discretize the structural domain and 25 000 control volumes discretize the fluid domain. For turbulence modelling the k-ε model with wall functions is employed. Details of the configuration can be found in [33]. Figure 7 shows a snapshot of the turbulent kinetic energy indicating the large structural movement.

Figure 8 shows the convergence of the FSI iterations within one time step with and without displacement prediction. An acceleration factor of more than 5 can be achieved in this case.

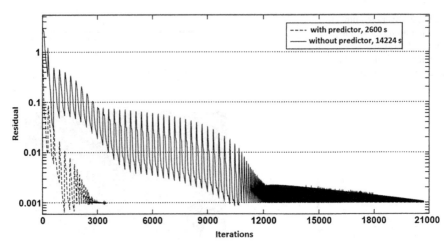

Fig. 8 Convergence of FSI iterations with and without displacement prediction.

6 Optimization

This chapter presents a shape optimization approach for fluid-structure interaction problems. Mathematically a general optimization problem can be written as follows:

$$\min \ J\left(\varphi\left(\mathbf{a}\right), \mathbf{a}\right) \quad \text{subject to} \quad F\left(\varphi\left(\mathbf{a}\right), \mathbf{a}\right) = 0 \tag{30}$$

with the objective functional J and the constraints F. The state variables $\varphi \in \mathbb{R}^{N}$ are internal values that can not be manipulated directly. They depend on the design variables $\mathbf{a} \in \mathbb{R}^{M}$, which can be controlled from the outside. Often, additional side constraints for the control variables have to be fulfilled:

$$\kappa_{i}^{l} \leq a_{i} \leq \kappa_{i}^{u} \quad \text{for} \quad i = 1, \ldots, M \tag{31}$$

with constants κ_{i}^{l} and κ_{i}^{u}.

In case of FSI problems the state variables can be flow and structure properties like velocity, pressure, temperature, turbulent kinetic energy, deformation, et cetera. For shape optimization the control variables are geometrical parameters, for instance, the coordinates of the control points of NURBS surfaces (see Sect. 6.2.1). An objective functional can be, for example, the minimization of pressure drop or structural deformation. The constraints are the partial differential equations describing the flow and structural problem. Typical side constraints are, e.g., restrictions to the inlet velocity to prevent transition to turbulence or limitations to the maximal deformation due to geometrical reasons.

In Fig. 9 a schematic view of a shape optimization for FSI problems is indicated. In the following we first give a description of the chosen NURBS surface-based approach for shape representation. Then, the theoretical background of the utilized

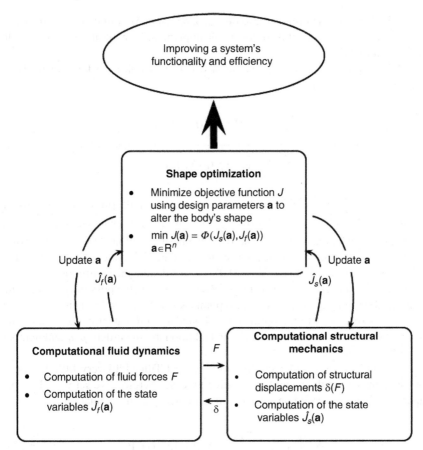

Fig. 9 Schematic shape optimization of a fluid-structure interaction problem.

optimization method is outlined. Finally, results for a test problem involving shape
optimization for a steady fluid-stucture interaction are presented.

6.1 Surface Representation

For the representation of surfaces NURBS (e.g., [20]) are employed. NURBS surfaces are a powerful technique to change the grid point position and therefore the
shape of the objects by adjusting the control points. Compared to the amount of grid
points the amount of control points is very small. NURBS are used in two ways
within this work. On the one hand to calculate an approximation of given surface
grid points, based on a predefined boundary condition enabling the easy definition of
an optimization surface. On the other hand during the optimization the coordinates

of the NURBS control points serve as design parameters. We briefly outline the basic concept. A detailed description of the whole implemented shape altering routine can be found in [4, 5, 25].

A NURBS surface is a B-spline surface with a nonuniform knot vector [29], which is defined as follows:

$$S(u, v) = \frac{\displaystyle\sum_{i=1}^{\hat{n}} \sum_{j=1}^{\hat{m}} w_{i,j} \mathbf{P}_{i,j} N_i^p(u) N_j^q(v)}{\displaystyle\sum_{r=1}^{\hat{n}} \sum_{s=1}^{\hat{m}} w_{r,s} N_r^p(u) N_s^q(v)}, \quad u, v \in [0, 1]. \tag{32}$$

$\mathbf{w} = (w_{i,j})_{i,j}$ denotes the weights. $\boldsymbol{\xi}, \boldsymbol{v}$ denote nonuniform knotvectors (their derivation is presented in Sect. 6.2.1). $N^p, N^q : [0, 1] \to \mathbb{R}$, with $N_i^p := N_{[\xi_i, \xi_{i+1}]}^p$ (N_j^q using $\boldsymbol{v}$ analogously) denote the B-Spline basis functions of orders p, q computed recursivley with the Cox-deBor recurrence (see [7]). Note, that we follow the convention of Farin et al. [8]: $0/0 := 0$. $\mathbf{P}_{i,j} \in \mathbb{R}^3$ denote the control points.

One of the main advantages using NURBS surfaces is the possibilty to compute them based on a given set of surface grid points $\mathbf{X} \in \mathbb{R}^n \times \mathbb{R}^m \times \mathbb{R}^3$. This allows to fit the NURBS surface onto the optimization boundary automatically (i.e. computing the control points $\mathbf{P}$) based on given $\mathbf{X}$ and additionally the orders p, q and number of control points $\hat{n}, \hat{m}$.

To simplify matters and according to Piegl and Tiller [20] very little has been published on setting the weights in a fitting process, therefore we set $\mathbf{w} = \underline{1}$ in this paper. This leads us, based on Eq. (32), to a simpler system of equations

$$S(u_k, v_l) = \sum_{i=1}^{\hat{n}} \sum_{j=1}^{\hat{m}} N_i^p(u_k) N_j^q(v_l) \mathbf{P}_{i,j}, \tag{33}$$

where u_k, v_l denotes the parameterized points of $\mathbf{X}$.

To finish the data fitting process Eq. (33) needs to be solved. Since there are notably less control points compared to surface grid points, we need to solve an overdetermined linear system of the form

$$\mathbf{A}\mathbf{x} = \mathbf{b}, \text{ with } \mathbf{A} \in \mathbb{R}^m \times \mathbb{R}^n, \mathbf{x} \in \mathbb{R}^n, \mathbf{b} \in \mathbb{R}^m, m \gg n. \tag{34}$$

According to [19] the corresponding linear regression problem can be defined as follows:

$$\|\mathbf{A}\mathbf{x}^* - \mathbf{b}\|_2 = \min_{\mathbf{x} \in \mathbb{R}^n} \|\mathbf{A}\mathbf{x} - \mathbf{b}\|_2. \tag{35}$$

A suitable method to solve this problem is the QR-decomposition with Householder reflection, which offers a stable algorithm. A detailed overview about this algorithm is given in Becker et al. [4]. According to Dahmen and Reusken [6] an important condition to achieve a unique solution is that $\mathbf{A}$ has a full rank, i.e., rank$(\mathbf{A}) = n$. Schoenberg and Whitney [24] stated that this is always fulfilled if at least one of n

parameterized grid points u_k exists between every pair of subsequent knots, i.e.,

$$\xi_i < u_k < \xi_{i+1}, \quad i = 1, \ldots, \hat{n} + p, \quad k = 1, \ldots, n, \tag{36}$$

and analogously for v_l. A more detailed description of the computation of the nonuniform knot vectors considering the condition (36) is given later in Sect. 6.2.1.

6.2 Shape Optimization in an FSI framework

To implement a shape optimization within our FSI framework, three basic components are needed: an initialization routine, an optimizer, and an evaluation routine (see Fig. 10).

6.2.1 Initialization

The first step in the shape optimization procedure is the identification of the optimization surface and the automatic approximation with a NURBS surface. This is done by marking the surface during the meshing. This set of points is the input for the NURBS surface fitting procedure as shown in Fig. 11.

To set up the system of equations (33) the surface grid points on the designated optimization surface must be identified. Their coordinates in $\mathbb{R}^3$ are stored in the array **X**. Although when the grid is blockstructured, as it is in FASTEST, the array

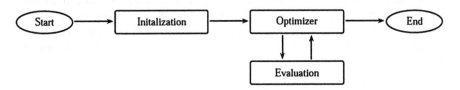

Fig. 10 Implementation of a shape optimization within our FSI framework.

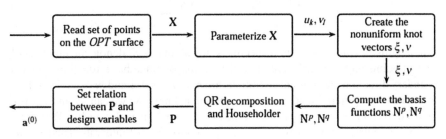

Fig. 11 Initialization routine of the optimization framework.

X can be build across block boundaries and regardless of the orientation of the local coordinate system of each block. The next step is the construction of a well chosen parameterization of **X** to derive $\mathbf{u} = (u_1, \ldots, u_n)$ and $\mathbf{v} = (v_1, \ldots, v_m)$. For this we use an approach based on an algorithm of Ma and Kruth [14], where, for instance, **u** (similar for **v**) is defined as

$$u_1 = 0, \quad u_k = \sum_{j=1}^{m} \hat{u}_{k,j}, \quad k \in \{2, \ldots, n\}, \tag{37}$$

with

$$\hat{u}_{1,l} = 0, \quad \hat{u}_{k,l} = \frac{\sum_{i=1}^{k-1}(\|\mathbf{X}_{i+1,l} - \mathbf{X}_{i,l}\|)^e}{\sum_{i=1}^{n-1}(\|\mathbf{X}_{i+1,l} - \mathbf{X}_{i,l}\|)^e}, \quad k \in \{2, \ldots, n\}, \ l \in \{1, \ldots, m\}. \tag{38}$$

The parameter $e \in [0, 1]$, introduced by Jung and Kim [13], offers the possibilty to influence the type of parameterization. For $e = 0$, a uniform parameterization, for $e = 1$, a chordlength parameterization and for $e = 0.5$, a centripedal parameterization is performed. However, e can be chosen within $[0, 1]$, dependent on the setting.

In order to calculate the knot vectors, several boundary conditions have to be considered. To obtain a unique set of control points, it is necessary that **A** has full rank. Hence, at least one grid point needs to be located between every pair of subsequent knots (Eq. (36)). Furthermore, the grid density should be considered and the corner points of the NURBS surface should coincide with the corresponding control points. For this the following algorithm is employed. Let us consider the discrete parameters u_k, $k \in \{1, \ldots, n\}$ and v_l, $l \in \{1, \ldots, m\}$ as a data basis to compute the knot vectors $\boldsymbol{\xi}, \boldsymbol{\nu}$ using

$$\boldsymbol{\xi} = (\underbrace{0, \ldots, 0}_{p}, \xi_{p+1}, \ldots, \xi_{\hat{n}}, \underbrace{1, \ldots, 1}_{p}),$$
$$\boldsymbol{\nu} = (\underbrace{0, \ldots, 0}_{q}, \nu_{q+1}, \ldots, \nu_{\hat{m}}, \underbrace{1, \ldots, 1}_{q}), \tag{39}$$

with lengths $p + \hat{n}$ and $q + \hat{m}$, respectively. Recall that $\hat{n}, \hat{m}$ denote the numbers of control points and $p, q \in \mathbb{N}$ are the orders of the B-spline basis functions in each boundary direction. Additional conditions are $n > \hat{n} - p + 1$ and $m > \hat{m} - q + 1$, to assure having a grid point between two subsequent knots. As we can see in Eq. (39), we need to calulate $(\hat{n} - p)$ entries for the knot vector $\boldsymbol{\xi}$ ($\boldsymbol{\nu}$ analogously). Therefore, we split all entries of $\mathbf{u} = (u_1, \ldots, u_k, \ldots, u_n)$ into $(\hat{n} - p)$ intervals. The $(\hat{n} - p)$ knot vector entries equal the mean average build within each of the $(\hat{n} - p)$ intervals.

With known knot vectors $\boldsymbol{\xi}$ and $\boldsymbol{v}$, as well as all discrete grid points u_k, $k = 1, \ldots, n$ and v_l, $l = 1, \ldots, m$ we are able to calculate the B-spline basis functions by the Cox-deBor recurrence (see [7]) and store them into the following matrices:

$$\mathbf{N}^p \in \mathbb{R}^{\hat{n}} \times \mathbb{R}^n, \text{ in } u\text{-direction and} \tag{40}$$

$$\mathbf{N}^q \in \mathbb{R}^{\hat{m}} \times \mathbb{R}^m, \text{ in } v\text{-direction.} \tag{41}$$

Now we can solve the system of equations (33) as described in Sect. 6.1 and find the initial positions of the control points $\mathbf{P}$.

To finish the preparation phase a relation between the set of control points $\mathbf{P}^{(g)}$, where $\mathbf{P}_{i,j}^{(g)} \in \mathbb{R}^3$ denotes the (i, j)-th entry at the g-th evaluation call, and the following set of design variables has to be established:

$$\mathbf{a}^{(g)} \in \mathbb{R}^M, \quad \text{subject to } \kappa_\tau^l \leq a_\tau^{(g)} \leq \kappa_\tau^u, \ \tau \in \{1, \ldots, M\}, \text{ and } \mathbf{a} = (a_\tau)_\tau, \tag{42}$$

where $M \in \mathbb{N}$ denotes the number of design variables. The initial set of design variables is always defined as $\mathbf{a}^{(0)} = 0$. The control points of the g-th evaluation call are derived as follows:

$$\mathbf{P}_{i,j}^{(g)} = \begin{pmatrix} \alpha_{i,j}^{(g)} \\ \beta_{i,j}^{(g)} \\ \gamma_{i,j}^{(g)} \end{pmatrix} + \mathbf{P}_{i,j}^{(g-1)}. \tag{43}$$

For each pair (i, j) we distinguish four cases for the usage of the control points:

- Case I: Fixed control points:

$$\alpha_{i,j}^{(g)} = \beta_{i,j}^{(g)} = \gamma_{i,j}^{(g)} = 0. \tag{44}$$

- Case II: One design variable per control point utilizing a direction vector: $\mathbf{s}^\tau = (s_i^\tau)_{i \in \{1,2,3\}}$

$$\alpha_{i,j}^{(g)} = s_1^\tau a_\tau^{(g)}, \quad \beta_{i,j}^{(g)} = s_2^\tau a_\tau^{(g)}, \quad \gamma_{i,j}^{(g)} = s_3^\tau a_\tau^{(g)}, \quad \|\mathbf{s}^\tau\|_2 = 1. \tag{45}$$

- Case III: Three design variables per control point in each spatial direction:

$$\alpha_{i,j}^{(g)} = a_\tau^{(g)}, \quad \beta_{i,j}^{(g)} = a_{\tau+1}^{(g)}, \quad \gamma_{i,j}^{(g)} = a_{\tau+2}^{(g)}. \tag{46}$$

- Case IV: The control point depends on another control point's design variable with a scaling factor $\mathbf{t}^\tau = (t_i^\tau)_{i \in \{1,2,3\}}$:

$$\alpha_{i,j}^{(g)} = t_1^\tau \alpha_{k,l}^{(g)}, \quad \beta_{i,j}^{(g)} = t_2^\tau \beta_{k,l}^{(g)}, \quad \gamma_{i,j}^{(g)} = t_3^\tau \gamma_{k,l}^{(g)}, \quad (i, j) \neq (k, l). \tag{47}$$

This is an important feature to reduce the number of design variables, e.g., in symmetric cases.

The above definitions allow to connect the design variables intuitively with the control points. In this approach, the design variable's value $a_\tau^{(g)} \in [\kappa_\tau^l, \kappa_\tau^u]$ represents the displacement along either a direction vector (case II) or a single direction in space (case III).

6.2.2 Optimizer

As optimizer the SIMPLEX algorithm according to Nelder and Mead [17] is chosen, which is a derivative-free optimization method for unconstrained minimization problems. The method is widespread due to the fact that it makes no assumptions about the objective functional except that it is continuous. Furthermore, it is quite numerically robust. Figure 12 shows the basic procedure of the algorithm. The general idea of the algorithm is to compare the objective functional values $E(a^{(1)}), \ldots, E(a^{(M+1)})$ at the $(M+1)$ vertices of an initial general simplex in the M-dimensional space of design parameters and to move it towards a local minimal point.

The following notations are used:

- $a^{(l)}$ is the set of design parameters resulting in the lowest evaluation value $E(a^{(l)})$
- $a^{(h)}$ is the set of design parameters resulting in the highest evaluation value $E(a^{(h)})$
- $a^{(s)}$ is the set of design parameters resulting in the second highest evaluation value $E(a^{(s)})$
- $a^{(t)}$ is the set of design parameters at the centroid of all sets without the highest and is derived with $a^{(t)} = \frac{1}{M} \sum_{i=1, i \neq h}^{M+1} a^{(i)}$
- $a^{(t)}$ is the reflected set of design parameters at the centroid $a^{(t)}$ and is derived with $a^{(r)} = a^{(t)} - (a^{(h)} - a^{(t)})$
- $a^{(e)}$ is the expanded set of design parameters at the reflected set $a^{(r)}$ and is derived with $a^{(e)} = a^{(r)} + (a^{(r)} - a^{(t)})$
- $a^{(c)}$ is the contracted set of design parameters and is derived with $a^{(c)} = a^{(t)} + 0, 5(a^{(h)} - a^{(t)})$

Three basic operations (reflection, contraction, and expansion) move the simplex by substituting the vertex with the highest objective functional value by a point with a lower value. Therefore, the centroid of the simplex is calculated and its functional value is compared with the other values at the vertices of the simplex. Depending on the centroid value the vertex with the highest functional value is exchanged, or alternatively the complete simplex becomes shrinked around the vertex with the lowest value with

$$a^{(i)} = a^{(l)} + 0, 5(a^{(i)} - a^{(l)}), \forall i. \tag{48}$$

The method represents a robust algorithm for unconstrained optimization problems. However, it is not able to handle side constraints, since it is possible that the simplex collapses and degenerates due to the boundaries. Afterwards the simplex is caught in a subspace and will fail to find the optimum within the complete parameter

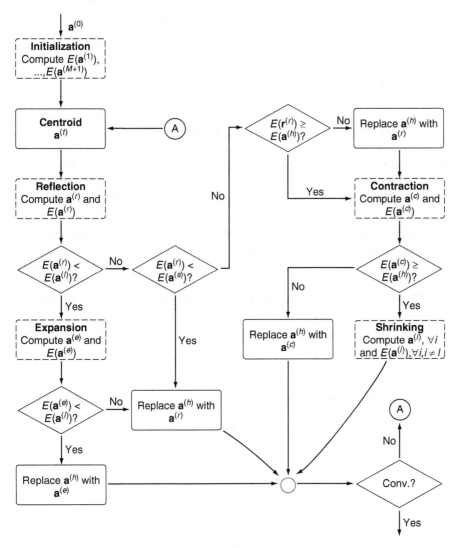

Fig. 12 SIMPLEX algorithm (Dashed boxes includes the call of the evaluation routine).

space. Thus, for constrained optimization it is necessary to ensure that the simplex does not collapse. This is done by transforming the constrained optimization problem into an unconstrained problem by introducing penalty functions. The penalty function can be defined as

$$P = \sum_{i=1}^{M} \theta(\max(a_i - \kappa_i^u, 0) + \max(-\kappa_i^l - a_i, 0)). \qquad (49)$$

where κ_i^{u} and κ_i^{l} are the upper and lower bounds of the i-th design parameter a_i and θ is a scaling factor. We set this factor usually to a value of $\theta = 1000$.

6.2.3 Evaluation

During the processing of the evaluation routine (illustrated in Fig. 13) a new set of control points $\mathbf{P}^{(g)} \in \mathbb{R}^{\hat{n}} \times \mathbb{R}^{\hat{m}} \times \mathbb{R}^3$ is calculated with Eq. (43) using a new set of design variables $\mathbf{a}^{(g)}$ as input. The B-spline basis functions already calculated during the initialization (see Sect. 6.2.1) can be reused in this routine. Now, a new set of surface grid points $\mathbf{X}^{(g)} \in \mathbb{R}^n \times \mathbb{R}^m \times \mathbb{R}^3$ can be calculated using the NURBS equation (33), where $X_{i,j,k}^{(g)}$ denotes the (i, j, k)th entry of the tensor $\mathbf{X}^{(g)}$.

Note that this method generates the new surface grid $\mathbf{X}^{(g)}$ only. To move the remaining computation grid within the flow domain according to the displaced optimization surface, grid generation techniques already used for FSI grid displacements are applied. Within the FSI approach introduced in Sect. 3, the FSI grid generation routine receives computed displacements of the coupling surface from the structural solver *FEAP* [21]. To utilize this routine for the purpose of shape

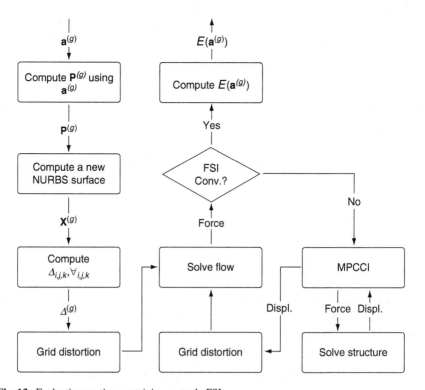

Fig. 13 Evaluation routine containing a steady FSI.

optimization we need to compute the displacements between $\mathbf{X}^{(g)}$ and the original optimization surface $\mathbf{X}^0$, i.e.

$$\Delta_{i,j,k} = |X_{i,j,k}^{(g)} - X_{i,j,k}^0|, \quad \forall\, i,j,k. \tag{50}$$

The grid generation routine uses $\boldsymbol{\Delta} = (\Delta_{i,j,k})_{i,j,k}$ to generate new grids within the optimization surface's neighboring blocks. This routine is explained in detail in Sect. 3.

Now the flow variables like velocity, pressure or temperature are solved and the forces at the FSI coupling interface are computed. These forces are send via the interpolation tool MpCCI to the structural solver, which computes based on the forces a structural displacment. This displacment serves as input to the grid distortion routine as described in Sect. 4. This computes us a new flow domain which can be solved again until a equilibrium between force and displacment is reached. Finally the evaluation value $E(\mathbf{a}^{(g)})$ is calculated an serves as input for the optimizer.

6.3 Optimization example

As test case we consider the shape optimization of a 2D steady FSI problem based on the benchmark configuration of Turek and Hron [31] with altered outer boundaries as shown in Fig. 14. The velocity at the inlet is defined as

$$U_{in} = 1.5\overline{U}\frac{y(H-y)}{(\frac{H}{2})^2}, \tag{51}$$

where $\overline{U} = 0.1$ m/s denotes the characteristic average velocity in x-direction and H denotes the channel height. The density of the fluid is $\rho_f = 1000$ kg/m^3 and its dynamic viscosity is $\mu_f = 1$ kg/(ms). Thus, the Reynolds number is $Re = 10$ based on the cylinder diameter. The fluid domain is discretized by 50000 hexahedral control volumes in the xy-plane.

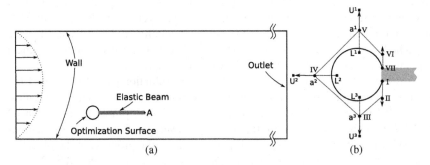

Fig. 14 (a) Setup of the steady FSI optimization example. (b) Relation between CP and DV.

The structural density is set to $\rho_s = 1000\,\text{kg/m}^3$, the Youngs-Modulus $E_s = 5000\,\text{N/mm}^2$ and the Possions Ratio to $\nu_s = 0.4$. The structural domain has the dimensions as set in the paper of Turek and Hron [31] and is disretized with 120 finite brick elements in the xy-plane.

The rigid cylinder is the shape optimization body and the objective is the minimization of the displacement of point A (cf. Fig. 14). Figure 14 shows the NURBS surface representing the optimization surface. The NURBS surface of order four consists of seven control points (roman numbers in Fig. 14) in the xy-plane. The control points I and VII are fixed at the end points, III to V are directly attached to three design variables $a_{i=\{1,2,3\}}$, $\kappa_i^l \leq a_\tau \leq \kappa_i^u$ with

$$\mathbf{P}_{III}^{(g+1)} = \begin{pmatrix} 0 \\ -1 \\ 0 \end{pmatrix} a_3^{(g+1)} + \mathbf{P}_{III}^{(g)}, \quad -0.05 \leq a_3 \leq 0.05 \tag{52}$$

$$\mathbf{P}_{IV}^{(g+1)} = \begin{pmatrix} -1 \\ 0 \\ 0 \end{pmatrix} a_2^{(g+1)} + \mathbf{P}_{IV}^{(g)}, \quad -0.04 \leq a_2 \leq 0.05 \tag{53}$$

$$\mathbf{P}_{V}^{(g+1)} = \begin{pmatrix} 0 \\ +1 \\ 0 \end{pmatrix} a_1^{(g+1)} + \mathbf{P}_{V}^{(g)}, \quad -0.05 \leq a_1 \leq 0.05. \tag{54}$$

Control points II and VI are dependent with following relation

$$\mathbf{P}_{II}^{(g+1)} = \begin{pmatrix} 0 \\ -0.3 \\ 0 \end{pmatrix} a_3^{(g+1)} + \mathbf{P}_{II}^{(g)} \tag{55}$$

$$\mathbf{P}_{VI}^{(g+1)} = \begin{pmatrix} 0 \\ +0.3 \\ 0 \end{pmatrix} a_1^{(g+1)} + \mathbf{P}_{VI}^{(g)}. \tag{56}$$

At first the results of the steady FSI simulation of the non-optimized cylinder are presented. Figure 15a and 15b show the development of the displacement of point A in x- and y-direction at the end of the elastic beam as well as the drag and lift force during the FSI iterations. After 30 fluid-structure coupling iterations a steady state is reached. At the converged solution the beam has experienced a final displacement at point A of 0.04216 m (equal to 5.2% of channel height H).

Next the optimization is considered. For each evaluation call of the SIMPLEX algorithm a complete steady fluid-structure coupling routine needs to be executed. The SIMPLEX required 115 evaluations to achieve the optimal result within 63 SIMPLEX loops. Thus, the algorithm needed an average of nearly two evaluations per SIMPLEX loop (recall Fig. 12). Figure 16 shows the progress of the objective function value for each SIMPLEX loop. The displacement was reduced by 29 % to a final value of 0.02981 m (equal to 3.7% of channel height H)

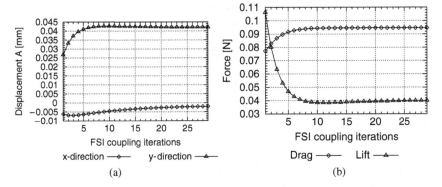

Fig. 15 (a) Displacement of point A in x- and y-direction. (b) Drag and lift force onto the elastic beam.

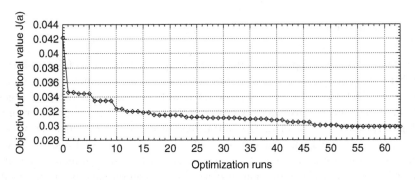

Fig. 16 Development of the objective function value during the optimization.

Fig. 17 Initial shape of the cylinder and its final displacement of the elastic beam (gray) as well as the final shape and displacement (black). Please note that the initial and deformed blockstructure is shown as well.

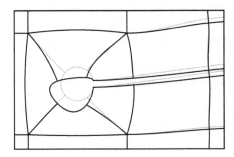

The final shape of the cylinder is illustrated in Fig. 17. The upper half of the cylinder got flattend; this increases the velocity above the elastic beam. The lower half got larger to decrease the velocity. The result is a smaller displacement of the elastic beam.

To support the above mentioned statements concerning the optimal result, Fig. 18a and Fig. 18b show the relative pressure distribution of the inital and final

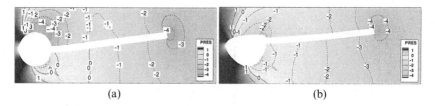

(a)   (b)

Fig. 18 (a) Relative pressure distribution of the initial configuration. (b) Relative pressure distribution of the final configuration.

configuration. One can easily see the reduction of the pressure below the elastic beam and the increase above.

7 Conclusions

A partitioned solution approach for the simulation and optimization of FSI problems has been presented. Special emphasis has been given to the grid moving techniques for which algebraic and elliptic approaches as well as special treatments of edges and faces have been considered. A proper combination of these methods ensures a high grid quality also for large deformations of the structure which, in particular, is of high importance for reliably simulating FSI involving turbulent flows. The possibilities for accelerating the computations by the usage of multigrid methods, adaptive underrelaxation, and displacement prediction have been discussed. The results indicate the high potential of such approaches. Finally, a concept for integrating the FSI solver into an optimization procedure for FSI problems has been presented which is based on NURBS surface representation and an unified treatment of grid movement and shape variation. Results for an exemplary test case illustrate the capabilities of the integrated optimization procedure.

Acknowledgements The financial support of the work by the *Deutsche Forschungsgemeinschaft* within the Research Unit 493 *Fluid-Structure Interaction: Modelling, Simulation, Optimization* is gratefully acknowledged. Thanks are addressed to Stephen Sachs and Michael Werner (grid movement techniques), Johannes Siegmann (NURBS implementation), and Julian Michaelis (SIMPLEX implementation) for their valuable contributions.

References

1. FASTEST – User Manual. Institute of Numerical Methods in Mechanical Engineering, Technische Universität Darmstadt (2004)
2. MpCCI – Mesh-Based Parallel Code Coupling Interface. User Guide V2.0. Fraunhofer, SCAI (2004)
3. Aitken, C.: Studies in practical mathematics. II. the evaluation of the latent roots and latent vectors of a matrix. Royal Societiy of Edinburgh **57**, 269–304 (1937)

4. Becker, G., Falk, U., Schäfer, M.: Shape optimization with higher-order surfaces in consideration of fluid-structure interaction. In: S. Hartmann, A. Meister, M. Schäfer, S. Turek (eds.) International Workshop on Fluid-Structure Interaction: Theory, Numerics and Application. Kassel Univ. Press, Kassel (2009)

5. Becker, G., Siegmann, J., Michaelis, J., Schäfer, M.: Comparison of a derivative-free and a derivative-based shape optimization method. In: H.C. Rodrigues, J.M. Guedes, P.R. Fernandes, M.M. Neves (eds.) Proceedings of the 8th World Congress on Structural and Multidisciplinary Optimization, Lisbon, Portugal (2009)

6. Dahmen, W., Reusken, A.: Numerik für Ingenieure und Naturwissenschaftler. Springer-Verlag Berlin Heidelberg, Berlin, Heidelberg (2008)

7. De Boor, C.: A practical guide to splines. Springer (2001)

8. Farin, G.E., Hoschek, J., Kim, M.S.: Handbook of computer aided geometric design. Elsevier, Amsterdam (2002)

9. Ferziger, J., Perić, M.: Computational Methods for Fluid Dynamics. Springer, Berlin (1996)

10. Gomes, J., Lienhart, H.: Experimental study on a fluid-structure interaction reference test case. In: H.J. Bungartz, M. Schäfer (eds.) Fluid-Structure Interaction - Modelling, Simulation, Optimisation, *LNCSE*, vol. 53, pp. 356–370. Springer (2006)

11. Hackbusch, W.: Multi-Grid Methods and Applications. Springer-Verlag, Berlin (1985)

12. Irons, B., Tuck, R.: A version of the aitken accelerator for computer iteration. International Journal of Numerical Methods in Engineering **1**, 275–27 (1969)

13. Jung, H., Kim, K.: A new parameterisation method for NURBS surface interpolation. The International Journal of Advanced Manufacturing Technology **16**(11), 784–790 (2000)

14. Ma, W., Kruth, J.P.: NURBS curve and surface fitting for reverse engineering. The International Journal of Advanced Manufacturing Technology **14**(12), 918–927 (1998)

15. Mohammadi, B., Pironneau, O.: Shape optimization in fluid mechanics. Annual Review of Fluid Mechanics **36**(1), 255–279 (2004)

16. Mok, D.: Partitionierte Lösungsansätze in der Strukturdynamik und der Fluid-Struktur-Interaktion. Ph.D. thesis, Institut für Baustatik, Universität Stuttgart (2001)

17. Nelder, J.A., Mead, R.: A simplex method for function minimization. The Computer Journal **7**(4), 308–313 (1965)

18. Ogden, R.W.: Non-Linear Elastic Deformations. Dover Publications (1997)

19. Peters, G., Wilkinson, J.H.: The least squares problem and pseudo-inverses. The Computer Journal **13**(3), 309–316 (1970)

20. Piegl, L., Tiller, W.: The NURBS book. Springer Verlag, New York (1996)

21. Schäfer, M., Heck, M., Yigit, S.: An implicit partitoned method for the numerical simulation of fluid-structure interaction. In: H.J. Bungartz, M. Schäfer (eds.) Fluid-Structure Interaction: Modelling, Simulation, Optimization, 53, pp. 171–194. Springer, Berlin, Heidelberg (2006)

22. Schäfer, M., Lange, H., Heck, M.: Implicit partitioned fluid-structure interaction coupling. In: Proceedings of 1er Colloque du GDR Interactions Fluide-Structure, pp. 31–88 (2005)

23. Schäfer, M., Meynen, S., Sieber, R., Teschauer, I.: Efficiency of multigrid methods for the numerical simulation of coupled fluid-solid problems. Scientific Computing and Applications, Advances in Computation: Theory and Practice pp. 257–266 (2001)

24. Schoenberg, I.J., Whitney, A.: On Polya frequency function. III. The positivity of translation determinants with an application to the interpolation problem by spline curves. Transactions of the American Mathematical Society **74**(2), 246–259 (1953)

25. Siegmann, J., Becker, G., Michaelis, J., Schäfer, M.: A general sensitivity based shape optimization approach for arbitrary surfaces. In: H.C. Rodrigues, J.M. Guedes, P.R. Fernandes, M.M. Neves (eds.) Proceedings of the 8th World Congress on Structural and Multidisciplinary Optimization, Lisbon, Portugal (2009)

26. Spekreijse, S.: Elliptic grid generation based on laplace equations and algebraic transformations. Journal of Computational Physics **118**, 38–61 (1995)

27. Sternel, D.C., Schäfer, M., Heck, M., Yigit, S.: Efficiency and accuracy of fluid-structure interaction simulations using an implicit partitioned approach. Computational Mechanics, **43**(1), 103–113 (2008)

28. Taylor., R.: FEAP – A Finite Element Analysis Programm. Version 7.4 User Manual. University of California at Berkeley (2002)
29. Tiller, W.: Rational B-splines for curve and surface representation. Computer Graphics and Applications, IEEE **3**(6), 61–69 (1983)
30. Truesdell, C., Noll, W.: The Non-Linear Field Theories of Mechanics, Third Edition. Springer (2004)
31. Turek, S., Hron, J.: Proposal for numerical benchmarking of fluid-structure interaction between an elastic object and laminar incompressible flow. In: H.J. Bungartz, M. Schäfer (eds.) Fluid-Structure Interaction: Modelling, Simulation, Optimization, 53, pp. 371–385. Springer, Berlin, Heidelberg (2006)
32. Yigit, S., Heck, M., Sternel, D.C., Schäfer, M.: Efficiency of Fluid-Structure Interaction Simulations with Adaptive Underrelaxation and Multigrid Acceleration. International Journal of Multiphysics, **1**, 85–99 (2007)
33. Yigit, S.: Phänomene der Fluid-Struktur-Wechselwirkung und deren numerische Berechnung. Ph.D. thesis, Institute of Numerical Methods in Mechanical Engineering, TU Darmstadt (2008)
34. Yigit, S., Schäfer, M., Heck, M.: Grid movement techniques and their influence on laminar fluid structure interaction computations. Journal of Fluids and Structures, **24**(6), 819–832 (2008)

An Adaptive Finite Element Method for Fluid-Structure Interaction Problems Based on a Fully Eulerian Formulation

R. Rannacher and T. Richter

Abstract In this article, we continue the investigation of the general variational framework for the adaptive finite element approximation of fluid-structure interaction problems proposed in Dunne [12] and Dunne & Rannacher [13]. The modeling is based on an Eulerian description of the (incompressible) fluid as well as the (elastic) structure dynamics. This approach uses a technique which is similar to the *Level Set* method in the simulation of two-phase flows in so far that it also tracks initial data and from this determines to which "phase" a point belongs. The advantage is that, in contrast to the common ALE approach, the computation takes place on fixed, though dynamically adapted, reference meshes what avoids the critical degeneration in case of large deformation or boundary contact of the structure. Based on this monolithic model of the fluid-structure interaction, we apply the *Dual Weighted Residual (DWR) Method* for goal-oriented a posteriori error estimation and mesh adaptation to fluid-structure interaction problems. Several test examples are presented in order to illustrate the potential of this approach.

1 Introduction

Computational fluid mechanics and structure mechanics are two major areas of numerical simulation of physical systems. With the introduction of high performance computing it has become possible to tackle coupled systems with fluid and structure dynamics. General examples of such fluid-structure interaction (FSI) problems are flow transporting elastic particles (particulate flow), flow around elastic structures (e.g., airfoils) and flow in elastic structures (e.g., heart valves). In all

R. Rannacher
Institute of Applied Mathematics, University of Heidelberg
e-mail: rannacher@iwr.uni-heidelberg.de

T. Richter
Institute of Applied Mathematics, University of Heidelberg
e-mail: thomas.richter@iwr.uni-heidelberg.de

H.-J. Bungartz et al. (eds.), *Fluid Structure Interaction II*, Lecture Notes in Computational Science and Engineering 73, DOI 10.1007/978-3-642-14206-2_7,
© Springer-Verlag Berlin Heidelberg 2010

these settings the dilemma in modeling the coupled dynamics is that the fluid model is normally based on an Eulerian perspective in contrast to the usual Lagrangian description of the solid model. This makes the setup of a common variational formulation difficult. However, such a variational formulation of FSI is needed as the basis of a consistent approach to residual-based a posteriori error estimation and mesh adaptation as well as to the solution of optimal control problems by the Euler-Lagrange method. The subject of the present paper is a novel monolithic finite element (FE) formulation for fluid-structure interaction problems in a fully Eulerian setting. This work is largely based on the doctoral dissertation of Dunne [12] and the survey articles Dunne & Rannacher [13], Bönisch, Dunne & Rannacher [7] and Dunne, Rannacher & Richter [14].

To achieve a closed variational formulation for the fluid-structure interaction problem, usually an auxiliary unknown coordinate transformation function T_f is introduced for the fluid domain. With its help the fluid problem is rewritten on the transformed domain. Then, all computations are done on the reference domain which is fixed in time. As part of the computation, the auxiliary transformation function T_f has to be determined at each time step. Such, so-called "arbitrary Lagrangian-Eulerian" (ALE) methods are described for instance in Dunne, Rannacher & Richter [14].

We however follow the alternative way of posing the fluid as well as the structure problem in a fully Eulerian-Eulerian (Eulerian) framework. A similar approach has been used by Lui & Walkington [22] in the context of the transport of visco-elastic bodies in a fluid. In the Eulerian setting a phase variable is employed on the fixed mesh to distinguish between the different phases liquid and solid. This approach to identifying the fluid-structure interface is generally referred to as "interface capturing", a method commonly used in the simulation of multiphase flows.

Our method has similarities to the Level Set (LS) method of Osher & Sethian [25]. In the classical LS approach the distance function has to continually be reinitialized, due to the smearing effect by the convection velocity in the fluid domain. This makes the use of the LS method delicate for modeling FSI problems particularly in the presence of cornered structures. To cope with this difficulty, in Dunne [11, 12] a variant of the LS method, the Initial Position (IP) method, has been proposed that makes reinitialization unnecessary and which easily copes with cornered structures. This approach does not depend on the specific structure model. In the present paper, we formulate a modification of the original fully Eulerian method avoiding the additional IP-set, which has been proposed by Richter & Wick [30].

The equations we use are based on the momentum and mass conservation equations for the flow of an incompressible Newtonian fluid and the deformation of a compressible St. Venant-Kirchhoff or likewise incompressible neo-Hookean solid. The spatial discretization is accomplished by a second-order finite element method with conforming equal-order (bilinear) trial functions using "local projection stabilization" as proposed by Becker & Braack [3, 4]. The time discretization is based on a Galerkin approach similar to the Crank-Nicolson scheme (see Schmich & Vexler [29]). This method allows for rigorous error estimation and time step control in a coupled space-time setting.

Based on the fully Eulerian variational formulation of the FSI system, we use the "dual weighted residual" (DWR) method, as described in Becker & Rannacher [5, 6] and Bangerth & Rannacher [1], to derive "goal-oriented" a posteriori error estimates. The evaluation of these error estimates requires the approximate solution of a linear dual variational problem. The resulting a posteriori error indicators are then used for automatic local mesh adaptation. The rigorous application of the DWR method to FSI problems requires a Galerkin discretization in space as well as in time, which was one of the motivations for introducing the monolithic fully Eulerian formulation.

The method for computing FSI described in this paper has been validated at several stationary model problems, for example a lid-driven cavity involving the interaction of an incompressible Stokes fluid with a linearized incompressible neo-Hookean solid (see Dunne & Rannacher [13]). Further, as a more challenging test the self-induced oscillation of a thin elastic bar immersed in an incompressible fluid has been treated (FLUSTRUK-A benchmark, Hron & Turek [21]). For this test problem, our method has also been compared against a standard "arbitrary Lagrange Eulerian" (ALE) approach. The potential of the fully Eulerian formulation of the FSI problems is indicated by its good behavior for large structure deformations. All computations and visualizations have been done using the flow-solver package GASCOIGNE [16] and the graphics package VISUSIMPLE [32]. The details on the software implementation can be found in Dunne [11, 12].

The outline of this paper is as follows. Sect. 2 ("Formulation") introduces the basic notation for the flow and structure model and also the fully Eulerian formulation of the FSI Problem, which will be used throughout this paper. Sect. 3 ("Discretization") describes the discretization in space and time as well as the techniques for solving the resulting algebraic systems. Sect. 4 ("Numerical Examples") contains the results obtained for different test-cases comprising large deformation and free movement of the structure in the fluid. For completeness also some of the results from Dunne & Rannacher [13] are recalled concerning the comparison of the fully Eulerian and the ALE approach at the FSI benchmark FLUSTRUK-A. The paper is closed by Sect. 5 ("Summary").

2 Formulation

In this section, we present the fully Eulerian formulation of the FSI problem. The Eulerian approach was first proposed in Dunne [11], then simplified by Richter & Wick [30].

2.1 Notation

We begin with introducing some notation which will be used throughout this paper. By $\Omega \subset \mathbb{R}^d$ ($d = 2$ or $d = 3$), we denote the domain of definition of the

FSI problem. The domain Ω is supposed to be *time independent* but to consist of two possibly time-dependent subdomains, the fluid domain $\Omega_f(t)$ and the structure domain $\Omega_s(t)$. Unless needed, the explicit time dependency will be skipped in this notation. The boundaries of Ω, Ω_f, and Ω_s are denote by $\partial\Omega$, $\partial\Omega_f$, and $\partial\Omega_s$, respectively. The common interface between Ω_f and Ω_s is $\Gamma_i(t)$, or simply Γ_i.

The initial structure domain is denoted by $\widehat{\Omega}_s$. Spaces, domains, coordinates, values (such as pressure, displacement, velocity) and operators associated to $\widehat{\Omega}_s$ (or $\widehat{\Omega}_f$) will likewise be indicated by a "hat".

Partial derivatives of a function f with respect to the i-th coordinate are denoted by $\partial_i f$, and the total time-derivative by $d_t f$. The divergence of a vector and tensor is written as $\operatorname{div} f = \sum_i \partial_i f_i$ and $(\operatorname{div} F)_i = \sum_j \partial_j F_{ij}$. The gradient of a vector valued function v is the tensor with components $(\nabla v)_{ij} = \partial_j v_i$.

By $[f]$, we denote the jump of a (possibly discontinuous) function f across an interior boundary, where n is always the unit vector n at points on that boundary.

For a Lebesgue measurable set X, we denote by $L^2(X)$ the Lebesque space of square-integrable functions on X equipped with the usual inner product and norm

$$(f,g)_X := \int_X fg \, dx, \quad \|f\|_X^2 = (f,f)_X,$$

respectively, and correspondingly for vector- and matrix-valued functions. Mostly the domain X will be Ω, in which case we will skip the domain index in products and norms. For Ω_f and Ω_s, we similarly indicate the associated spaces, products, and norms by a corresponding index "f" or "s".

We will generally use roman letters, V, for denoting spaces of functions depending only on spatial variables and calligraphic letters, $\mathscr{V}$, for spaces of functions depending additionally on time. Let $L_X := L^2(X)$ and $L_X^0 := L^2(X)/\mathbb{R}$. The functions in L_X (with $X = \Omega$, $X = \Omega_f(t)$, or $X = \Omega_s(t)$) with first-order distributional derivatives in L_X make up the Sobolev space $H^1(X)$. Further, $H_0^1(X) = \{v \in H^1(X) : v_{|\partial X_D} = 0\}$, where ∂X_D is that part of the boundary ∂X at which Dirichlet boundary conditions are imposed. We will use the function spaces $V_X := H^1(X)^d$, $V_X^0 := H_0^1(X)^d$, and for time-dependent functions

$$\mathscr{L}_X := \mathscr{L}^2[0,T;L_X], \quad \mathscr{V}_X := \mathscr{L}^2[0,T;V_X] \cap \mathscr{H}^1[0,T;V_X^*],$$
$$\mathscr{L}_X^0 := \mathscr{L}^2[0,T;L_X^0], \quad \mathscr{V}_X^0 := \mathscr{L}^2[0,T;V_X^0] \cap \mathscr{H}^1[0,T;V_X^*],$$

where V_X^* is the dual of V_X^0, and $\mathscr{L}^2$ and $\mathscr{H}^1$ indicate the corresponding properties in time. Again, the X-index will be skipped in the case of $X = \Omega$, and for $X = \Omega_f$ and $X = \Omega_s$ a corresponding index "f" or "s" will be used.

2.2 Fluid

For the liquid part, we assume Newtonian incompressible flow governed by the usual Navier-Stokes equations, i.e., the equations describing conservation of mass and momentum. The (constant) density and kinematic viscosity of the fluid are ρ_f and ν_f, respectively.

The equations are written in an Eulerian framework in the time-dependent domain $\Omega_f(t)$. The physical unknowns are the scalar pressure field $p_f \in \mathscr{L}_f$ and the vector velocity field $v_f \in v_f^D + \mathscr{V}_f^0$. Here, v_f^D is a suitable extension of the prescribed Dirichlet data on the boundaries (both moving or stationary) of Ω_f, and g_1 is a suitable extension to all of $\partial\Omega_f$ of the Neumann data for $\sigma_f \cdot n$ on the boundaries. We have "hidden" the fluid-structure interface conditions of continuity of velocity and normal stress in parts of the boundary data v_f^D and g_1.

The variational form of the Navier-Stokes equations in an Eulerian framework is obtained by multiplying them with suitable test functions from the test space V_f^0 for the momentum equations and L_f for the mass conservation equation.

Problem 1 (Fluid model in Eulerian formulation).
Find $\{v_f, p_f\} \in \{v_f^D + \mathscr{V}_f^0\} \times \mathscr{L}_f$, such that $v_f(0) = v_f^0$, and

$$
(\rho_f(\partial_t + v_f \cdot \nabla)v_f, \psi^v)_f + (\sigma_f, \varepsilon(\psi^v))_f = (g_1, \psi^v)_{\partial\Omega_f} + (f, \psi^v)_f,
$$
$$
(\mathrm{div}\, v_f, \psi^P)_f = 0,
\tag{1}
$$

for all $\{\psi^v, \psi^P\} \in V_f^0 \times L_f$, where

$$
\sigma_f := -p_f I + 2\rho_f \nu_f \varepsilon(v_f), \quad \varepsilon(v) := \tfrac{1}{2}(\nabla v + \nabla v^T).
$$

2.3 Structure

In the examples in Sect. 4, we consider two different types of materials, an *incompressible neo-Hookean* (INH) material and a compressible elastic material described by the *St. Venant-Kirchhoff* (STVK) model. These two models will be described by a set of two parameters, the Poisson ratio ν_s and the Young modulus E_s, or alternatively, the Lamé coefficients λ_s and μ_s. These parameters satisfy the following relations:

$$
\nu_s = \frac{\lambda_s}{2(\lambda_s + \mu_s)}, \quad E_s = \mu_s \frac{3\lambda_s + 2\mu_s}{\lambda_s + \mu_s},
$$
$$
\mu_s = \frac{E_s}{2(1 + \nu_s)}, \quad \lambda_s = \frac{\nu_s E_s}{(1 + \nu_s)(1 - 2\nu_s)},
$$

where $\nu_s = \frac{1}{2}$ for incompressible and $\nu_s < \frac{1}{2}$ for compressible material. $\hat{u}^D$ and $\hat{v}^D$ are suitable extensions of the prescribed Dirichlet data on the boundaries (usually $\hat{u}^D = \hat{v}^D = 0$) and $\hat{g}_2$ is the extension of the Neumann data to the complete boundary $\partial\hat{\Omega}_s$.

2.3.1   Incompressible neo-Hookean (INH) material

Problem 2 (INH structure model in Lagrangian formulation).
Find $\{\hat{u}_s, \hat{v}_s, \hat{p}_s\} \in \{\hat{u}^D + \hat{\mathcal{V}}_s^0\} \times \{\hat{v}_s^D + \hat{\mathcal{V}}_s^0\} \times \mathcal{L}_s$, such that $\hat{u}_s(0) = \hat{u}_s^0$, $\hat{v}_s(0) = \hat{v}_s^0$, and

$$(\rho_s d_t \hat{v}_s, \hat{\psi}^u)_{\hat{s}} + (\hat{\sigma}_s \hat{F}^{-T}, \hat{\varepsilon}(\hat{\psi}^u))_{\hat{s}} = (\hat{g}_2, \hat{\psi}^u)_{\partial\hat{\Omega}_s} + (\hat{f}_2, \hat{\psi}^u)_{\hat{s}},$$

$$(d_t \hat{u}_s - \hat{v}_s, \hat{\psi}^v)_{\hat{s}} = 0, \qquad\qquad (2)$$

$$(\det \hat{F}, \hat{\psi}^P)_{\hat{s}} = (1, \hat{\psi}^P)_{\hat{s}},$$

for all $\{\hat{\psi}^u, \hat{\psi}^v, \hat{\psi}^P\} \in \hat{V}_s^0 \times \hat{V}_s^0 \times \hat{L}_s$, where $\hat{F} := I + \hat{\nabla}\hat{u}_s$ and

$$\hat{\sigma}_s := -\hat{p}_s I + \mu_s(\hat{F}\hat{F}^T - I), \quad \hat{\varepsilon}(\hat{\psi}^u) := \tfrac{1}{2}(\hat{\nabla}\hat{\psi}^u + \hat{\nabla}\hat{\psi}^{uT}).$$

2.3.2   Compressible St. Venant-Kirchhoff (STVK) material

Problem 3 (STVK structure model in Lagrangian formulation).
Find $\{\hat{u}_s, \hat{v}_s\} \in \{\hat{u}^D + \hat{\mathcal{V}}_s^0\} \times \{\hat{v}_s^D + \hat{\mathcal{V}}_s^0\}$, such that $\hat{u}_s(0) = \hat{u}_s^0$, $\hat{v}_s(0) = \hat{v}_s^0$, and

$$(\rho_s d_t \hat{v}_s, \hat{\psi}^u)_{\hat{s}} + (\hat{J}\,\hat{\sigma}_s\,\hat{F}^{-T}, \hat{\varepsilon}(\hat{\psi}^u))_{\hat{s}} = (\hat{g}_2, \hat{\psi}^u)_{\partial\hat{\Omega}_s} + (\hat{f}_2, \hat{\psi}^u)_{\hat{s}},$$

$$(d_t \hat{u}_s - \hat{v}_s, \hat{\psi}^v)_{\hat{s}} = 0, \qquad\qquad (3)$$

for all $\{\hat{\psi}^u, \hat{\psi}^v\} \in \hat{V}_s^0 \times \hat{V}_s^0$, where $\hat{\varepsilon}(\hat{\psi}^u) := \tfrac{1}{2}(\hat{\nabla}\hat{\psi}^u + \hat{\nabla}\hat{\psi}^{uT})$.

2.3.3   Conversion of the structure models to the Eulerian framework

To rewrite the above conservation equations in an Eulerian frame, we need the pressure $\hat{p}_s$, the displacement $\hat{u}_s$ and its gradient $\hat{\nabla}\hat{u}_s$ in the Eulerian sense, which are denoted by p_s, u_s and ∇u_s, respectively. There holds $p_s(x) = \hat{p}_s(\hat{x})$ and $u_s(x) = \hat{u}_s(\hat{x})$, or more precisely

$$p_s(x) = \hat{p}_s(T_s(x)) = \hat{p}_s(\hat{x}), \quad u_s(x) = \hat{u}_s(T_s(x)) = \hat{u}_s(\hat{x}),$$

where T_s is the (inverse) displacement function of points in the deformed domain Ω_s back to points in the initial domain $\hat{\Omega}_s$. The corresponding displacement function is $\hat{T}_s(\hat{x})$:

$$\hat{T}_s : \hat{\Omega}_s \to \Omega_s, \quad \hat{T}_s(\hat{x}) = \hat{x} + \hat{u}_s = x,$$
$$T_s : \Omega_s \to \Omega_s, \quad T_s(x) = x - u_s = \hat{x}. \tag{4}$$

Since $\det \hat{\nabla} \hat{T}_s = \det \hat{F} \neq 0$, the displacements T_s and $\hat{T}_s$ are well defined.

To access the Eulerian deformation, we use the identity $T_s(\hat{T}_s(\hat{x})) = \hat{x}$ for the displacement functions obtained from (4). Differentiating this yields:

$$(I - \nabla u)(I + \hat{\nabla} \hat{u}) = I \quad \Leftrightarrow \quad \hat{\nabla} \hat{u} = (I - \nabla u)^{-1} - I.$$

Thus, the Cauchy stress tensor σ_s can be written for INH and STVK materials in an Eulerian framework as follows:

$$\sigma_s = \begin{cases} -p_s I + \nu_s (FF^T - I) & \text{(INH material)}, \\ J^{-1} F(\lambda_s (\text{tr} E) I + 2\mu_s E) F^T & \text{(STVK material)}, \end{cases}$$

$$F = I + \hat{\nabla} \hat{u} = (I - \nabla u)^{-1}, \quad J = \det F, \quad E = \frac{1}{2}(F^T F - I).$$

Finally, we write the structure equations in Eulerian framework for both types of material, INH and STVK.

Problem 4 (INH structure model in Eulerian formulation).

Find $\{u_s, v_s\} \in \{u_s^D + \mathscr{V}_s^0\} \times \{v_s^D + \mathscr{V}_s^0\}$ and $p_s \in \mathscr{L}_s$, such that $u_s(0) = u_s^0$, $v_s(0) = v_s^0$, and

$$(\hat{\rho}_s J_s \partial_t v_s, \psi^v)_s + (\hat{\rho}_s J_s v_s \cdot \nabla v_s, \psi^v)_s + (\sigma_s, \nabla \psi^v)_s$$
$$= (g_2, \psi^v)_{\Gamma_{sN}} + (\sigma_s n_s, \psi^v)_{\Gamma_i} + (\hat{\rho}_s J_s f_s, \psi^v)_s,$$
$$(\partial_t u_s + v_s \cdot \nabla u_s - v_s, \psi^u)_s = 0, \tag{5}$$
$$(1 - \det F_s, \psi^p)_s = 0,$$

for all $\{\psi^u, \psi^v\} \in V_s^0 \times V_s^0$ and $\psi^p \in L_s$ where $F_s := I - \nabla u_s$, $J_s := \det F_s$, $E := \frac{1}{2}(F_s^{-T} F_s^{-1} - I)$, and

$$\sigma_s := -p_s I + \mu_s (F_s^{-1} F_s^{-T} - I).$$

Problem 5 (STVK structure model in Eulerian formulation).

Find $\{u_s, v_s\} \in \{u_s^D + \mathscr{V}_s^0\} \times \{v_s^D + \mathscr{V}_s^0\}$, such that $u_s(0) = u_s^0$, $v_s(0) = v_s^0$, and

$$(\hat{\rho}_s J_s \partial_t v_s, \psi^v)_s + (\hat{\rho}_s J_s v_s \cdot \nabla v_s, \psi^v)_s + (\sigma_s, \nabla \psi^v)_s,$$
$$= (g_2, \psi^v)_{\Gamma_{sN}} + (\sigma_s n_s, \psi^v)_{\Gamma_i} + (\hat{\rho}_s J_s f_s, \psi^v)_s,$$
$$(\partial_t u_s + v_s \cdot \nabla u_s - v_s, \psi^u)_s = 0, \tag{6}$$
$$(1 - \det F_s, \psi^p)_s = 0,$$

for all $\{\psi^u, \psi^v\} \in V_s^0 \times V_s^0$, where $F_s := I - \nabla u_s$, $J_s := \det F_s$, $E := \frac{1}{2}(F_s^{-T} F_s^{-1} - I)$, and

$$\sigma_s := J_s F_s^{-1} (\lambda_s \text{tr} E\, I + 2\mu_s E) F_s^{-T}.$$

2.4   The FSI problem in Eulerian formulation

Now, we can combine the Eulerian formulations of the flow and the structure part of
the problem into a complete variational formulation of the nonstationary as well as
the stationary FSI problem in Eulerian framework. Here, the continuity of velocity
across the fluid-structure interface Γ_i is strongly enforced by requiring one common
continuous field for the velocity $v \in \mathcal{V}$ in Ω. In the case of STVK material the
(non-physical) pressure p_s in the structure subdomain is determined as harmonic
extension of the flow pressure p_f, together denoted by $p \in \mathcal{L}$ in Ω. The Dirichlet
boundary data v_f^D and v_s^D on parts of $\partial\Omega$ are merged into a suitable velocity field
$v^D \in V$. The deformation u_s is defined on all of Ω by a continuous extension
to the flow domain. We denote the deformation by $u \in \mathcal{V}$ and will describe this
(non-physical) extension later.

The balance of forces on the interface $\sigma_f \cdot n = \sigma_s \cdot n$ on Γ_i now implicitly
appears as part of the combined variational formulation in form of a boundary
integral:

$$([\sigma \cdot n], \psi^v)_{\Gamma_i} = \int_{\Gamma_i} (\sigma_f - \sigma_s) \cdot n_f \psi^v \, do.$$

The remaining parts of the Neumann boundary data g_1 and g_2 on $\partial\Omega$ are combined
to g_3. For ease of notation, we introduce the characteristic functions χ_f of Ω_f
and χ_s of Ω_s respectively. We write the Cauchy stress tensor and the density in the
whole domain as follows:

$$\sigma := \chi_f \sigma_f + \chi_s \sigma_s, \quad \rho := \chi_f \hat{\rho}_f + \chi_s \hat{\rho}_s. \tag{7}$$

The characteristic functions depend on the time t and - in case of the solid part -
on the deformation u. Let $\hat{\chi}_f$ and $\hat{\chi}_s$ be the characteristic functions of the reference
domains $\hat{\Omega}_f$ and $\hat{\Omega}_s$ respectively. Then, on the moving Eulerian domains $\Omega_f(t)$
and $\Omega_s(t)$ the characteristic functions can be expressed by

$$\chi_f(t, x) = \hat{\chi}_f(\hat{x}), \quad \chi_s(t, x) = \hat{\chi}_s(\hat{x}).$$

Since the layout of the Eulerian subdomains is not known a priori, the characteristic
functions must be given implicitly by the solution. To close this gap, Dunne [11]
introduces the *set of initial points* (IP-Set) $\varphi(\Omega)$ for all points of Ω at time t. This
function indicates the origin of a point with coordinate $x \in \Omega$ at time t in the
reference domain $\hat{x} \in \hat{\Omega}$ by $\hat{x} = \varphi(t, x)$. The IP-Set is transported in the full domain
with a certain velocity w. While in the structure domain the structure's velocity
v_s itself is used to transport the IP-Set, using the fluid's velocity would lead to
very entangled displacements and breakdown of the scheme. Instead, a harmonic
continuation of the structure's velocity into the flow domain is used.

Here, a different approach, more simular to the ALE coordinates is used. No
further velocity field w is necessary. In the structure domain $\Omega_s(t)$ the deformation
has exactly the meaning of the IP-Set, by

$$\hat{x} = \Phi(t, x)|_{\Omega_s(t)} := x - u_s(t, x),$$

the mapping between reference and deformed coordinate is given. Using the harmonic extension of u_s to the whole domain, the characteristic functions on the moving domains are given by:

$$\chi_s(t, x) = \hat{\chi}_s(\hat{x}) = \hat{\chi}_s(x - u(t, x)), \quad \chi_f(t, x) = \hat{\chi}_f(\hat{x}) = \hat{\chi}_f(x - u(t, x)).$$

With u being available on all of Ω, we can decide for every t and x the correspondence to the subdomain by

$$x \in \Omega \Rightarrow \begin{cases} x - u(t, x) \in \hat{\Omega}_s & \Rightarrow x \in \Omega_s(t) \\ x - u(t, x) \notin \hat{\Omega}_s & \Rightarrow x \in \Omega_f(t). \end{cases} \tag{8}$$

The extension of u to the flow domain does not necessarily define a mapping $T_f(t) : \Omega_f(t) \to \hat{\Omega}_f$ from the moving flow domain to the reference domain, only the information $T_f(t, x) \notin \hat{\Omega}_s$ is used. Hence, we have more freedom to specify the boundary values of the extension of u on $\partial\Omega/\partial\hat{\Omega}_s$, e.g. by choosing:

$$-\Delta u = 0 \text{ in } \Omega_f(t), \quad u = u_s \text{ on } \Gamma_i(t), \quad n \cdot u = 0 \text{ on } \partial\Omega_f(t)/\Gamma_i(t).$$

This allows for arbitrary large deformation and movement of the structure in the flow domain.

Now, we can combine formulas (1), (5) and (6) to obtain a complete variational formulation of the FSI problem in Eulerian coordinates.

In case of the incompressible material (INH), we use one continuous pressure variable for both subproblems. Since the "physical pressure" does not need to be continuous across the interface one could instead decouple the pressure variables and harmonically extend each one to the other domain:

Problem 6 (FSI Problem in Eulerian formulation, INH material).
Find $\{u, v, p\} \in \{u^D + \mathcal{V}^0\} \times \{v^D + \mathcal{V}^0 \times \mathcal{L}\}$, such that $u(0) = u^0$, $v(0) = v^0$, and

$$(\rho(\partial_t v + v \cdot \nabla v), \psi) + (\sigma, \psi) = (g_3, \psi)_{\partial\Omega} + (f_3, \psi),$$
$$(\text{div } v, \xi) = 0,$$
$$(\chi_s(\partial_t u + v \cdot \nabla u - v), \varphi) \tag{9}$$
$$+\alpha_u\{(\chi_f \alpha_u \nabla u, \nabla\varphi) - (\partial_n u, \varphi)_{\Gamma_i}\} = 0,$$

for all $\{\psi, \varphi, \xi\} \in V^0 \times V^0 \times L$, and with

$$\sigma_f = -p_f I + v_f \rho_f (\nabla v + \nabla v^T),$$

$\psi^P \in L_s$ where $F_s := I - \nabla u_s$, $J_s := \det F_s$, $E := \frac{1}{2}(F_s^{-T} F_s^{-1} - I)$, and

$$\sigma_s := -p_s I + \mu_s(F_s^{-1} F_s^{-T} - I).$$

The parameter $\alpha_u \sim \alpha_0 h^{-2}$ needs to be chosen large enough to impose the important condition $d_t u = v$. The boundary integral is necessary to prevent feedback from the continuation. When dealing with the compressible STVK material, the pressure is harmonically extended to the structure domain.

Problem 7 (FSI Problem in Eulerian formulation, STVK material).
Find $\{u, v, p\} \in \{u^D + \mathscr{V}^0\} \times \{v^D + \mathscr{V}^0 \times \mathscr{L}\}$, such that $u(0) = u^0$, $v(0) = v^0$, and

$$(\rho(\partial_t v + v \cdot \nabla v), \psi) + (\sigma, \psi) = (g_3, \psi)_{\partial \Omega} + (f_3, \psi),$$
$$(\chi_f \operatorname{div} v, \xi) + \alpha_p \{(\chi_s \nabla p, \nabla \xi) - (\partial_n p, \xi)_{\Gamma_i}\} = 0,$$
$$(\chi_s (\partial_t u + v \cdot \nabla u - v), \varphi) \tag{10}$$
$$+ \alpha_u \{(\chi_f \alpha_u \nabla u, \nabla \varphi) - (\partial_n u, \varphi)_{\Gamma_i}\} = 0,$$

for all $\{\psi^u, \psi^v\} \in V^0 \times V^0$ and $\psi^p \in L$ where $F_s := I - \nabla u_s$, $J_s := \det F_s$, $E := \frac{1}{2}(F_s^{-T} F_s^{-1} - I)$, and

$$\sigma_s := J_s F_s^{-1} (\lambda_s \operatorname{tr} E \, I + 2\mu_s E) F_s^{-T}.$$

In these variational formulations the location of the interface Γ_i is given implicitly by the deformation:

$$\Gamma_i(t) := \{x \in \Omega, \ x - u(t, x) \in \hat{\Gamma}_i\}.$$

The resulting system is nonlinear even if linear models are used for the two subproblems, e.g. a Stokes fluid and a linear elastic structure.

In some situations the solution of an FSI problem may tend to a "steady state" as $t \to \infty$. The equations for stationary fluid structure interaction in Eulerian formulation are given in Dunne [11]. Corresponding stationary equations without the IP-Set are developed in Richter & Wick [30].

2.5 Comparison to ALE formulations

In the classical ALE formulation for FSI problems the structure problem is formulated on the static reference domain $\hat{\Omega}_s$. The flow problem is transformed back to the reference flow domain $\hat{\Omega}_f$ by an artificial transformation $\hat{T}_f : \hat{\Omega}_f \to \Omega$ usually derived by the continuation of the deformation to the flow domain. The situation in Eulerian coordinates is similar: here, the flow problem resides in the natural coordinate system while the structure problem needs to be transformed. This transformation $T_s : \Omega_s \to \hat{\Omega}_s$ however is a "natural transformation" since it is given by the deformation itself. If the structure solver is capable of handling a certain deformation, the transformation is well behaved and cannot deteriorate in the context of FSI. The deformation field is extended to the flow domain in both formulations, but while the extended (and artificial) field is used to transform the flow variables in the ALE formulation, it is only necessary to lookup the domain of influence in the Eulerian formulation. Here, no gradient evaluation is required and the regularity is of lower importance. In the fully Eulerian formulation, the extension of the deformation to the flow domain does not have to define a mapping between the reference and the deformed fluid domain. Hence, more freedom is possible in modeling this extension.

Both formulations contain equations for velocity, deformation and pressure, a total of five solution variables in two spatial dimensions. The Eulerian formulation is strongly nonlinear due to the implicit dependence of the domain on the deformation. In the ALE formulation the transformation of the flow domain imposes strong nonlinearities which can prohibit large deformation. The Eulerian formulation tends to be slightly more costly due to the implicit definition of the moving fluid-structure interface $\Gamma_i(t)$.

3 Discretization

In this section, we detail the discretization in space and time of the FSI problem based on its different variational formulations. Our method of choice is the Galerkin finite element (FE) method with "conforming" finite elements, both in space and time. For a general introduction to the FE method, we refer to Carey and Oden [10], Girault & Raviart [17], Brenner & Scott [9], or Braess [8]. Space-time Galerkin methods are described in Eriksson, Estep, Hansbo & Johnson [15], Schmich & Vexler [29] and Meidner & Vexler [23] and Schmich [28] for the Navier-Stokes equations. Having a Galerkin method for the completely discretized scheme at hand rigorous error estimation is accessible.

First, we introduce a semi-discrete formulation incorporating a continuous or discontinuous Galerkin method in time. The resulting methods will be variants of the implicit Euler method and of the Crank-Nicolson scheme.

Then, we provide the framework for the spatial finite element method and we describe the complete variational formulations which are the basis of the Galerkin discretization. Since we are using a so-called "equal-order" approximation of all physical quantities, additional pressure stabilization has to be incorporated which is done here by the "local projection" technique of Becker and Braack [3].

At each time step a nonlinear algebraic problem is solved using a Newton-like method. This relies on solving the linear defect-correction problem, which requires the evaluation of the corresponding Jacobi matrix.

3.1 Galerkin formulation

For the discretization of Problem 6 (INH) and Problem 7 (STVK) we choose a Galerkin approach in space and time using finite dimensional FE subspaces

$$Ł_{kh} \subset Ł := L^2(I, L), \quad \mathcal{V}_{kh} \subset \mathcal{V} := C(I, V).$$

Within the present abstract setting the discretization in time is likewise by a Galerkin method such as the dG(r) ("discontinuous" Galerkin) or the cG(r) ("continuous" Galerkin) method of degree $r \geq 0$. The dG(0) method is closely related to the first-order backward Euler scheme and the cG(1) method to the second-order Crank-Nicolson scheme (indeed even algebraically identical for autonomous problems). The full space-time Galerkin framework is mainly introduced as the basis for a systematic approach to residual-based a posteriori error estimation.

At first, we introduce a compact form of the variational formulation of the FSI problem. For arguments $U = \{v, u, p\}$ and $\Psi = \{\psi^v, \psi^u, \psi^p\} \in \mathcal{W} := \mathcal{V} \times \mathcal{V} \times Ł$, we introduce the space-time semilinear form

$$
\begin{aligned}
A(U)(\Psi) := \int_0^T &\big\{ (\rho(\partial_t v + v \cdot \nabla v), \psi^v) + (\sigma(U), \varepsilon(\psi^v)) \\
&+ \begin{cases} (\operatorname{div} v, \psi^p) & \text{(INH material)} \\ (\chi_f \operatorname{div} v, \psi^p) + (\chi_s \alpha_p \nabla p, \nabla \psi^p) & \text{(STVK material)} \end{cases} \\
&+ (\chi_s \alpha_v (\partial_t u - v + v \cdot \nabla u), \psi^u) + (\chi_f \nabla u, \nabla \psi^u) \\
&- (g_3, \psi^v)_{\partial\Omega} - (f_3, \psi^v) \big\} \, dt.
\end{aligned}
\tag{11}
$$

With this notation, we can write the variational problems 6 and 7 in compact form as follows.

Problem 8 (Compact Eulerian formulation of the FSI problems).
Find $U \in U^D + \mathcal{W}$, such that

$$A(U)(\Psi) = 0 \quad \forall \Psi \in \mathcal{W}, \tag{12}$$

where U^D is an appropriate extension of the Dirichlet boundary and initial data and the space $\mathcal{W}$ is defined by

$$\mathcal{W} := \big\{ \Psi = \{\psi^v, \psi^u, \psi^p\} \in \mathcal{V} \times \mathcal{V} \times Ł, \ \psi^u(0) = \psi^v(0) = 0 \big\}.$$

3.2 Semidiscretization in time

For time discretization, we consider a mixed *continuous Galerkin* (cG(1)) method for velocity and deformation and a *discontinuous Galerkin* (dG(0)) method for the pressure. This Galerkin method can be interpreted as a variant of the Crank-Nicolson scheme. A detailed introduction to Galerkin methods for time discretization is given in the textbook of Eriksson, Estep, Hansbo & Johnson [15] or in the thesis of Schmich [28]. We partition the time interval $\bar{I} = [0, T]$ into $M \in \mathbb{N}$ subintervals,

$$\bar{I} = \{0\} \cup I_1 \cup I_2 \cup \cdots \cup I_M,$$

where $I_m = (t_{m-1}, t_m]$, $t_m > t_{m-1}$, and $k_m := t_m - t_{m-1}$. For the velocity and deformation, we consider the following two time-discrete spaces:

$$\mathcal{V}_k^{cG} := \{v_k \in C(\bar{I}, V) \big| v_k|_{I_m} \in P_1(I_m, V), m = 1, 2, \ldots, M\},$$

$$\mathcal{V}_k^{dG} := \{v_k \in L^2(I, V) \big| v_k|_{I_m} \in P_0(I_m; V), m = 1, 2, \ldots, M, v_k(0) \in V_k\},$$

where $P_1(I, V)$ and $P_0(I, V)$ are the spaces of linear and constant functions, respectively, on I with values in V. The first space $\mathcal{V}_k^{cG}$ contains functions which are continuous and piecewise linear in time, the second one functions which are discontinuous and constant on every subinterval I_m. The choice of $\mathcal{V}_k^{cG}$ as trial and $\mathcal{V}_k^{dG}$ as test space leads to a variant of the trapezoidal rule, using $\mathcal{V}_k^{dG}$ for both test and trial spaces a variant of the implicit Euler scheme can be derived. For the pressure, we choose piecewise constant functions in time from the space

$$Ł_k^{dG} := \{p_k \in L^2(I, L) \big| p_k|_{I_m} \in P_0(I_m; L), m = 1, 2, \ldots, M\}.$$

Then, by $\mathcal{W}_k^{cG} := \mathcal{V}_k^{cG} \times \mathcal{V}_k^{cG} \times Ł_k^{dG}$ and likewise by $\mathcal{W}_k^{dG} := \mathcal{V}_k^{dG} \times \mathcal{V}_k^{dG} \times Ł_k^{dG}$, we denote the time discrete trial and test spaces. In compact notation, the time discretized Galerkin formulation of our problem reads: find $U_k = (v_k, u_K, p_k) \in U_K^D + \mathcal{W}_k^{cG}$ such that

$$A(U_k)(\Psi_k) = 0 \quad \forall \Psi_k \in W_k^{dG},$$

with the space-time semilinear form as given in (11).

Since the test function from $\mathcal{W}_k^{dG}$ are discontinuous in time, this Galerkin scheme splits into a time-stepping method. Setting

$$v_k^m := v_k(t_m), \quad u_k^m := u_k(t_m), \quad p_k^m := p_k|_{I_m}, \quad \Psi_k^m := \Psi_k|_{I_m},$$

and $U_k^m := (v_k^m, u_k^m, p_k^m)$, we can approximate the integrals in the semilinear form (11) with the trapezoidal rule and solve for the approximation U^m with U^{m-1} given from the preceding time step. Here, one has to be careful with the implicit association of points $x \in \Omega$ to the specific subdomain. A fluid point $x \in \Omega_f(t_{m-1})$ can change to a structure point $x \in \Omega_s(t_m)$ within one time interval I_m. We thus define corresponding to (7)

$$\chi_f^m := \chi_f(t_m), \quad \chi_s^m := \chi_s(t_m),$$
$$\sigma^m := \chi_f^m \sigma_f(U^m) + \chi_s^m \sigma_s(U^m), \quad \rho^m := \chi_f^m \hat{\rho}_f + \chi_s^m \hat{\rho}_s,$$

and, for example, the momentum part in (11) approximated on I_m takes the form

$$\int_{I_m} (\rho(\partial_t v + v \cdot \nabla v), \psi^v) dt \approx \left(\frac{\rho^{m-1} + \rho^m}{2}(v^m - v^{m-1}), \varphi^v \right)$$
$$+ \frac{k_m}{2} \left(\rho^{m-1} v^{m-1} \cdot \nabla v^{m-1} + \rho^m v^m \cdot \nabla v^m, \varphi^v \right).$$

3.3 *Spatial discretization*

The spatial discretization is by a conforming finite element Galerkin method on meshes Ω_h consisting of cells denoted by K,

$$\bar{\Omega} = \bigcup_{i=1,\dots,N} \bar{K}_i,$$

which are (convex) quadrilaterals in 2d or hexahedrals in 3d. Such a decomposition Ω_h is referred to as "regular" if any cell edge is either a subset of the domain boundary components Γ_D, Γ_N, or a complete face or edge of another cell. However, to facilitate mesh refinement and coarsening, we allow the cells to have a certain number of nodes that are at the midpoint of sides or faces of neighboring cells. These "hanging nodes" do not carry degrees of freedom and the corresponding function values are determined by linear or bilinear interpolation of neighboring "regular" nodal points. For more details on this construction, we refer to Carey & Oden [10] or Bangerth & Rannacher [1].

The mesh parameter h is a scalar cellwise constant function defined by $h|_K :=$ $h_K = \mathrm{diam}(K)$. We set $h_{\max} := \max_{K \in \Omega_h} h_K$. For a cell K, we denote by ρ_K the diameter of the maximal inscribed ball in K. To ensure proper approximation properties of the finite element spaces which are constructed based on the meshes Ω_h, we require the "uniform-shape" and the "uniform-size" conditions to be fulfilled:

Mesh regularity condition: *Each cell $K \in \mathscr{T}_h$ is the image of the reference unit cube $\hat{K} = [0,1]^d$ under some d-linear mapping $\sigma_K : \hat{K} \to K$. This mapping is uniquely described by the 2^d coordinate values of the corners of K, if the ordering of the corners is preserved, see Fig. 1. The Jacobian tensors σ'_K of these mappings are invertible and satisfy the uniform bounds*

$$\sup_{h>0} \max_{K \in \mathscr{T}_h} \|\sigma'_K\| \le c, \qquad \sup_{h>0} \max_{K \in \mathscr{T}_h} \|(\sigma'_K)^{-1}\| \le c. \tag{13}$$

This condition is satisfied if the cells $K \in \mathscr{T}_h$ possess the usual structural properties of uniform "non-degeneracy", "uniform shape", and "uniform size property".

To increase the number of cells in a decomposition Ω_h, we employ "mesh refinement", which consists of subdividing a cell into 2^d subcells. Cell subdivision is done by connecting the midpoints of opposing edges or faces of a cell. A refinement is

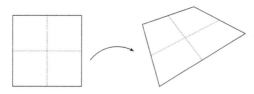

Fig. 1 Reference mapping $\sigma_K : \hat{K} \to K$.

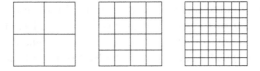

Fig. 2 A regular mesh after two cycles of global refinement.

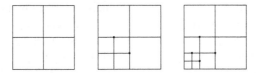

Fig. 3 A regular mesh after two cycles of local refinement.

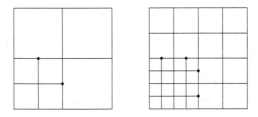

Fig. 4 A regular mesh after two cycles of patchwise local refinement.

"global" if this is done for every cell. An example of a regular mesh and two global refinements is shown in Fig. 2. Each of the resulting meshes after refinement is also regular. "Coarsening" of 2^d cells is possible if they were generated by prior refinement of some "parent cell". A group of 2^d such cells is referred to as a "(cell) patch".

In addition to "global" refinement, we will also use "local" refinement. This consists of only subdividing some cells in a given decomposition. Such refinement leads to nodes that are placed on the middle of the neighboring cells' edges or faces, i.e., to "hanging nodes", where only one hanging node is allowed per edge or face. In Fig. 3 local refinement is applied twice leading to hanging nodes indicated by "bullets".

Sometimes, we will require that a decomposition Ω_h is organized in a patchwise manner. This means that Ω_h is the result of global refinement of the coarser decomposition Ω_{2h}, as shown in Fig. 4.

Given a function space V the decomposition Ω_h and the cellwise space of polynomial functions $Q(K)$, we construct the corresponding finite element subspace $V_h \subset V$ by

$$V_h := \{\varphi \in V, \varphi|_K \in Q(K), K \in \Omega_h\}.$$

Each polynomial function space $Q(K)$ is actually defined on a "reference cell" $\hat{K} := (0,1)^d$ as the reference function space $\hat{Q}(\hat{K})$. We use the function space of polynomial degree $p \geq 0$ on $\hat{K}$ defined by

$$\hat{Q}^p(\hat{K}) := \mathrm{span}\{\hat{x}^\alpha, \hat{x} = (\hat{x}_1, \ldots, \hat{x}_d) \in \hat{K}, \alpha = (\alpha_1, \ldots, \alpha_d), \alpha_i \in \{0, \ldots, p\}\},$$

with the usual multi-index notation. In the numerical tests presented below, only finite elements with $p = 1$ ("d-linear elements") are used. Therefore, we omit the superscript p and simply refer to $\hat{Q}(\hat{K})$. The reference function space $\hat{Q}(\hat{K})$ is mapped to the corresponding cell K with the help of the d-linear mapping $\sigma_K :$ $\hat{K} \to K$,

$$Q(K) = \{\varphi(x) = \hat{\varphi}(\sigma_K^{-1}(x)), \ \hat{\varphi} \in \hat{Q}(\hat{K}), \ x \in K\}.$$

The d-linear mapping $\sigma_K : \hat{K} \to K$ is uniquely described by the 2^d coordinate values of the corners of K, if the ordering of the corners is preserved, see Fig. 1. Hence, in this case, the reference function space and the mapping function space are the same. Accordingly, the resulting finite elements are called "isoparametric".

To allow for dynamically changing meshes, we associate with Ω_{mh} the decomposition of the domain Ω used in the time interval I_m. The corresponding finite element subspace is denoted by $V_{mh} \subset V$. The fully discrete trial and test spaces are then defined as

$$\mathscr{V}_{kh}^{cG} := \{v_{kh} \in C(\bar{I}, V) \big| v_{kh}|_{I_m} \in P_1(I_m, V_{mh}), \ m = 1, 2, \dots, M\},$$

$$\mathscr{V}_{kh}^{dG} := \{v_{kh} \in L^2(I, V) \big| v_{kh}|_{I_m} \in P_0(I_m, V_{mh}), m = 1, 2, \dots, M, \ v_{kh}(0) \in V_{0h}\},$$

$$\mathsf{L}_{kh}^{dG} := \{p_{kh} \in L^2(I, L) \big| p_{kh}|_{I_m} \in P_0(I_m, V_{mh}), \ m = 1, 2, \dots, M\}.$$

The spatial discretization by "equal-order" finite elements for velocity and pressure needs stabilization in order to compensate for the missing "inf-sup stability". We use the so-called *"local projection stabilization"* (LPS) introduced by Becker and Braack [3,4]. An analogous approach is also employed for stabilizing the convection in the flow model as well as in the transport equation for the displacement u. To this end, we define the mesh-dependent bilinear form

$$(\varphi, \psi)_\delta := \sum_{K \in \Omega_h} \delta_K \, (\varphi, \psi)_K,$$

where the parameter δ_K is adaptively determined by

$$\delta_K := \frac{\alpha h_K^2}{\chi_f \rho_f \nu_f + \chi_s \mu_s + \beta \rho |v_h|_{\infty; K} h_K}.$$

Further, we introduce the "fluctuation operator" $\pi_h : V_h \to V_{2h}$ on the finest mesh level Ω_h by $\pi_h := I - P_{2h}$, where $P_{2h} : V_h \to V_{2h}$ denotes the L^2-projection. The operator π_h measures the fluctuation of a function in V_h with respect to its projection into the next coarser space V_{2h}. With this notation, we define the stabilization form

$$S_\delta(U_h)(\Psi_h) := \int_0^T \{(\nabla \pi_h p_h, \nabla \pi_h \psi_h^p)_\delta + (\rho v_h \cdot \nabla \pi_h v_h, v_h \cdot \nabla \pi_h \psi_h^v)_\delta$$

$$+ (\chi_s v_h \cdot \nabla \pi_h u_h, v_h \cdot \nabla \pi_h \psi_h^u)_\delta\} \, dt,$$

where the first term stabilizes the fluid pressure, the second one the INH structure pressure, the third one the transport in the flow model, and the fourth one the transport of the displacement u_h. The LP stabilization has the important property that it acts only on the diagonal terms of the coupled system and that it does not contain any second-order derivatives. However, it is only "weakly" consistent, as it does not vanish for the continuous solution, but it tends to zero with the right order as $h \to 0$. The choice of the numbers α, β in the stabilization parameter δ_K is, based on practical experience, in our computations $\alpha = 1/2$, and $\beta = 1/6$.

With this notation the stabilized Galerkin finite element approximation of Problem 8 reads as follows.

Problem 9 (Space-Time discrete Galerkin approximation of the FSI problem in Eulerian framework).
Find $U_{kh} = \{v_{kh}, u_{kh}, p_{kh}\} \in U_{kh}^D + \mathcal{W}_{kh}^{cG}$, such that

$$A_\delta(U_{kh})(\Psi_{kh}) := A(U_{kh}, \Psi_{kh}) + S_\delta(U_{kh})(\Psi_{kh}) = 0 \quad \forall \Psi_h \in \mathcal{W}_h^{dG}, \qquad (14)$$

where the "discrete" finite element spaces $\mathcal{W}_{kh}^{cG}$ and W_{kh}^{dG} are defined analogously as their "continuous" counterparts.

As on the spatially continuous level the existence of solutions to this semi-discrete problem is not guaranteed and has to be justified separately for each particular situation.

3.4 Error estimation and mesh adaptation

The main issues of this article is the automatic mesh adaptation within the Eulerian finite element approximation of the FSI problem. The computations presented below have been done on three different types of meshes:

- globally refined meshes obtained using several steps of uniform refinement of a coarse initial mesh,
- locally refined meshes obtained using a purely geometry-based criterion by marking all cells for refinement which have certain prescribed distances from the fluid-structure interface,
- locally refined meshes obtained using a systematic residual-based criteria by marking all cells for refinement which have error indicators above a certain threshold.

The goal is to employ the so-called *"Dual Weighted Residual Method"* (DWR method) for the adaptive solution of FSI problems. This method has been developed in Becker & Rannacher [5, 6] (see also Bangerth & Rannacher [1]) as an extension of the duality technique for a posteriori error estimation described in Eriksson, Estep, Hansbo & Johnson [15]. The DWR method provides a general framework for the derivation of "goal-oriented" a posteriori error estimates together with criteria of mesh adaptation for the Galerkin discretization of general linear and nonlinear variational problems, including optimization problems. It is based on a complete *variational* formulation of the problem, such as (12) for the FSI problem. In fact, this was one of the driving factors for deriving the Eulerian formulation underlying (12).

We begin with a brief outline of the DWR method for the special case of an FSI problem governed by an abstract variational equation posed on a space-time domain such as (12). We restrict us to a simplified version of the DWR method, which suffices for the present purposes. For a more elaborated version, which is particularly useful in the context of optimization problems, we refer to the literature stated above. For notational simplicity, we think the nonhomogeneous boundary and initial data U^D to be incorporated into a linear forcing term $F(\cdot)$, or to be exactly representable in the approximating space $\mathscr{W}_h$. Then, we seek $U \in U^D + \mathscr{W}^0$ such that

$$A(U)(\Psi) = F(\Psi) \quad \forall \Psi \in \mathscr{W}^0. \tag{15}$$

The corresponding (stabilized) Galerkin approximation seeks $U_h \in U_h^D + \mathscr{W}_h^0$ such that

$$A(U_h)(\Psi_h) + S_\delta(U_h)(\Psi_h) = F(\Psi) \quad \forall \Psi_h \in \mathscr{W}_h^0. \tag{16}$$

Suppose that the goal of the computation is the evaluation of the value $J(U)$ for some functional $J(\cdot)$ (for simplicity assumed to be linear) which is defined on $\mathscr{W}$. We want to control the quality of the discretization in terms of the error

$$J(U - U_h) = J(U) - J(U_h).$$

To this end, we introduce the directional derivative $A'(U)(\Phi, \cdot)$ the existence of which is assumed. With the above notation, we introduce the bilinear form

$$L(U, U_h)(\Phi, \Psi) := \int_0^1 A'(U_h + s(U - U_h))(\Phi, \Psi) \, ds,$$

and formulate the "dual problem"

$$L(U, U_h)(\Phi, Z) = J(\Phi) \quad \forall \Phi \in \mathscr{W}^0. \tag{17}$$

In the present abstract setting the existence of a solution $Z \in \mathscr{W}^0$ of the dual problem (17) has to be assumed. Now, taking $\Phi = U - U_h \in \mathscr{W}^0$ in (17) and using the Galerkin orthogonality property

$$A(U)(\Psi_h) - A(U_h)(\Psi_h) = S_\delta(U_h)(\Psi_h), \quad \Psi \in \mathscr{W}_h^0,$$

yields the error representation

$$
\begin{aligned}
J(U - U_h) &= L(U, U_h)(U - U_h, Z) \\
&= \int_0^1 A'(U_h + s(U - U_h))(U - U_h, Z) \, ds \\
&= A(U)(Z) - A(U_h)(Z) \\
&= F(Z - \Psi_h) - A(U_h)(Z - \Psi_h) - S_\delta(U_h)(\Psi_h) \\
&=: \rho(U_h)(Z - \Psi_h) - S_\delta(U_h)(\Psi_h),
\end{aligned}
$$

where $\Psi_h \in \mathscr{W}^0$ is an arbitrary element, usually taken as the generic nodal inter-polant $I_h Z \in \mathscr{W}_h^0$ of Z. For the evaluation of the terms on the right-hand side, we split the integrals in the residual term $\rho(U_h)(Z - \Psi_h)$ into their contributions from the single time intervals I_m and spatial mesh cells $K \in \mathscr{T}_h^m$ and integrate by parts. This results in an estimate of the error $|J(U - U_h)|$ in terms of computable local residual terms $\rho_K(U_h)$ multiplied by certain weight factors $\omega_K(Z)$ which depend on the dual solution Z,

$$
|J(U - U_h)| \le \sum_{m=1}^{M} \sum_{K \in \mathscr{T}_h^m} \rho_K^m(U_h) \, \omega_K^m(Z) + |S_\delta(U_h)(\Psi_h)|. \tag{18}
$$

The explicit form of the terms in the sum on the right-hand side is rather compli-cated and will not be repeated here. The second term due to the regularization is assumed to be small and is therefore neglected. For details, we refer to Dunne [12] and Dunne & Rannacher [13].

Since the dual solution Z is unknown, the evaluation of the weights $\omega_K(Z)$ requires further approximation. First, we linearize by assuming

$$
L(U, U_h)(\Phi, \Psi) \approx L(U_h, U_h)(\Phi, \Psi) = A'(U_h)(\Phi, \Psi),
$$

and use the approximate "discrete" dual solution $Z_h \in \mathscr{W}_h^0$ defined by

$$
A'(U_h)(\Phi, Z_h) = J(\Phi_h) \quad \forall \Phi_h \in \mathscr{W}_h^0. \tag{19}
$$

From Z_h, we generate improved approximations to Z in a post-processing step by patchwise higher-order interpolation. For example in 2d on 2×2-patches of cells in $\mathscr{T}_h$ the 9 nodal values of the piecewise bilinear Z_h are used to construct a patchwise biquadratic function $\tilde{Z}$ as indicated in Fig. 5. This is then used to obtain the approximate error estimate

$$
|J(U - U_h)| \approx \eta := \sum_{m=1}^{M} \sum_{K \in \mathscr{T}_h^m} \rho_K^m(U_h) \, \omega_K^m(\tilde{Z}), \tag{20}
$$

which is the basis of automatic mesh adaptation.

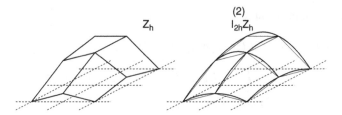

Fig. 5 Local postprocessing by patchwise "biquadratic" interpolation, $I_{2h}^{(2)} Z_h$, of the "bilinear" discrete solution Z_h in 2D.

Remark 1. The *dual solution* Z has the features of a "generalized" Green function $G(K, K')$, as it describes the dependence of the target error quantity $J(U - U_h)$, which may be concentrated at some cell K, on local properties of the data, i.e. in this case the residuals $\rho_{K'}$ on cells K'.

Remark 2. The solvability of the primal and dual problems (15), (16) and (17), (19), respectively, is not for granted. This is a difficult task regarding the rather few existence results in the literature for general FSI problems. Further, the assumption of differentiability cause concerns in treating the FSI problems in the Eulerian framework since the dependence of the characteristic function $\chi_f(x, u) = \hat{\chi}_f(x - u)$ on the deflection u is generally not differentiable (only Lipschitzian). However, this non-differentiability can be resolved by the "Hadamard structure theorem", on the assumption that the interface between fluid and structure forms a lower dimensional manifold and the differentiation is done in a weak variational sense. In essence this has the same effect as discretizing along the interface and replacing the directional derivative by a mesh-size dependent difference quotient, a pragmatic approach that has proven itself in similar situations, e.g. for Hencky elasto-plasticity in Rannacher and Suttmeier [27].

For the *primal* problem the directional (Gâteaux) derivative of the complete FSI problem does not need to be exact, it only needs to be "good enough" for Newton iteration to ensure convergence, leading to a reduction of the residuals of the non-linear system. Thus, for the primal problem the nonlinear system is used to measure the "quality" of the approximation. For the *dual* problem though things may initially seem less clear, since the dual problem is simply a *linear* problem directly based on the Gâteaux derivative. Of course, an immediate "measure of quality" of the discrete dual solution is the residual of the linear system. But there is no immediate measure for the quality of the discrete dual solution in relation to the *continuous* dual solution. This uncertainty stems from the highly nonlinear influence of the displacement u in the Gâteaux derivative, which is seen in the transformed fluid equations in the ALE framework. For the Eulerian framework, additional boundary Dirac integrals, which stem from the shape derivatives are causing these nonlinearities. This seemingly lack of clarity though is not typical to FSI problems. It is only more obvious in such problems since everything *visible* depends on the position of the interface. Generally though this uncertainty concerning the discrete dual solution is present in *all* nonlinear problems, since in such problems the Gâteaux derivatives depend on the primal solution and can only be approximated by using the discrete primal solutions.

In the case of FSI problems, we assume that the interface obtained on the current mesh is already in good agreement with the correct one, $\Gamma_{hi} \approx \Gamma_i$, and set up the dual problem formally with Γ_{hi} as a fixed interface. In all test computations, we did not encounter difficulties in obtaining the discrete solutions. In fact the performance of the error estimator for a given goal functional was always good for both the ALE and the Eulerian framework.

3.5 Mesh adaptation algorithm

The approach which is used for the adaptive refinement of the spatial mesh is straightforward. Let an error tolerance TOL be given. Then, on the basis of the (approximate) a posteriori error estimate (20), the mesh adaptation proceeds as follows:

1. Compute the primal solution U_h from (16) on the current mesh, starting from some initial state, e.g., that with zero deformation.
2. Compute the solution $\tilde{Z}_h$ of the approximate discrete dual problem (19).
3. Evaluate the cell-error indicators $\eta_K^m := \rho_K^m(U_h)\,\omega_K^m(\tilde{Z}_h)$.
4. If $\eta < TOL$ then accept U_h and evaluate $J(U_h)$, otherwise proceed to the next step.
5. Determine the 30% cells with largest and the 10% cells with smallest values of η_K. The cells of the first group are refined and those of the second group coarsened. Then, continue with Step 1. (Coarsening usually means canceling of an earlier refinement. Further refinement may be necessary to prevent the occurrence of too many hanging nodes. In two dimensions this strategy leads to about a doubling of the number of cells in each refinement cycle. By a similar strategy it can be achieved that the number of cells stays about constant during the adaptation process within the time stepping procedure.

Remark 3. The error representation (18) has been derived assuming the error functional $J(\cdot)$ as linear. In many applications nonlinear, most often quadratic, error functionals occur. An example is the spatial L^2-norm error

$$J(U_h) := \|(U - U_h)(T)\|,$$

at the end time T. For nonlinear (differentiable) error functionals the DWR approach can be extended to yield an error representation of the form (18); see Becker & Rannacher [6], Bangerth & Rannacher [1].

3.6 Evaluation of directional derivatives and solution of the algebraic problems

The time-discretized scheme as derived in Sect. 3.2 can be formulated as a time-stepping scheme, similar to the Crank-Nicolson scheme. The nonlinear problem in every time-step is solved by a Newton-like method. The resulting linear subproblems are then solved by the "Generalized Minimal Residual (GMRES)" method with preconditioning by a geometric multigrid method with block-ILU smoothing. Since such an approach is rather standard nowadays, we omit its details and refer to some relevant literature, e.g., Turek [31], Rannacher [26], or Hron and Turek [20]. For the implementational details of using the multigrid method on locally refined meshes, we refer to Becker and Braack [2].

The linear operators are essentially (if time-stepping parts and factors stemming from the approximation of the temporal derivatives are neglected) the directional derivatives of the governing semilinear form of the variational formulation, i.e. the form $A(U)(\Psi)$ in the Eulerian framework,

$$A'(U)(\Phi, \Psi) := \frac{d}{d\varepsilon} A(U + \varepsilon\Phi)(\Psi)|_{\varepsilon=0}. \tag{21}$$

For only "weakly nonlinear" systems such as the original Navier-Stokes equations in Eulerian framework obtaining the directional derivative is a straight forward task and can be done analytically "by hand". For structure mechanical systems (for example based on the St. Vernant-Kirchhoff material law) though writing down the explicit directional derivative can become cumbersome. For example in the Lagrangian case the scalar product $(\hat{J}\hat{F}^{-T}, \widehat{\nabla}\hat{\varphi}^v)$ is strongly nonlinear in $\hat{u}$,

$$\hat{J}\hat{\sigma}\hat{F}^{-T} = \hat{F}(\lambda_s \mathrm{tr}\hat{E}\,I + 2\mu_s\hat{E}), \quad \hat{F} = I + \widehat{\nabla}\hat{u}, \quad \hat{E} = \tfrac{1}{2}(\hat{F}^T\hat{F} - I).$$

However, in the Eulerian framework the corresponding scalar product takes the form $(\sigma, \nabla\varphi)$, which does not become any easier since the Cauchy stress tensor σ is based on the inverse of the "reverse deformation gradient" $I - \nabla u$,

$$\sigma = JF^{-1}(\lambda_s \mathrm{tr}E\,I + 2\mu_s E)F^{-T}, \quad F = I - \nabla u, \quad E = \tfrac{1}{2}(F^{-T}F^{-1} - I).$$

To alleviate this problem one may use a method that is the basis of "Automatic Differentiation" such as described in Rall [24] and Griewank [18]. The method is used to determine the derivative of a function at a given position. It is based on the technique of mechanically applying the basic rules of differentiation to the "serialized evaluation" of a function. This is achieved by breaking down the evaluation of the function for a given value into a sequence of basic elementary evaluations. Consequently, since evaluation is done in a sequence the resulting values from one evaluation are used in a later evaluation. To these elementary parts the rules of differentiation (i.e. the chain rule, the sum rule and the product rule) are applied. For the details of the application of this techniques within the simulation of FSI problems, we refer to Dunne [12].

4 Numerical examples

For the verification of the Eulerian formulation of fluid-structure interaction problems different numerical test-cases have been studied by Dunne [12], Dunne & Rannacher [13] and Richter & Wick [30]. Being slightly more expensive, the Eulerian formulation yields the same results (and the same convergence order) as the well-established ALE approach. In Dunne & Rannacher [13] the FLUSTRUK-A benchmark by Hron & Turek [21] is studied. These results are compared with different approaches, both monolithic and partitioned. For completeness, we recall some of the most important observations obtained by these test calculation. In Richter & Wick [30] at several numerical test-cases the behavior of the Eulerian and the ALE formulation for large deformations is investigated, finding that the Eulerian formulation is capable to model large deformations easier without the need of remeshing or reinitialization. In this section, we present numerical examples which particularly demonstrate the capability of the Eulerian formulation to allow for large deformation and free movement of the structure.

4.1 The FSI benchmark FLUSTRUK-A revisited

From Dunne & Rannacher [13], we recall the essential results for the FSI benchmark FLUSTRUK-A described in Hron & Turek [21]. A thin elastic bar immersed in an incompressible fluid develops self-induced time-periodic oscillations of different amplitude depending on the material properties assumed. The configuration of this benchmark is shown in Fig. 6. The test cases FSI-2* and FSI-3* are slight variations from the original settings FSI-2 and FSI-3 using different material parameters for a better demonstration of special properties using the Eulerian approach. This benchmark has been defined to validate and compare the different computational approaches and software implementations for solving FSI problems. In order to have a fair comparison of our Eulerian-based method with the traditional Eulerian-Lagrangian approach, also an ALE method has been implemented for this benchmark problem.

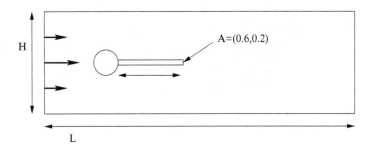

Fig. 6 Configuration of the FSI benchmark FLUSTRUK-A.

Configuration: The computational domain has length $L = 2.5$, height $H = 0.41$, and left bottom corner at $(0, 0)$. The center of the circle is positioned at $C = (0.2, 0.2)$ with radius $r = 0.05$. The elastic bar has length $l = 0.35$ and height $h = 0.02$. Its right lower end is positioned at $(0.6, 0.19)$ and its left end is clamped to the circle. Control points are $A(t)$ fixed at the trailing edge of the structure with $A(0) = (0.6, 0.20)$, and $B = (0.15, 0.2)$ fixed at the cylinder (stagnation point).

Boundary and initial conditions: The boundary conditions are as follows: Along the upper and lower boundary the usual "no-slip" condition is used for the velocity. At the (left) inlet a constant parabolic inflow profile,

$$v(0, y) = 1.5 \, \bar{U} \, \frac{4y(H - y)}{H^2},$$

is prescribed which drives the flow, and at the (right) outlet zero-stress $\sigma \cdot n = 0$ is realized by using the "do-nothing" approach in the variational formulation (see Heywood, Rannacher & Turek [19] and Rannacher [26]). This implicitly forces the pressure to have zero mean value at the outlet. The initial condition is zero flow velocity and structure displacement.

Material properties: The fluid is assumed as incompressible and Newtonian, the cylinder as fixed and rigid, and the structure as (compressible) St. Venant-Kirchhoff (STVK) material.

Discretization: The first set of computations is done on globally refined meshes for validating the proposed method and its software implementation. Then, for the same configuration adaptive meshes are used where the refinement criteria are either purely heuristic, i.e., based on the cell distance from the interface, or are based on a simplified stationary version of the DWR approach. In all cases a uniform time-step size of $0.005 \, s$ is used. The curved cylinder boundary is approximated to second order by polygonal mesh boundaries.

Below, we recall the results obtained for three of the most important test cases within the benchmark FLUSTRUK-A. The corresponding characteristic problem parameters are listed in Table 1. These tests are intended to demonstrate that the fully Eulerian-based simulation method developed in this project is comparable and

Table 1 Parameter settings for the FSI test cases.

parameter	FSI-2*	FSI-3	FSI-3*
structure model	STVK	STVK	INH
$\rho_f [10^3 kg \, m^{-3}]$	1	1	1
$\nu_f [10^{-3} m^2 s^{-1}]$	1	1	1
ν_s	0.4	0.4	0.5
$\rho_s [10^3 kg \, m^{-3}]$	20	1	1
$\mu_s [10^6 kg \, m^{-1} s^{-2}]$	0.5	2	2
$\bar{U} [m \, s^{-1}]$	0	2	2

competitive with the common ALE approach with respect to accuracy including mass conservation, while being sufficiently flexible in dealing with large deformation.

4.1.1 Moderate deformation: ALE versus Eulerian Method

We begin with the FSI-3 test case. Some snapshots of the results of these simulations are shown in Fig. 7. The corresponding time-dependent behavior of the displacements are shown in Fig. 8. With the Eulerian approach we estimated an oscillation with a frequency of $5.48\,s^{-1}$ and a lower frequency of $5.04\,s^{-1}$ using ALE coordinates. Hron and Turek [20] identified a frequency of $5.3\,s^{-1}$ as reference value. As maximum amplitude of the vertical displacement, the Eulerian approach gave the

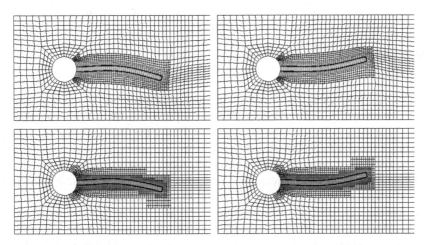

Fig. 7 FSI-3 Test: Some snapshots of results obtained by the ALE (top two) and the Eulerian (bottom two) approaches.

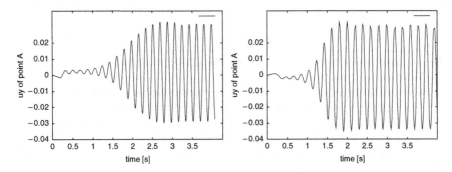

Fig. 8 FSI-3 Test: Vertical Displacement of the control point A, obtained by the Eulerian approach (left, $N = 3876$ cells) with maximum amplitude $3.01 \cdot 10^{-2}$ and frequency $5.48\,s^{-1}$, and by the ALE approach (right, $N = 2082$ cells) with maximum amplitude $3.19 \cdot 10^{-2}$ and frequency $5.04\,s^{-1}$.

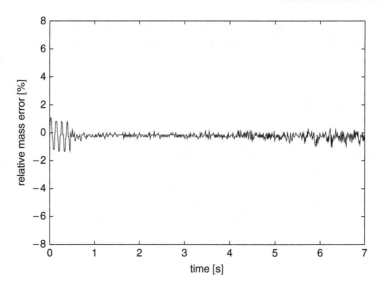

Fig. 9 FSI-3*: Relative mass error of the bar.

value $3.01 \cdot 10^{-2}$ versus $3.19 \cdot 10^{-2}$ for ALE coordinates. Here, the corresponding reference value are given by Hron and Turek [20] as $3.438 \cdot 10^{-2}$. While our results differ from these reference values by less than 5% for the frequency, the maximum amplitude shows a larger deviation of up to 10%.

The FSI-3* test case is used to illustrate the mass conservation properties of the Eulerian solution approach. Here, the *incompressible neo-Hookean* (INH) material is used instead of the compressible *St. Venant-Kirchhoff* material. Since in the Eulerian approach the structure deformations are not in a Lagrangian framework, it is not immediately clear, due to the coupling with the fluid, how well the mass of the structure is conserved in an Eulerian approach, especially in the course of an instationary simulation comprising hundreds of time steps. In Fig. 9, we display the bar's relative mass error as a function of time. Except for certain initial jitters, the relative error is less than 1%.

4.1.2 Large deformation: local versus zonal mesh refinement

In the test case FSI-2* the fluid is initially at rest and the bar is subjected to a vertical (gravitational) force. This causes the bar to bend downward until it touches the bottom wall. The density of the bar has been chosen larger than in the original FSI-2 benchmark to yield a bigger deformation of the bar. We compare the efficiency of zonal refinement versus local refinement by the DWR method for this test case within the Eulerian methods. Figure 10 shows corresponding sequences of snapshots, while the position of the trailing tip A is displayed in Fig. 11. We see that by sensitivity-driven local refinement within the DWR method on only 1 900 cells

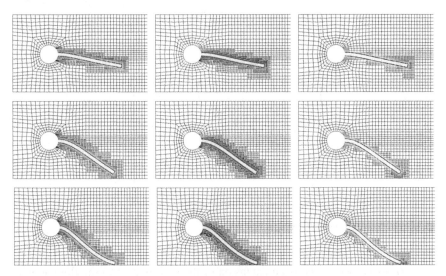

Fig. 10 A sequence of snap-shots of the bar's large deformation under gravitational loading obtained by the Eulerian approach using zonal refinement with $N \sim 3\,000$ and $N \sim 12\,000$ cells (left and middle), and local mesh refinement by the DWR method (right) with only $N \sim 1\,900$ cells.

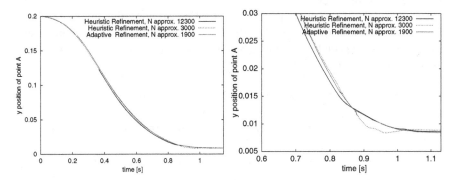

Fig. 11 Time varying position of the point $x_A(t)$ over the full time interval $[0, 1.1]$ (left) and over a zoomed interval $[0.6, 1.1]$ (right).

almost the same accuracy can be achieved as by zonal refinement on 12 300 cells. The gain in CPU time needed is almost 85 % (about 30 h for the zonal versus about 4 h for the local refinement).

4.2 Free movement

Limiting factor for large deformation using the ALE framework is the implicit transformation of the fluid mesh to artificial coordinates, usually defined via

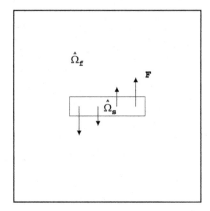

Fig. 12 Rectangular elastic structure in unit square.

$\hat{T}_f = id + \hat{u}_f$, where $\hat{u}_f$ is an extension of the deformation to the flow domain $\hat{\Omega}_f$. In the Eulerian formulation, the flow problem is not transformed but instead given in natural coordinates. Here, the artificial extension of the deformation field u to Ω_f is not used as transformation, but just needed to trace the domain of influence in the definition of the characteristic functions (8).

In the middle of the unit square $\Omega = (0,1)^2$, we prescribe an elastic object $\hat{\Omega}_s = \{(x, y) : x \in (0.3, 0.7), y \in (0.45, 0.55)\}$ (see Fig. 12). On the boundary $\partial\Omega$, the *do-nothing* condition is given for the velocity and a homogenous Neumann condition for the deformation:

$$-\nu_f \partial_n v + p \cdot n = 0 \text{ on } \partial\Omega, \quad \partial_n u = 0 \text{ on } \partial\Omega.$$

The dynamics of the system is driven by a force acting on the elastic structure:

$$(\rho_s(\partial_t v + v \cdot \nabla v), \varphi)_{\Omega_s(t)} + (\sigma_s, \nabla\varphi)_{\Omega_s(t)} = (\rho_s J_s F, \varphi),$$

with the volume force F given by

$$F = \begin{pmatrix} -y + \frac{1}{2} \\ x - \frac{1}{2} \end{pmatrix}.$$

This force results in a rotation of the rectangle and a rotational flow evolving in the domain Ω. By the standard ALE approach (without remeshing and reinitialization) this problem is not computable, since the extension of the deformation u to the flow domain would lead to an increasingly large distortion of the meshes. In Fig. 13 we show snapshots of the solution for different time points.

For comparison we show corresponding results obtained using an ALE formulation. For the construction of the mapping $\hat{T}_f : \hat{\Omega}_f \rightarrow \Omega_f(t)$ a bi-harmonic extension of the deformation u to the flow domain is used. This usually yields the

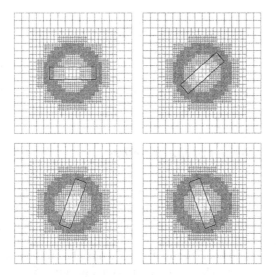

Fig. 13 Rotation of the elastic rectangle by a volume force.

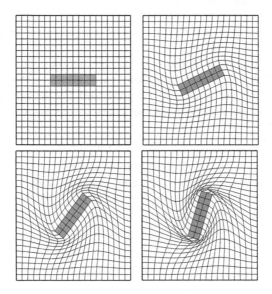

Fig. 14 Implicit mesh deformation using ALE coordinates for the rotating rectangular structure.

best results and allows for the largest possible deformation, see Dunne [12] for details. In Fig. 14 we show the deformation and the resulting distorted meshes for different time-steps. The last snapshot indicates the largest possible deformation of the flow domain before breakdown of the solution scheme in the next time-step. For the ALE comparison a very coarse mesh is used for a better visualization of the effects.

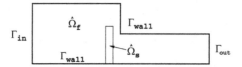

Fig. 15 Narrowing channel with elastic obstacle. Flow driven by pressure drop: $p = 0$ on Γ_{in} and $p = -1$ on Γ_{out}.

4.3 Boundary contact of the structure

The ALE formulation has problems when large deformation appears close to the boundary of the domain. In particular the contact of the elastic structure with the boundary is not possible within a monolithic formulation using simple ALE coordinates without remeshing techniques. The transformation $\hat{T}_f(t) : \hat{\Omega}_f \rightarrow \Omega_f(t)$ of the flow domain would degenerate. Here, we model a pressure induced flow in a narrowing channel, blocked by an elastic structure, see Fig. 15.

At the inflow and outflow boundary the *do-nothing* condition is prescribed with a pressure drop causing a flow to the right:

$$-\nu_f \partial_n v + p \cdot n = 0 \text{ on } \Gamma_{\text{in}}, \quad -\nu_f \partial_n v + p \cdot n = -1 \text{ on } \Gamma_{\text{out}}.$$

Prescribing a pressure drop instead of a Dirichlet inflow condition, the problem stays well-posed even if the channel is closed. Along the remaining boundary Γ_{wall} homogenous Dirichlet condition $v = 0$ is prescribed. For the extension of the displacement u to the flow domain a slip-like boundary condition is enforced on the outer boundary,

$$u \cdot n = 0 \text{ on } \Gamma_{\text{wall}}.$$

In Fig. 16, we show snapshots of the deformed structure $\Omega_s(t)$ for different time points. First, by the evolving flow the structure is bended to the right. "Contact" of the structure with the wall is realized up to one layer of mesh elements. After mesh refinement, the structure gets closer to the boundary of the domain.

5 Summary

In this paper, we presented a monolithic, fully Eulerian variational formulation for "fluid-structure interaction (FSI)" problems. This approach allows us to treat FSI problems with free bodies and large deformations. This is the main advantage of this method compared to interface tracking methods such as the arbitrary Lagrangian-Eulerian (ALE) method. For the FSI benchmark FLUSTRUK-A the Eulerian approach turns out to yield results which are in good agreement with those obtained by the ALE approach. In order to have a "fair" comparison both methods have been implemented using the same numerical components and software library

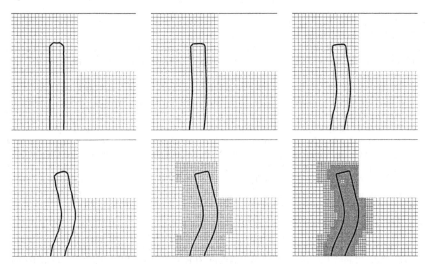

Fig. 16 Near-contact of the elastic structure with the wall. "Contact" is possible up to one layer of mesh elements.

GASCOIGNE [16]. The method based on the Eulerian approach is inherently more expensive than the ALE method, by about a factor of two, but it allows to treat also large deformations and boundary contact of the structure. This potential has been investigated for several model configurations.

The monolithic variational formulation of the FSI problem forms the basis for the application of the "Dual Weighted Residual (DWR)" method for "goal-oriented" a posteriori error estimation and mesh adaptation. In this method inherent sensitivities of the FSI problem are utilized by solving linear "dual" problems, similar as in the Euler-Lagrange approach to solving optimal control problems. The feasibility of this approach, in a simplified quasi-stationary version, has been demonstrated for the FSI benchmark FLUSTRUK-A.

Acknowledgements We gratefully acknowledge the financial support of this project by the Deutsche Forschungsgemeinschaft within the Research Unit 493 Fluid-Structure Interaction: Modelling, Simulation, Optimization.

References

1. W. Bangerth and R. Rannacher, *Adaptive Finite Element Methods for Differential Equations*, Birkhäuser, 2003.
2. R. Becker and M. Braack, *Multigrid techniques for finite elements on locally refined meshes*. Numer. Linear Algebra Appl. 7, 363–379, 2000.
3. R. Becker and M. Braack, *A finite element pressure gradient stabilization for the Stokes equations based on local projections*, Calcolo 38, 173-199, 2001.

4. R. Becker and M. Braack, A two-level stabilization scheme for the Navier-Stokes equations, *Proc. ENUMATH-03*, pp. 123-130, Springer, 2003.
5. R. Becker and R. Rannacher, Weighted a-posteriori error estimates in FE methods, Lecture ENUMATH-95, Paris, Sept. 18-22, 1995, in: *Proc. ENUMATH-97*, (H.G. Bock, et al., eds), pp. 621-637, World Scientific Publ., Singapore, 1998.
6. R. Becker and R. Rannacher, An optimal control approach to error estimation and mesh adaption in finite element methods, *Acta Numerica 2000* (A. Iserles, ed.), pp. 1-101, Cambridge University Press, 2001.
7. S. Bönisch, Th. Dunne, and R. Rannacher, Lecture on numerical simulation of liquid-structure interaction, in Lecture Notes of Oberwolfach Seminar *Hemodynamical Flows: Aspects of Modeling, Analysis and Simulation*, Nov. 20-26, 2005, Mathematical Resarch Institute Oberwolfach (G.P. Galdi, R. Rannacher, A.M. Robertson, S. Turek, eds), Birkhäuser, Basel, 2007.
8. D. Braess, *Finite Elements*, Cambridge University Press, Cambridge, United Kingdom, 1997.
9. S. Brenner and R. L. Scott, *The Mathematical Theory of Finite Element Methods*, Springer, Berlin Heidelberg New York, 1994.
10. G. Carey and J. Oden, *Finite Elements, Computational Aspects*, volume III. Prentice-Hall, 1984.
11. Th. Dunne, *An Eulerian approach to fluid-structure interaction and goal-oriented mesh refinement*, Proc. "Finite Elements for Flow Problems (FEF05)", IACM Special Interest Conference supported by ECCOMAS, April 4-6, 2005, Swansea, Wales, UK, Int. J. Numer. Methods Fluids 51, 1017–1039, 2006.
12. Th. Dunne, *Adaptive Finite Element Simulation of Fluid Structure Interaction Based on an Eulerian Formulation*, Doctoral dissertation, Institute of Applied Mathematics, University of Heidelberg, 2007.
13. Th. Dunne and R. Rannacher, *Adaptive finite element simulation of fluid-structure interaction based on an Eulerian variational formulation*, in "Fluid-Structure Interaction: Modeling, Simulation, Optimization" (H.-J. Bungartz and M. Schäfer, eds), pp. 371–386, Springer, Berlin-Heidelberg, 2006.
14. Th. Dunne, R. Rannacher and T. Richter, *Numerical simulation of fluid-structure interaction based on monolithic variational formulations*, in Series "Comtemporary Challenges in Mathematical Fluid Mechanics" (G.P. Galdi, R. Rannacher, eds.), World Scientific, Singapore, to appear 2010.
15. K. Eriksson, D. Estep, P. Hansbo, and C. Johnson, *Computational differential equations*, Cambridge University Press, Cambridge, 1996.
16. GASCOIGNE, A C++ numerics library for scientific computing, Institute of Applied Mathematics, University of Heidelberg, URL http://www.gascoigne.uni-P2.de/.
17. V. Girault and P.-A. Raviart, *Finite Element Methods for the Navier-Stokes Equations*, Springer: Berlin-Heidelberg-New York, 1986.
18. A. Griewank, *On automatic differentiation in mathematical programming: recent developments and applications*, (M. Iri and K. Tanabe, eds), pp. 83–108, Kluwer Academic Publishers, 1989.
19. J. Heywood, R. Rannacher, and S. Turek, *Artificial boundaries and flux and pressure conditions for the incompressible Navier-Stokes equations*, Int. J. Numer. Math. Fluids 22, 325–352, 1992.
20. J. Hron and S. Turek, *Proposal for numerical benchmarking of fluid-structure interaction between an elastic object and laminar incompressible flow*, in "Fluid-Structure Interaction: Modelling, Simulation, Optimisation" (H.-J. Bungartz and M. Schäfer, eds), Springer, Berlin-Heidelberg, 2006.
21. J. Hron and S. Turek, *A monolithic FEM/multigrid solver for an ALE formulation of fluid-structure interaction with applications in biomechanics*, in "Fluid-Structure Interaction: Modelling, Simulation, Optimisation" (H.-J. Bungartz and M. Schäfer, eds), pp. 146–170, Springer, Berlin-Heidelberg, 2006.
22. C. Liu and N. J. Walkington, An Eulerian description of fluids containing visco-elastic particles, *Arch. Rat. Mech. Anal.* 159, 229–252, 2001.
23. D. Meidner und B. Vexler, Adaptive Space-Time Finite Element Methods for Parabolic Optimization Problems, *SIAM Journal on Control and Optimization* 46(1), S. 116 142, 2007.

24. L. B. Rall, *Automatic Differentiation - Techniques and Application*, Springer, New York, 1981.
25. S. Osher and J. A. Sethian, Propagation of fronts with curvature based speed: algorithms based on Hamilton-Jacobi formulations, *Journal of Computational Physics* 79, 12, 1988.
26. R. Rannacher, Finite element methods for the incompressible Navier-Stokes equations, in *Fundamental Directions in Mathematical Fluid Mechanics* (G. P. Galdi, J. Heywood, R. Rannacher, eds), pp. 191–293, Birkhäuser, Basel-Boston-Berlin, 2000.
27. R. Rannacher and F.-T. Suttmeier Error estimation and adaptive mesh design for FE models in elasto-plasticity, in *Error-Controlled Adaptive FEMs in Solid Mechanics* (E. Stein, ed.), pp. 5–52, John Wiley, Chichster, 2002.
28. M. Schmich, *Adaptive Finite Element Methods for computing nonstationary incompressible flows*, Doctoral dissertation, Institute of Applied Mathematics, University of Heidelberg, 2009.
29. M. Schmich and B. Vexler, Adaptivity with Dynamic Meshes for Space-Time Finite Element Discretizations of Parabolic Equations, *SIAM J. Sci. Comput., Vol. 30(1), pp. 369-393*, 2008.
30. T. Richter and T. Wick, Finite Elements for Fluid-Structure Interaction in ALE and Fully Eulerian Coordinates, submitted to *Computer Methods in Applied Mechanics and Engineering*, 2009.
31. S. Turek, *Efficient solvers for incompressible flow problems: an algorithmic and computational approach*, Springer, Heidelberg-Berlin-New York, 1999.
32. VISUSIMPLE, An open source interactive visualization utility for scientific computing, Institute of Applied Mathematics, University of Heidelberg, URL http://visusimple.uni-P2.de/.

Numerical Simulation and Benchmarking of a Monolithic Multigrid Solver for Fluid-Structure Interaction Problems with Application to Hemodynamics

S. Turek, J. Hron, M. Mádlík, M. Razzaq, H. Wobker, and J.F. Acker

Abstract An Arbitrary Lagrangian-Eulerian (ALE) formulation is applied in a fully coupled monolithic way, considering the fluid-structure interaction (FSI) problem as one continuum. The mathematical description and the numerical schemes are designed in such a way that general constitutive relations (which are realistic for biomechanics applications) for the fluid as well as for the structural part can be easily incorporated. We utilize the LBB-stable finite element pairs $Q_2 P_1$ and $P_2^+ P_1$ for discretization in space to gain high accuracy and perform as time-stepping the 2nd order Crank-Nicholson, respectively, a new modified Fractional-Step-θ-scheme for both solid and fluid parts. The resulting discretized nonlinear algebraic system is solved by a Newton method which approximates the Jacobian matrices by a divided differences approach, and the resulting linear systems are solved by direct or iterative solvers, preferably of Krylov-multigrid type.

For validation and evaluation of the accuracy and performance of the proposed methodology, we present corresponding results for a new set of FSI benchmark configurations which describe the self-induced elastic deformation of a beam attached to a cylinder in laminar channel flow, allowing stationary as well as periodically oscillating deformations. Then, as an example of FSI in biomedical problems, the influence of endovascular stent implantation on cerebral aneurysm hemodynamics is numerically investigated. The aim is to study the interaction of the elastic walls of the aneurysm with the geometrical shape of the implanted stent structure for prototypical 2D configurations. This study can be seen as a basic step towards the understanding of the resulting complex flow phenomena so that in future aneurysm rupture shall be suppressed by an optimal setting of the implanted stent geometry.

S. Turek, M. Razzaq, H. Wobker, and J. F. Acker
Institute for Applied Mathematics, TU Dortmund, Vogelpothsweg 87, 44227 Dortmund, Germany
e-mail: stefan.turek@mathematik.tu-dortmund.de

J. Hron and M. Mádlík
Mathematical Institute, Charles University Prague, Sokolovska 83, 18675 Prague, Czech republic
e-mail: hron@karlin.mff.cuni.cz

H.-J. Bungartz et al. (eds.), *Fluid Structure Interaction II*, Lecture Notes in Computational Science and Engineering 73, DOI 10.1007/978-3-642-14206-2_8, © Springer-Verlag Berlin Heidelberg 2010

1 Introduction

In this paper, we consider the general problem of viscous flow interacting with an
elastic body which is being deformed by the fluid action. Such a problem is of great
importance in many real life applications, typical examples are the areas of biomedical fluids which include the influence of hemodynamic factors in blood vessels,
cerebral aneurysm hemodynamics, joint lubrication and deformable cartilage, and
blood flow interaction with elastic veins [2, 9, 25, 26, 34]. The theoretical investigation of fluid-structure interaction problems is complicated by the need of a mixed
description for both parts: While for the solid part the natural view is the material
(Lagrangian) description, for the fluid it is usually the spatial (Eulerian) description.
In the case of their combination some kind of mixed description (usually referred
to as the Arbitrary Lagrangian-Eulerian description or ALE) has to be used which
brings additional nonlinearity into the resulting equations (see [17]).

The numerical solution of the resulting equations of the fluid-structure interaction
problem poses great challenges since it includes the features of structural mechanics, fluid dynamics, and their coupling. The most straightforward solution strategy,
mostly used in the available software packages (see for instance [16]), is to decouple
the problem into the fluid part and solid part, for each of those parts using some well
established solution method; then the interaction process is introduced as external
boundary conditions in each of the subproblems. This has the advantage that there
are many well tested numerical methods for both separate problems of fluid flow
and elastic deformation, while on the other hand the treatment of the interface and
the interaction is problematic due to high stiffness and sensitivity. In contrast, the
monolithic approach discussed here treats the problem as a single continuum with
the coupling automatically taken care of as internal interface. This on the other hand
requires more robust nonlinear and linear solvers for the global problem.

Besides a short description of the underlying numerical aspects regarding discretization and solution procedure for this monolithic approach (see [17, 24]), we
present corresponding results for a set of FSI benchmarking test cases ('channel
flow around cylinder with attached elastic beam', see [30]), and we concentrate
on prototypical numerical studies for 2D aneurysm configurations and first steps
towards full 3D models. The corresponding parameterization is based on abstrac-
tions of biomedical data (i.e., cutplanes of 3D specimens from New Zealand white
rabbits as well as computer tomographic and magnetic resonance imaging data of
human neurocrania). In our studies, we allow the walls of the aneurysm to be elastic
and hence deforming with the flow field in the vessel. Moreover, we examine several
configurations for stent geometries which clearly influence the flow behavior inside
the aneurysm such that a very different elastic displacement of the walls is observed,
too. We demonstrate that both the elastic modeling of the aneurysm walls as well
as the proper description of the geometrical details of the shape of the aneurysm
and particularly of the stents are of great importance for a quantitative analysis of
the complex interaction between structure and fluid. This is especially true in view
of more realistic blood flow models and anisotropic constitutive laws for the elastic
walls.

2 Fluid-structure interaction problem formulation

The general fluid-structure interaction problem consists of the description of the fluid and solid parts, appropriate interface conditions at the interface and conditions for the remaining boundaries, respectively. Here, we consider the flow of an incompressible Newtonian fluid interacting with an elastic solid. We denote the domain occupied by the fluid by Ω_t^f and the solid part by Ω_t^s at the time $t \in [0, T]$. Let $\Gamma_t^0 = \bar{\Omega}_t^f \cap \bar{\Omega}_t^s$ be the part of the boundary where the elastic solid interacts with the fluid. In the following, the description for both fields and the interface conditions are introduced. Furthermore, discretization aspects and computational methods are described.

2.1 Fluid mechanics

The fluid flow is assumed to be laminar. It can be described by the Navier-Stokes equations for incompressible flows

$$\rho^f \left(\frac{\partial \mathbf{v}^f}{\partial t} + \mathbf{v} \cdot \nabla \mathbf{v} \right) - \nabla \cdot \sigma^f = \mathbf{0}, \quad \nabla \cdot \mathbf{v} = 0 \quad \text{in} \quad \Omega_t^f, \tag{1}$$

where ρ^f is the constant density. The state of the flow is described by the velocity and pressure fields $\mathbf{v}^f$, p^f, respectively. The external forces, for example due to gravity or human motion, are assumed to be not significant and are neglected. Although the blood is known to be non-Newtonian in general, we assume it to be Newtonian in this study. This is because we consider large arteries with radii of more than 2 mm, where the velocity and shear rate are high and the kinematic viscosity ν^f is nearly constant [20], such that the non-Newtonian effects can be neglected. The constitutive relation for the stress tensor reads

$$\sigma^f = -p^f \mathsf{I} + 2\mu \mathsf{D}(\mathbf{v}^f), \tag{2}$$

where μ is the dynamic viscosity of the fluid, p^f is the Lagrange multiplier corresponding to the incompressibility constraint in (1), and $\mathsf{D}(\mathbf{v}^f)$ is the strain-rate tensor:

$$\mathsf{D}(\mathbf{v}^f) = \frac{1}{2}(\nabla \mathbf{v}^f + (\nabla \mathbf{v}^f)^T). \tag{3}$$

For the fluid-structure interaction we use the ALE form of the balance equations. The corresponding discretization techniques are discussed in Sect. 3. Let us remark that also non-Newtonian flow models can be used for modeling blood flow, for instance of Power Law type or even including viscoelastic effects (see [7]) which is planned for future extensions.

2.2 *Structural mechanics*

The governing equations for the structural mechanics are the balance equations

$$\rho^s(\frac{\partial \mathbf{v}^s}{\partial t} + (\nabla \mathbf{v}^s)\mathbf{v}^s - \mathbf{g}) - \nabla \cdot \sigma^s = \mathbf{0}, \quad \text{in} \quad \Omega_t^s, \tag{4}$$

where the superscript s denotes the structure, ρ^s is the density of the material, $\mathbf{g}^s$ represents the external body forces acting on the structure, and σ^s is the Cauchy stress tensor. The deformation of the structure is described by the displacement $\mathbf{u}^s$, with velocity field $\mathbf{v}^s = \frac{\partial \mathbf{u}^s}{\partial t}$. Written in the more common Lagrangian description, i.e. with respect to some fixed reference (for example initial) state Ω^s, we have

$$\rho_0^s(\frac{\partial^2 \mathbf{u}^s}{\partial t^2} - \mathbf{g}) - \nabla \cdot \Sigma^s = \mathbf{0}, \quad \text{in} \quad \Omega^s, \tag{5}$$

where $\Sigma^s = J\sigma^s F^{-T}$ is the first Piola-Kirchhoff stress tensor. J denotes the determinant of the deformation gradient tensor F, defined as $F = I + \nabla \mathbf{u}^s$. Unlike the Cauchy stress tensor σ^s, Σ^s is non-symmetric. Since constitutive relations are often expressed in terms of symmetric stress tensor, it is natural to introduce the second Piola-Kirchhoff tensor S^s

$$S^s = F^{-T} \Sigma^s = JF^{-1}\sigma^s F^{-T}, \tag{6}$$

which is symmetric. For elastic material the stress is a function of the deformation (and possibly of thermodynamic variables such as the temperature) but it is independent of deformation history and thus of time. The material characteristics may still vary in space. In a homogeneous material mechanical properties do not vary, the strain energy function depends only on the deformation. A material is mechanically isotropic if its response to deformation is the same in all directions. The constitutive equation is then a function of F. More precisely, it is usually written in terms of the Green-Lagrange strain tensor, as

$$E = \frac{1}{2}(C - I), \tag{7}$$

where I is the identity tensor and $C = F^T F$ is the left Cauchy-Green strain tensor.

For the subsequent FSI benchmark calculations we employ the St. Venant-Kirchhoff material model as an example for homogeneous isotropic material whose reference configuration is the natural state (i.e. where the Cauchy stress tensor is zero everywhere). The St. Venant-Kirchhoff material model is specified by the following constitutive law

$$\sigma^s = \frac{1}{J}F(\lambda^s(\text{tr } E)I + 2\mu^s E)F^T, \quad S^s = \lambda^s(\text{tr } E)I + 2\mu^s E, \tag{8}$$

where λ^s denotes the first Lamé coefficient, and μ^s the shear modulus. More complex constitutive relations for hyperelastic materials may be found in [14], and particular models for biological tissues and blood vessels are reported in [11, 15]. The material elasticity is characterized by a set of two parameters, the Poisson ratio ν^s and the Young modulus E. These parameters satisfy the relations

$$\nu^s = \frac{\lambda^s}{2(\lambda^s + \mu^s)}, \qquad E = \frac{\mu^s(3\lambda^s + 2\mu^2)}{(\lambda^s + \mu^s)}, \tag{9}$$

$$\mu^s = \frac{E}{2(1 + \nu^s)}, \qquad \lambda^s = \frac{\nu^s E}{(1 + \nu^s)(1 - 2\nu^s)}, \tag{10}$$

where $\nu^s = 1/2$ for incompressible and $\nu^s < 1/2$ for compressible material. In the large deformation case it is common to describe the constitutive equation using a stress-strain relation based on the Green Lagrange strain tensor $\mathbf{E}$ and the second Piola-Kirchhoff stress tensor $\mathbf{S}(\mathbf{E})$ as a function of $\mathbf{E}$. However, also incompressible materials can be handled in the same way (see [17]).

For the hemodynamic applications, a Neo-Hooke material model is taken which can be used for compressible or incompressible (for $\nu^s \to 1/2 \Rightarrow \lambda^s \to \infty$) material and which is described by the constitutive laws:

$$\sigma^s = -p^s \mathbf{I} + \frac{\mu^s}{J}(\mathbf{F}\mathbf{F}^T - \mathbf{I}), \tag{11}$$

$$0 = -p^s + \frac{\lambda^s}{2}(J - \frac{1}{J}). \tag{12}$$

Both models, the St. Venant-Kirchhoff and the Neo-Hooke material model, share the isotropic and homogenous properties, and both can be used for the computation of large deformations. However, the St. Venant-Kirchhoff model does not allow for large strain computation, while the Neo-Hooke model is also valid for large strains. In the case of small strains and small deformations, both material laws yield the same linearized material model. We implemented the St. Venant-Kirchhoff material model as the standard model for the compressible case, since the setup of the benchmark does not involve large strains in the oscillating beam structure. Its implementation is simpler and, therefore, the FSI benchmark will hopefully be adopted by a wider group of researchers. If someone wants or has to use the Neo-Hooke material, the results for a given set of E and ν or λ and μ are comparable, if the standard Neo-Hooke material model as in (11), (12) is used. Similarly as in the case of more complex blood flow models, also more realistic constitutive relations for the anisotropic behavior of the walls of aneurysms can be included which however is beyond the scope of this paper.

2.3 Interaction conditions

The boundary conditions on the fluid-solid interface are assumed to be

$$\sigma^f \mathbf{n} = \sigma^s \mathbf{n}, \quad \mathbf{v}^f = \mathbf{v}^s, \quad \text{on} \quad \Gamma_t^0, \tag{13}$$

where $\mathbf{n}$ is a unit normal vector to the interface Γ_t^0. This implies the no-slip condition for the flow and that the forces on the interface are in balance.

3 Discretization and solution techniques

For the moment, we restrict our considerations to two dimensions which allows systematic tests of the proposed methods for biomedical applications in a very efficient way such that the qualitatitive behavior can be carefully analyzed. The corresponding fully implicit, monolithic treatment of the fluid-structure interaction problem suggests that an A-stable second order time stepping scheme and that the same finite elements for both the solid part and the fluid region should be utilized. Moreover, to handle the fluid incompressibility constraints, we have to choose a stable finite element pair. For that reason, the conforming biquadratic, discontinuous linear $Q_2 P_1$ pair is used. Let us define the usual finite dimensional spaces U for displacement, V for velocity, P for pressure approximation as follows

$$U = \{\mathbf{u} \in L^\infty(I, [W^{1,2}(\Omega)]^2), \mathbf{u} = \mathbf{0} \text{ on } \partial\Omega\},$$
$$V = \{\mathbf{v} \in L^2(I, [W^{1,2}(\Omega_t)]^2) \cap L^\infty(I, [L^2(\Omega_t)]^2), \mathbf{v} = \mathbf{0} \text{ on } \partial\Omega\},$$
$$P = \{p \in L^2(I, L^2(\Omega))\}.$$

Then the variational formulation of the fluid-structure interaction problem is to find $(\mathbf{u}, \mathbf{v}, p) \in U \times V \times P$ that satisfy the corresponding weak form of the balance equations including appropriate initial conditions. The spaces U, V, P on an interval $[t^n, t^{n+1}]$ would be approximated in the case of the Q_2, P_1 pair as

$$U_h = \{\mathbf{u}_h \in [C(\Omega_h)]^2, \mathbf{u}_h|_T \in [Q_2(T)]^2 \quad \forall T \in \mathscr{T}_h, \mathbf{u}_h = \mathbf{0} \text{ on } \partial\Omega_h\},$$
$$V_h = \{\mathbf{v}_h \in [C(\Omega_h)]^2, \mathbf{v}_h|_T \in [Q_2(T)]^2 \quad \forall T \in \mathscr{T}_h, \mathbf{v}_h = \mathbf{0} \text{ on } \partial\Omega_h\},$$
$$P_h = \{p_h \in L^2(\Omega_h), p_h|_T \in P_1(T) \quad \forall T \in \mathscr{T}_h\}.$$

Let us denote by $\mathbf{u}_h^n$ the approximation of $\mathbf{u}(t^n)$, $\mathbf{v}_h^n$ the approximation of $\mathbf{v}(t^n)$ and p_h^n the approximation of $p(t^n)$. Consider for each $T \in \mathscr{T}_h$ the bilinear transformation $\psi_T : \hat{T} \to T$, where $\hat{T}$ is the unit square. Then, $Q_2(T)$ is defined as

$$Q_2(T) = \{q \circ \psi_T^{-1} : q \in \text{span} < 1, x, y, xy, x^2, y^2, x^2y, y^2x, x^2y^2 >\}, \tag{14}$$

with nine local degrees of freedom located at the vertices, midpoints of the edges and in the center of the quadrilateral. The space $P_1(T)$ consists of linear functions defined by

$$P_1(T) = \left\{ q \circ \psi_T^{-1} : q \in \text{span} < 1, x, y > \right\}, \tag{15}$$

with the function value and both partial derivatives located in the center of the quadrilateral, as its three local degrees of freedom, which leads to a discontinuous pressure. The inf-sup condition is satisfied (see [5]); however, the combination of the bilinear transformation ψ with a linear function on the reference square $P_1(\widehat{T})$ would imply that the basis on the reference square did not contain the full basis. So, the method can at most be first order accurate on general meshes (see [3,5])

$$\|p - p_h\|_0 = O(h). \tag{16}$$

The standard remedy is to consider a local coordinate system (ξ, η) obtained by joining the midpoints of the opposing faces of T (see [3, 22, 29]). Then, we set on each element T

$$P_1(T) := \text{span} < 1, \xi, \eta > . \tag{17}$$

For this case, the inf-sup condition is also satisfied and the second order approximation is recovered for the pressure as well as for the velocity gradient (see [5, 13])

$$\|p - p_h\|_0 = O(h^2) \quad \text{and} \quad \|\nabla(\mathbf{u} - \mathbf{u}_h)\|_0 = O(h^2). \tag{18}$$

For a smooth solution, the approximation error for the velocity in the L_2-norm is of order $O(h^3)$ which can easily be demonstrated for prescribed polynomials or for smooth data on appropriate domains.

In the last section we present first results for 3-dimensional computation where we use the $P_2^+ P_1$ pair which also satisfies the Babuška–Brezzi stability condition and yields a stable discretization of the incompressible problems (see [10]).

3.1 Time discretization

In view of a more compact presentation, the applied time discretization approach is described only for the fluid part (see [23] for more details). In the following, we restrict to the (standard) incompressible Navier-Stokes equations

$$\mathbf{v}_t - \nu \Delta \mathbf{v} + \mathbf{v} \cdot \nabla \mathbf{v} + \nabla p = \mathbf{f}, \quad \nabla \cdot \mathbf{v} = 0, \quad \text{in} \quad \Omega \times (0, T], \tag{19}$$

for given force $\mathbf{f}$ and viscosity ν, with prescribed boundary values on the boundary $\partial \Omega$ and an initial condition at $t = 0$.

3.1.1 Basic-θ-scheme

The basic θ-scheme for time discretization reads:

Given $\mathbf{v}^n$ and $\Delta t = t_{n+1} - t_n$, then solve for $\mathbf{v} = \mathbf{v}^{n+1}$ and $p = p^{n+1}$

$$\frac{\mathbf{v} - \mathbf{v}^n}{\Delta t} + \theta[-\nu\Delta\mathbf{v} + \mathbf{v}\cdot\nabla\mathbf{v}] + \nabla p = \mathbf{g}^{n+1}, \quad \mathbf{div\,v} = 0, \quad \text{in} \quad \Omega \tag{20}$$

with right hand side $\mathbf{g}^{n+1} := \theta\mathbf{f}^{n+1} + (1-\theta)\mathbf{f}^n - (1-\theta)[-\nu\Delta\mathbf{v}^n + \mathbf{v}^n\cdot\nabla\mathbf{v}^n]$. The parameter θ has to be chosen depending on the time-stepping scheme, e.g., $\theta = 1$ for the Backward Euler (BE), or $\theta = 1/2$ for the Crank-Nicholson-scheme (CN) which we prefer. The pressure term $\nabla p = \nabla p^{n+1}$ may be replaced by $\theta\nabla p^{n+1} + (1-\theta)\nabla p^n$, but with appropriate postprocessing, both strategies lead to solutions of the same accuracy. In all cases, we end up with the task of solving, at each time step, a nonlinear saddle point problem of given type which has then to be discretized in space as described above.

These two methods, CN and BE, belong to the group of *One-Step-θ-schemes*. The CN scheme can occasionally suffer from numerical instabilities because of its only weak damping property (not strongly A-stable), while the BE-scheme is of first order accuracy only (however: it is a good candidate for steady-state simulations). Another method which has proven to have the potential to excel in this competition is the Fractional-Step-θ-scheme (FS). It uses three different values for θ and for the time step Δt at each time level. In [24, 31] we additionally introduced a modified Fractional-Step-θ-scheme which particularly for fluid-structure interaction problems seems to be advantageous. A brief description is given below and a detailed description will appear in the thesis [23].

In the following, we use the more compact form for the diffusive and advective part:

$$N(\mathbf{v})\mathbf{v} = -\nu\Delta\mathbf{v} + \mathbf{v}\cdot\nabla\mathbf{v}. \tag{21}$$

3.1.2 Fractional-Step-θ-scheme

For the Fractional-Step-θ-scheme we proceed as follows. Choosing $\theta = 1 - \frac{\sqrt{2}}{2}$ $\theta' = 1 - 2\theta$, and $\alpha = \frac{1-2\theta}{1-\theta}$, $\beta = 1 - \alpha$, the macro time step $t_n \to t_{n+1} = t_n + \Delta t$ is split into the three following consecutive substeps (with $\tilde{\theta} := \alpha\theta\Delta t = \beta\theta'\Delta t$):

$$[I + \tilde{\theta}N(\mathbf{v}^{n+\theta})]\mathbf{v}^{n+\theta} + \nabla p^{n+\theta} = [I - \beta\theta\Delta t N(\mathbf{v}^n)]\mathbf{v}^n + \theta\Delta t\mathbf{f}^n,$$
$$\mathbf{div\,v}^{n+\theta} = 0,$$

$$[I + \tilde{\theta}N(\mathbf{v}^{n+1-\theta})]\mathbf{v}^{n+1-\theta} + \nabla p^{n+1-\theta} = [I - \alpha\theta'\Delta t N(\mathbf{v}^{n+\theta})]\mathbf{v}^{n+\theta}$$
$$+\theta'\Delta t\mathbf{f}^{n+1-\theta},$$

$$\mathbf{div\,v}^{n+1-\theta} = 0,$$

$$[I + \tilde{\theta} N(\mathbf{v}^{n+1})]\mathbf{v}^{n+1} + \nabla p^{n+1} = [I - \beta\theta \Delta t N(\mathbf{v}^{n+1-\theta})]\mathbf{v}^{n+1-\theta}$$
$$+\theta \Delta t \mathbf{f}^{n+1-\theta},$$

$$\mathbf{div}\mathbf{v}^{n+1} = 0.$$

3.1.3 A modified Fractional-Step-θ-scheme

Consider an initial value problem of the following form, with $X(t) \in \mathbf{R}^d, d \geq 1$:

$$\begin{cases} \dfrac{dX}{dt} &= f(X,t) \quad \forall t > 0, \\ X(0) &= X_0. \end{cases} \tag{22}$$

Then, a modified θ-scheme (see [31]) with macro time step Δt can be written again as three consecutive substeps, where $\theta = 1 - 1/\sqrt{2}$, $X^0 = X_0$, $n \geq 0$ and X^n is known:

$$\frac{X^{n+\theta} - X^n}{\theta \Delta t} = f\left(X^{n+\theta}, t^{n+\theta}\right),$$

$$X^{n+1-\theta} = \frac{1-\theta}{\theta} X^{n+\theta} + \frac{2\theta - 1}{\theta} X^n,$$

$$\frac{X^{n+1} - X^{n+1-\theta}}{\theta \Delta t} = f\left(X^{n+1}, t^{n+1}\right).$$

As shown in [31], the 'classical' and the modified Fractional-Step-θ-schemes are

- fully implicit,
- strongly A-stable, and
- second order accurate (in fact, they are 'nearly' third order accurate [31]).

These important properties promise some advantageous behavior, particularly in implicit CFD simulations for nonstationary incompressible flow problems. Applying one step of this scheme to the Navier-Stokes equations, we obtain the following variant of the scheme:

1. $$\begin{cases} \dfrac{\mathbf{v}^{n+\theta} - \mathbf{v}^n}{\theta \Delta t} + N(\mathbf{v}^{n+\theta})\mathbf{v}^{n+\theta} + \nabla p^{n+\theta} = \mathbf{f}^{n+\theta}, \\ \\ \mathbf{div}\mathbf{v}^{n+\theta} = 0. \end{cases}$$

2. $$\mathbf{v}^{n+1-\theta} = \tfrac{1-\theta}{\theta}\mathbf{v}^{n+\theta} + \tfrac{2\theta-1}{\theta}\mathbf{v}^n.$$

3. $$\begin{cases} \dfrac{\mathbf{v}^{n+1} - \mathbf{v}^{n+1-\theta}}{\theta \Delta t} + N(\mathbf{v}^{n+1})\mathbf{v}^{n+1} + \nabla \tilde{p}^{n+1} = \mathbf{f}^{n+1}, \\ \\ \mathbf{div}\mathbf{v}^{n+1} = 0. \end{cases}$$

4. $$p^{n+1} = (1 - \theta)p^{n+\theta} + \theta\tilde{p}^{n+1}.$$

These 3 substeps build one macro time step and have to be compared with the previous description of the Backward Euler, Crank-Nicholson and the classical Fractional-Step-θ-scheme which all can be formulated in terms of a macro time step with 3 substeps, too. Then, the resulting accuracy and numerical cost are better comparable and the rating is fair. The main difference to the previous 'classical' Fractional-Step-θ-scheme is that substeps 1. and 3. look like a Backward Euler step while substep 2. is an extrapolation step only for previously computed data such that no operator evaluations at previous time steps are required.

Substep 3b. can be viewed as postprocessing step for updating the new pressure which however is not mandatory. In fact, in our numerical tests [31] we omitted this substep 3b. and accepted the pressure from substep 3. as final pressure approximation, that means $p^{n+1} = \widetilde{p}^{n+1}$.

Summarizing, the numerical effort of the modified scheme for each substep is cheaper at least for 'small' time steps (treatment of the nonlinearity) and complex right hand side evaluations, while the resulting accuracy is similar. The modified θ-scheme is a *Runge-Kutta* scheme; it has been derived in [12] as a particular case of the Fractional-Step-θ-scheme.

3.2 Solution algorithms

After applying the standard finite element method, the system of nonlinear algebraic equations arising from the governing equations described in Sects. 2.1 and 2.2 reads (for incompressible solid material)

$$
\begin{pmatrix}
S_{uu} & S_{uv} & 0 \\
S_{vu} & S_{vv} & kB \\
c_u B_s^T & c_v B_f^T & 0
\end{pmatrix}
\begin{pmatrix}
\mathbf{u} \\
\mathbf{v} \\
p
\end{pmatrix}
=
\begin{pmatrix}
\mathbf{f_u} \\
\mathbf{f_v} \\
f_p
\end{pmatrix},
\tag{23}
$$

which is a typical saddle point problem, where S describes the diffusive and convective terms from the governing equations. The above system of nonlinear algebraic equations (23) is solved using the Newton method as basic iteration which can exhibit quadratic convergence provided that the initial guess is sufficiently close to the solution. The basic idea of the Newton iteration is to find a root of a function, $\mathbf{R(X)} = \mathbf{0}$, using the available known function value and its first derivative. One step of the Newton iteration can be written as

$$
\mathbf{X}^{n+1} = \mathbf{X}^n + \omega^n \left[\frac{\partial \mathbf{R(X^n)}}{\partial \mathbf{X}} \right]^{-1} \mathbf{R(X^n)},
\tag{24}
$$

where $\mathbf{X} = (\mathbf{u}_h, \mathbf{v}_h, p_h)$ and $\frac{\partial \mathbf{R(X^n)}}{\partial \mathbf{X}}$ is the Jacobian matrix. To ensure the convergence globally, some improvements of this basic iteration are used. The damped Newton method with line search improves the chance of convergence by adaptively changing the length of the correction vector (see [17, 29] for more details). The damping parameter $\omega^n \in (-1, 0)$ is chosen such that

$$\mathbf{R}(\mathbf{X}^{n+1}) \cdot \mathbf{X}^{n+1} \leq \mathbf{R}(\mathbf{X}^n) \cdot \mathbf{X}^n. \tag{25}$$

The damping greatly improves the robustness of the Newton iteration in the case when the current approximation $\mathbf{X}^n$ is not close enough to the final solution. The Jacobian matrix $\frac{\partial \mathbf{R}(\mathbf{X}^n)}{\partial \mathbf{X}}$ can be computed by finite differences from the residual vector $\mathbf{R}(\mathbf{X})$

$$\left[\frac{\partial \mathbf{R}(\mathbf{X}^n)}{\partial \mathbf{X}} \right]_{ij} \approx \frac{[\mathbf{R}]_i (\mathbf{X}^n + \alpha_j \mathbf{e}_j) - [\mathbf{R}]_i (\mathbf{X}^n - \alpha_j \mathbf{e}_j)}{2\alpha_j}, \tag{26}$$

where $\mathbf{e}_j$ are the unit basis vectors in $\mathbf{R}^n$ and the coefficients α_j are adaptively chosen according to the change in the solution in the previous time step. Since we know the sparsity pattern of the Jacobian matrix in advance, which is given by the used finite element method, this computation can be done in an efficient way such that the linear solver remains the dominant part in terms of the CPU time (see [29, 33]). A good candidate, at least in 2D, seems to be a direct solver for sparse systems like UMFPACK [8] or MUMPS [1]; while this choice provides very robust linear solvers, its memory and CPU time requirements are too high for larger systems (i.e. more than 20 000 unknowns). Large linear problems can be solved by Krylov-space methods (BiCGStab, GMRes [4]) with suitable preconditioners. One possibility is the ILU preconditioner with special treatment of the saddle point character of our system, where we allow certain fill-in for the zero diagonal blocks, see [6]. As an alternative, we also utilize a standard geometric multigrid approach based on a hierarchy of grids obtained by successive regular refinement of a given coarse mesh. The complete multigrid iteration is performed in the standard defect-correction setup with the V or F-type cycle. While a direct sparse solver [8] is used for the coarse grid solution, on finer levels a fixed number (2 or 4) of iterations by local MPSC schemes (Vanka-like smoother) [17, 29, 35] is performed. Such iterations can be written as

$$\begin{pmatrix} \mathbf{u}^{l+1} \\ \mathbf{v}^{l+1} \\ p^{l+1} \end{pmatrix} = \begin{pmatrix} \mathbf{u}^l \\ \mathbf{v}^l \\ p^l \end{pmatrix} - \omega \sum_{\text{element}\Omega_i} \begin{pmatrix} S_{\mathbf{uu}|\Omega_i} & S_{\mathbf{uv}|\Omega_i} & 0 \\ S_{\mathbf{vu}|\Omega_i} & S_{\mathbf{vv}|\Omega_i} & kB_{|\Omega_i} \\ c_{\mathbf{u}}B^T_{s|\Omega_i} & c_{\mathbf{v}}B^T_{f|\Omega_i} & 0 \end{pmatrix}^{-1} \begin{pmatrix} \mathbf{def}^l_{\mathbf{u}} \\ \mathbf{def}^l_{\mathbf{v}} \\ def^l_p \end{pmatrix}.$$

The inverse of the local (39×39) systems can be computed by hardware optimized direct solvers. The full nodal interpolation is used as the prolongation operator P with its transposed operator used as the restriction $\mathsf{R} = \mathsf{P}^T$, see [16, 29] for more details.

4 FSI benchmarking

In order to validate and to analyze different techniques to solve such FSI problems, also in a quantitative way, a set of benchmark configurations has been proposed in [30] (also see the contribution in this volume). The configurations consist of laminar incompressible channel flow around an elastic object which results in self-induced oscillations of the structure. Moreover, characteristic flow quantities and corresponding plots are provided for a quantitative comparison.

The domain is based on the 2D version of the well-known CFD benchmark in [32] and by omitting the elastic bar behind the cylinder one can easily recover the setup of the 'classical' *flow around cylinder* configuration which allows for validation of the flow part by comparing the results with the older flow benchmark. The setting is intentionally nonsymmetric to prevent the dependence of the onset of any possible oscillation on the precision of the computation. The mesh used for the computations is shown in Fig. 1. A parabolic velocity profile is prescribed at the left channel inflow

$$v^f(0, y) = 1.5\bar{U}\frac{y(H-y)}{\left(\frac{H}{2}\right)^2} = 1.5\bar{U}\frac{4.0}{0.1681}y(0.41 - y), \qquad (27)$$

such that the mean inflow velocity is $\bar{U}$ and the maximum of the inflow velocity profile is $1.5\bar{U}$. The *no-slip* condition is prescribed for the fluid on the other boundary parts. i.e. top and bottom wall, circle and fluid-structure interface Γ_t^0. The outflow condition can be chosen by the user, for example *stress free* or *do nothing* conditions. The outflow condition effectively prescribes some reference value for the

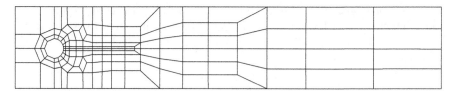

level	#refine	#el	#dof
0	0	62	1338
1	1	248	5032
2	2	992	19488
3	3	3968	76672
4	4	15872	304128
5	5	63488	1211392
6	6	253952	4835328
7	7	1015808	19320832

Fig. 1 Coarse mesh with number of degrees of freedom for refined levels.

pressure variable p. While this value could be arbitrarily set in the incompressible case, in the case of compressible structure this will have influence on the stress and consequently the deformation of the solid. In this description, we set the reference pressure at the outflow to have *zero mean value*. Suggested starting procedure for the non-steady tests is to use a smooth increase of the velocity profile in time as

$$v^f(t,0,y) = \begin{cases} v^f(0,y)\frac{1-\cos(\frac{\pi}{2}t)}{2} & \text{if } t < 2.0, \\ v^f(0,y) & \text{otherwise,} \end{cases} \tag{28}$$

where $v^f(0,y)$ is the velocity profile given in (27). The following FSI tests are performed for three different inflow speeds. FSI1 is resulting in a steady state solution, while FSI2 and FSI3 result in periodic solutions. The parameter values for the FSI1, FSI2 and FSI3 are given in the Table 1. Here, the computed values are summarized in Table 2 for the steady state test FSI1. In Figs. 2 and 3, plots of the resulting x- and y-displacements of the trailing edge point A (see [30]) of the elastic bar and plots of the forces (lift, drag) acting on the cylinder and the bar are drawn. Furthermore, computed values for three different mesh refinement levels and two different time steps for the nonsteady tests FSI2 and FSI3 are presented respectively, which show the (almost) grid independent solution behavior (for more details see [30]).

Table 1 Parameter settings for the FSI benchmarks.

parameter	FSI1	FSI2	FSI3
ρ^s [$10^3 \frac{kg}{m^3}$]	1	10	1
ν^s	0.4	0.4	0.4
μ^s [$10^6 \frac{kg}{ms^2}$]	0.5	0.5	2.0
ρ^f [$10^3 \frac{kg}{m^3}$]	1	1	1
ν^f [$10^{-3} \frac{m^2}{s}$]	1	1	1
$\bar{U}$ [$\frac{m}{s}$]	0.2	1	2

parameter	FSI1	FSI2	FSI3
$\beta = \frac{\rho^s}{\rho^f}$	1	10	1
ν^s	0.4	0.4	0.4
$Ae = \frac{E^s}{\rho^f \bar{U}^2}$	3.5×10^4	1.4×10^3	1.4×10^3
$Re = \frac{\bar{U}d}{\nu^f}$	20	100	200
$\bar{U}$	0.2	1	2

Table 2 Results for **FSI1**.

level	nel	ndof	ux of A [$\times 10^{-3}$ m]	uy of A [$\times 10^{-3}$ m]	drag [N]	lift [N]
2	992	19488	0.02287080	0.8193038	14.27359	0.7617550
3	3968	76672	0.02277423	0.8204231	14.29177	0.7630484
4	15872	304128	0.02273175	0.8207084	14.29484	0.7635608
5	63488	1211392	0.02271553	0.8208126	14.29486	0.7636992
6	253952	4835328	0.02270838	0.8208548	14.29451	0.7637359
7	1015808	19320832	0.02270493	0.8208773	14.29426	0.7637460
ref.			0.0227	0.8209	14.294	0.7637

FSI2: x & y displacement of the point A [m].

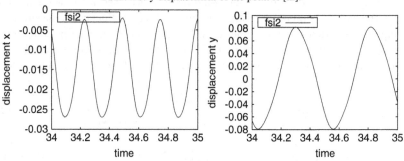

FSI2: lift and drag force [N] on the cylinder+elastic bar.

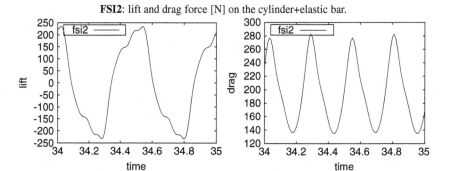

lev.	ux of A [$\times 10^{-3}$ m]	uy of A [$\times 10^{-3}$ m]	drag [N]	lift [N]
2	$-14.02 \pm 12.03[3.85]$	$1.25 \pm 79.3[1.93]$	$210.10 \pm 72.62[3.85]$	$0.25 \pm 227.9[1.93]$
3	$-14.54 \pm 12.50[3.86]$	$1.25 \pm 80.7[1.93]$	$212.83 \pm 75.89[3.86]$	$0.92 \pm 234.3[1.93]$
4	$-14.88 \pm 12.75[3.86]$	$1.24 \pm 81.7[1.93]$	$215.06 \pm 77.76[3.86]$	$0.82 \pm 237.1[1.93]$
lev.	ux of A [$\times 10^{-3}$ m]	uy of A [$\times 10^{-3}$ m]	drag [N]	lift [N]
2	$-14.01 \pm 12.04[3.86]$	$1.25 \pm 79.3[1.93]$	$210.09 \pm 72.82[3.86]$	$0.52 \pm 228.6[1.93]$
3	$-14.54 \pm 12.48[3.86]$	$1.25 \pm 80.7[1.93]$	$213.06 \pm 75.76[3.86]$	$0.85 \pm 234.4[1.93]$
4	$-14.87 \pm 12.73[3.86]$	$1.24 \pm 81.7[1.93]$	$215.18 \pm 77.78[3.86]$	$0.87 \pm 238.0[1.93]$
lev.	ux of A [$\times 10^{-3}$ m]	uy of A [$\times 10^{-3}$ m]	drag [N]	lift [N]
2	$-14.01 \pm 12.04[3.86]$	$1.28 \pm 79.2[1.93]$	$210.14 \pm 72.86[3.86]$	$0.49 \pm 228.7[1.93]$
3	$-14.48 \pm 12.45[3.86]$	$1.24 \pm 80.7[1.93]$	$213.05 \pm 75.74[3.86]$	$0.84 \pm 234.8[1.93]$
4	$-14.85 \pm 12.70[3.86]$	$1.30 \pm 81.6[1.93]$	$215.06 \pm 77.65[3.86]$	$0.61 \pm 237.8[1.93]$
ref.	$-14.85 \pm 12.70[3.86]$	$1.30 \pm 81.7[1.93]$	$215.06 \pm 77.65[3.86]$	$0.61 \pm 237.8[1.93]$

Fig. 2 Results for **FSI2** with time step $\Delta t = 0.002$, $\Delta t = 0.001$, $\Delta t = 0.0005$ [s].

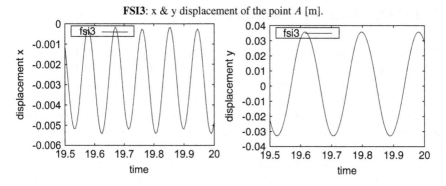

FSI3: x & y displacement of the point A [m].

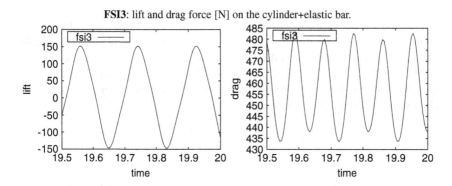

FSI3: lift and drag force [N] on the cylinder+elastic bar.

lev.	ux of A [$\times 10^{-3}$ m]	uy of A [$\times 10^{-3}$ m]	drag [N]	lift [N]
2	$-3.02 \pm 2.83[10.75]$	$1.41 \pm 35.47[5.37]$	$458.2 \pm 28.32[10.75]$	$2.41 \pm 145.58[5.37]$
3	$-2.78 \pm 2.62[10.93]$	$1.44 \pm 34.36[5.46]$	$459.1 \pm 26.63[10.93]$	$2.41 \pm 151.26[5.46]$
4	$-2.86 \pm 2.70[10.95]$	$1.45 \pm 34.93[5.47]$	$460.2 \pm 27.65[10.95]$	$2.47 \pm 154.87[5.47]$

lev.	ux of A [$\times 10^{-3}$ m]	uy of A [$\times 10^{-3}$ m]	drag [N]	lift [N]
2	$-3.02 \pm 2.85[10.75]$	$1.42 \pm 35.63[5.37]$	$458.7 \pm 28.78[10.75]$	$2.23 \pm 146.02[5.37]$
3	$-2.78 \pm 2.62[10.92]$	$1.44 \pm 34.35[5.46]$	$459.1 \pm 26.62[10.92]$	$2.39 \pm 150.68[5.46]$
4	$-2.86 \pm 2.70[10.92]$	$1.45 \pm 34.90[5.46]$	$460.2 \pm 27.47[10.92]$	$2.37 \pm 153.75[5.46]$

lev.	ux of A [$\times 10^{-3}$ m]	uy of A [$\times 10^{-3}$ m]	drag [N]	lift [N]
2	$-3.02 \pm 2.85[10.74]$	$1.32 \pm 35.73[5.36]$	$458.7 \pm 28.80[10.74]$	$2.23 \pm 146.00[5.36]$
3	$-2.77 \pm 2.61[10.93]$	$1.43 \pm 34.43[5.46]$	$459.1 \pm 26.50[10.93]$	$2.36 \pm 149.91[5.46]$
4	$-2.88 \pm 2.72[10.93]$	$1.47 \pm 34.99[5.46]$	$460.5 \pm 27.74[10.93]$	$2.50 \pm 153.91[5.46]$
ref.	$-2.88 \pm 2.72[10.93]$	$1.47 \pm 34.99[5.46]$	$460.5 \pm 27.74[10.93]$	$2.50 \pm 153.91[5.46]$

Fig. 3 Results for **FSI3** with time step $\Delta t = 0.001$, $\Delta t = 0.0005$, $\Delta t = 0.00025$ [s].

5 FSI Optimization benchmarking

The objective of the following benchmarking scenario is to extend the validated
FSI benchmark configurations towards optimization problems such that minimal
drag/lift values of the elastic object, minimal pressure loss or minimal nonstationary
oscillations through boundary control of the inflow, change of geometry or optimal
control of volume forces can be achieved. The main design aim for the subsequent
fluid structure interaction optimization problem is to minimize the lift on the beam
with the help of boundary control of the inflow data. Here, the simulation is based on
the described FSI1 configuration. Further extensions of this optimization problem
will be to control minimal pressure loss or minimal nonstationary oscillations of the
elastic beam through boundary control of the inflow section, change of the geometry (elastic channel walls or length/thickness of elastic beam) or optimal control of
volume forces.

As described before, the position of the beam is not symmetric such that the lift
is not zero. To minimize the lift value, we allow additional parabolic inflow at the
top and additional parabolic outflow at the bottom of the domain. The location of
the additional inlet and outlet is shown in the schematic diagram of the geometry in
Fig. 4. With V_1 and V_2 we denote the magnitude of the additional inflow and outflow
velocity, respectively. Then, the aim is to

$$\text{minimize } \left(\text{lift}^2 + \alpha(V_1^2 + V_2^2)\right)$$
$$\text{w.r.t.} \quad V_1, V_2.$$

The arbitrary but fixed parameter $\alpha > 0$ reflects the 'costs' of the additional in-
and outflow and has to be prescribed. The domain is based on the 2D version of the
described FSI benchmark, shown in Fig. 4; however, the thickness of the beam is
increased from 0.02 to 0.04. Fluid and structural parameter values are based on the
FSI1 benchmark. Table 3 provides an overview of the geometry parameters.

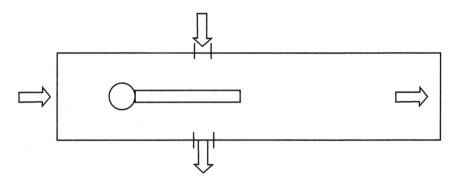

Fig. 4 Geometry and configuration of the problem.

Table 3 Overview of the geometry parameters.

geometry parameters		value [m]
channel length	L	2.5
channel width	H	0.41
cylinder center position	C	(0.2, 0.2)
cylinder radius	r	0.05
elastic structure length	l	0.35
elastic structure thickness	h	0.04

Table 4 Results for FSI-Optimization.

		level 1				level 2	
α	iter	optimal values (V_1, V_2)	lift	iter	optimal values (V_1, V_2)	lift
1e+0	57	(3.74e-1, 3.88e-1)	8.1904e-1	59	(3.66e-1, 3.79e-1)	7.8497e-1
1e-2	60	(1.04e+0, 1.06e+0)	2.2684e-2	59	(1.02e+0, 1.04e+0)	2.1755e-2
1e-4	73	(1.06e+0, 1.08e+1)	2.3092e-4	71	(1.04e+0, 1.05e+01)	2.2147e-4
1e-6	81	(1.06e+0, 1.08e+1)	2.3096e-6	86	(1.04e+0, 1.05e+01)	2.2151e-6

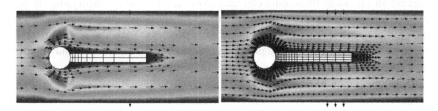

Fig. 5 No displacement is visible of the beam due to optimal boundary flow control: Level 1 (left), Level 2 (right).

Optimal points are those (V_1, V_2) values which result in minimal lift values on the beam depending on the parameter α. As α decreases we get the reduction of the lift on the beam, and the optimal values read (1.06, 10.8) for level 1 and (1.04, 10.5) for level 2 using the simplex algorithm proposed by Nelder and Mead [21]. Results are given in Table 4 which show the optimal values for the velocities V_1 and V_2 providing also the resulting lift on the beam as compared with the FSI1 benchmark values in Table 2. In Fig. 5, it is visible that the displacement of the beam decreases with decreasing α, as well as the lift value decreases due to the boundary control. Keep in mind that the original lift on the beam is approximately 7.6e-1 for the FSI1 benchmark while in the case of FSI1-Optimization, it reduces to approximately 2.3e-6 for α=1e-6. Interested readers are referred to [18] for a comprehensive survey of the original Nelder-Mead simplex algorithm and for its advantages and disadvantages.

6 Applications to hemodynamics

In the following, we consider the numerical simulation of special problems encountered in the area of cardiovascular hemodynamics, namely flow interaction with
thick-walled deformable material (here: the arterial walls) and rigid parts (here:
stents), which can become a useful tool for deeper understanding of the onset
of diseases of the human circulatory system, as for example blood cell and inti-
mal damages in stenosis, aneurysm rupture, evaluation of new surgery techniques
of the heart, arteries and veins (see [2, 19, 34] and the literature cited therein).
In this contribution, prototypical studies are performed for brain aneurysms. The
word 'aneurysm' comes from the latin word *aneurysma* which means dilatation. An
aneurysm is a local dilatation in the wall of a blood vessel, usually an artery, due
to a defect, disease or injury. Typically, as the aneurysm enlarges, the arterial wall
becomes thinner and eventually leaks or ruptures, causing subarachnoid hemorrhage
(SAH) (bleeding into brain fluid) or formation of a blood clot within the brain. In
the case of a vessel rupture, there is a hemorrhage, which is particularly rapid and
intense in case of an artery. In arteries the wall thickness can be up to 30% of the
diameter and its local thickening can lead to the creation of an aneurysm. The aim
of numerical simulations is to relate the aneurysm state (unrupture or rupture) with
wall pressure, wall deformation and effective wall stress. Such a relationship would
provide information for the diagnosis and treatment of unruptured and ruptured
aneurysms by elucidating the risk of bleeding or rebleeding, respectively.

As a typical example for the related CFD simulations, a real view is provided
in Fig. 6 which also contains the automatically extracted computational domain and
(coarse) mesh in 2D, however without stents. In order to use the proposed numerical
methods for aneurysm hemodynamics, in a first step only simplified two-dimensional examples, which however include the interaction of the flow with the deformable
material, are considered. Flow through a deformable vein with elastic walls of a
brain aneurysm is simulated to analyze qualitatively the described methods; here,
the flow is driven by prescribing the flow velocity at the inflow section (Poiseuille
flow) while the solid part of the boundary is either fixed or stress-free. Both ends of
the walls are fixed, and the flow is driven by a periodical change of the inflow at the
right end.

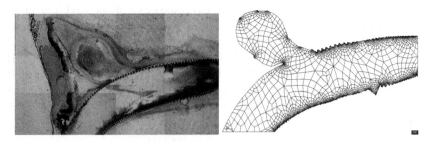

Fig. 6 Left: Real view of aneurysm. Right: Schematic drawing of the mesh.

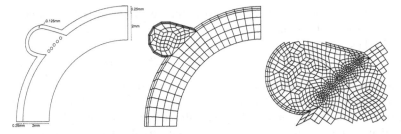

Fig. 7 Left: Schematic drawing of the measurement section. Middle: Mesh without stents (776 elements). Right: Mesh with stents (1431 elements) which are part of the simulations.

6.1 Geometry of the problem

For convenience, the geometry of the fluid domain under consideration is currently based on simplified 2D models, see Fig. 7, which allows to concentrate on the detailed qualitative evaluation of our approach based on the described monolithic ALE formulation. The underlying construction of the (2D) shape of the aneurysm can be explained as follows:

- The bent blood vessel is approximated by quarter circles around the origin.
- The innermost circle has the radius 6*mm*, the next has 8*mm*, and the last one has 8.25*mm*.
- This results in one rigid inner wall and an elastic wall between 8*mm* and 8.25*mm* of thickness 0.25*mm*.

The aneurysm shape is approximated by two arcs and lines intersecting the arcs tangentially. The midpoints of the arcs are the same $(-6.75; 6)$, they have the radius 1.125*mm* and 1.25*mm*. They are intersected tangentially by lines at angular value 1.3 radians. This results in a wall thickness of 0.125*mm* for the elastic aneurysm walls (see Fig. 7). The examined stents are of circular shape, placed on the neck of the aneurysm, and we use three, respectively five stents (simplified 'circles' in 2D as cutplanes from 3D configurations) of different size and position. Such stents (in real life) are typically used to keep arteries open and are located on the vessel wall while 'our' (2D) stent is located only near the aneurysm (Fig. 7). The purpose of this device is to reduce the flux into and within the aneurysm in order to occlude it by a clot or rupture.

6.2 Boundary and initial conditions

The (steady) velocity profile, to flow from the right to the left part of the channel, is defined as parabolic inflow, namely

$$\mathbf{v}^f(0, y) = \bar{U}(y - 6)(y - 8). \tag{29}$$

Correspondingly, the pulsatile inflow profile for the nonsteady tests for which peak systole and diastole occur for $\Delta t = 0.25s$ and $\Delta t = 0.75s$ respectively, is prescribed as

$$\mathbf{v}^f(t, 0, y) = \mathbf{v}^f(0, y)(1 + 0.75sin(2\pi t)). \tag{30}$$

The natural outflow condition at the lower left part effectively prescribes some reference value for the pressure variable p, here $p = 0$. While this value could be arbitrarily set in the incompressible case, in the case of a compressible structure this might have influence on the stress and consequently the deformation of the solid. The *no-slip* condition is prescribed for the fluid on the other boundary parts, i.e. top and bottom wall, stents and fluid-structure interface.

6.3 Numerical results

The Newtonian fluid used in the tests has a density $\rho^f = 1.035 \times 10^{-6}kg/mm^3$ and a kinematic viscosity $\nu^f = 3.38mm^2/s$ which is similar to the properties of blood. If we prescribe the inflow speed $\bar{U} = -50mm/s$, this results in a Reynolds number Re ≈ 120 based on the prescribed peak systole inflow velocity and the width of the veins which is $2mm$ such that the resulting flow is within the laminar region. Parameter values for the elastic vein in the described model are as follows: The density of the upper elastic wall is $\rho^s = 1.12 \times 10^{-6}kg/mm^3$, solid shear modulus is $\mu^s = 42.85kg/mms^2$, Poisson ratio is $\nu^s = 0.4$, Young modulus is $E = 120kN/mm^2$. As described before, the constitutive relations used for the materials are the incompressible Newtonian model (2) for the fluid and a hyperelastic Neo-Hooke material for the solid. This choice includes most of the typical difficulties the numerical method has to deal with, namely the incompressibility and significant deformations.

From a medical point of view, the use of stents provides an efficient treatment for managing the difficult entity of intracranial aneurysms. Here, the thickness of the aneurysm wall is attenuated and the aneurysm hemodynamics changes significantly. Since the purpose of this device is to control the flux within the aneurysm in order to occlude it by a clot or rupture, the resulting flow behavior into and within the aneurysm is the main objective, particularly in view of the different stent geometries. Therefore, we decided for the 2D studies to locate the stents only in direct connection to the aneurysm.

Comparing our studies with the CFD literature (see [2, 9, 27, 28, 34]), several research groups focus on CFD simulations with realistic 3D geometries, but typically assuming rigid walls. In contrast, we concentrate on the complex interaction between elastic deformations and flow perturbations induced by the stents. At the moment, we are only able to perform these simulations in 2D. However, with these studies we should be able to analyze qualitatively the influence of geometrical details onto the elastic material behavior, particularly in view of more complex blood models and constitutive equations for the structure. Therefore, the aims of our current studies can be described as follows:

1. What is the influence of the elasticity of the walls onto the flow behavior inside the aneurysm, particularly w.r.t. the resulting shape of the aneurysm?
2. What is the influence of the geometrical details of the (2D) stents, that means shape, size, position, on the flow into and inside the aneurysm?
3. Do both aspects, small-scale geometrical details as well as elastic fluid-structure interaction, have to be considered simultaneously or is one of them negligible in first order approximation?
4. Are modern numerical methods and corresponding CFD simulation tools able to simulate qualitatively the multiphysics behavior of such biomedical configurations?

In the following, we show some corresponding results for the described prototypical aneurysm geometry, first for the steady state inflow profile, followed by nonsteady tests for the pulsatile inflow, both with rigid and elastic walls, respectively.

6.3.1 Steady configurations

Due to the given inflow profile, which is not time-dependent, and due to the low Reynolds numbers, the flow behavior leads to a steady state which only depends on the elasticity and the shape of the stents. Moreover, for the following simulations, we only treat the aneurysm wall as elastic structure. Then, the aneurysm undergoes some slight deformations which can hardly be seen in the following figures. However, they result in a different volume of the flow domain (see Fig. 9) and lead to a significantly different local flow behavior since the spacing between stents and elastic walls may change (see Fig. 8).

In Fig. 11, we visualize the different flow behavior by coloring corresponding to the velocity magnitude and by showing corresponding vector plots inside the aneurysm. Particularly the influence of the number of stents on the complete fluid flow through the channel including the aneurysm can be clearly seen.

Summarizing these results for steady inflow, the simulations show that the stent implantation across the neck of the aneurysm prevents blood penetration into the aneurysm fundus. Moreover, the elastic geometrical deformation of the wall is slightly reduced by implanting the stents while the local flow behavior inside the aneurysm is more significantly influenced by the elastic properties of the outer wall, particularly due to different width between stents and walls of the aneurysm. In the next section, we will consider the behaviour of more realistic flow configurations with time-dependent pulsatile inflow which will be analyzed for the case of elastic behaviour of the aneurysm walls.

6.3.2 Pulsatile configurations

For the following pulsatile test case, we have taken again the aneurysm part as elastic while the other parts of the walls belonging to the channel are rigid. First of all, we show again (see Fig. 10) the resulting volume of the flow domain for five, three and

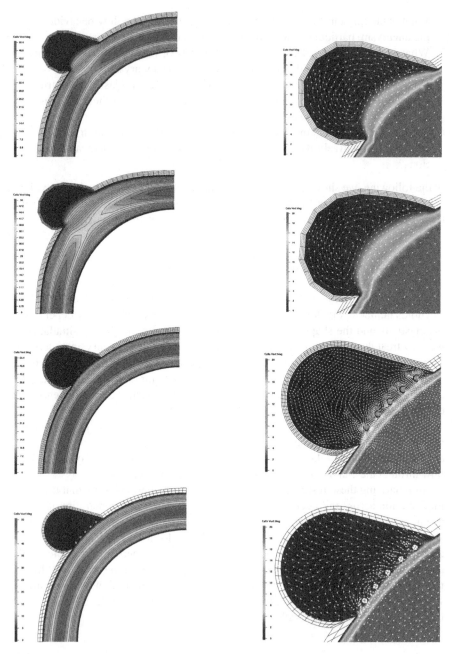

Fig. 8 Magnitude of the blood flow velocity for four configurations. Top to bottom: Rigid walls without stents, elastic walls without stents, rigid walls with stents, elastic walls with stents. Left: Overall view. Right: Scaled view of the aneurysm.

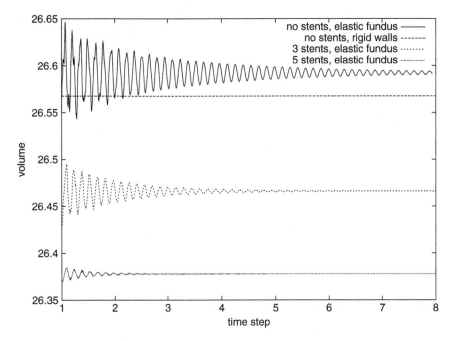

Fig. 9 Resulting volume of the fluid domain for different configurations.

no stents. In all cases, the oscillating behavior due to the pulsatile inflow is visible which also leads to different volume sizes. Looking carefully at the resulting flow behavior, we see differences w.r.t. the channel flow near the aneurysm, namely due to the different flow rate into the aneurysm, and significant local differences inside the aneurysm.

6.3.3 Extension to 3D

Finally, we show first results of extending the monolithic formulation to 3D. A similar problem of pulsatile flow in an elastic tube with an aneurysm-like cavity is solved. The material parameters are the same as in the previous section and the resulting deformation and flow field at different times are shown in Fig. 12.

7 Summary and future developments

We presented a monolithic ALE formulation of fluid-structure interaction problems suitable for applications with large structural deformations and laminar viscous flows, particularly arising in biomechanics. The corresponding discrete nonlinear

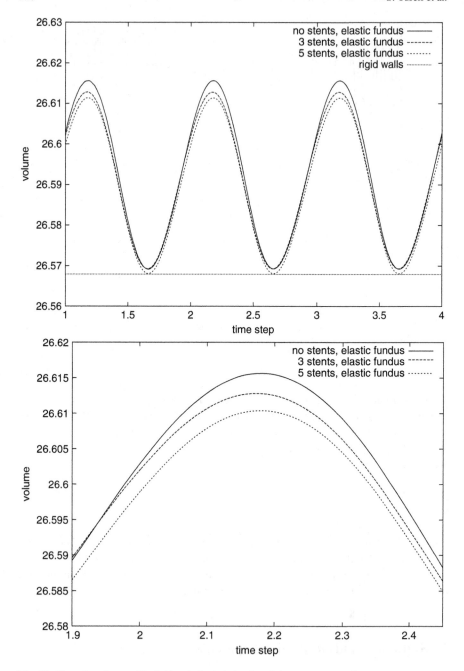

Fig. 10 Domain volume with rigid and elastic behavior of the aneurysm wall.

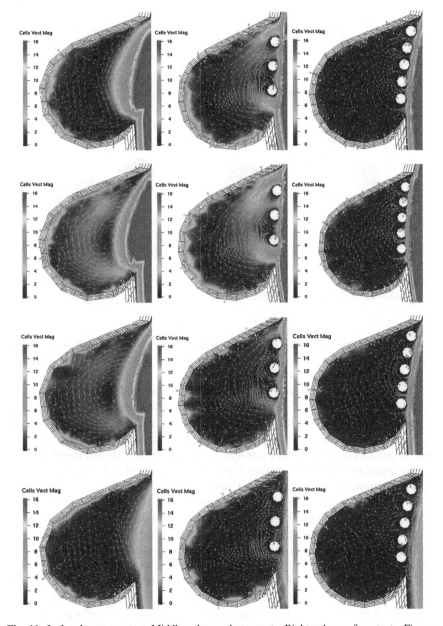

Fig. 11 Left column: no stent. Middle column: three stents. Right column: five stents. Figures demonstrate the local behavior of the fluid flow inside the aneurysm during one cycle.

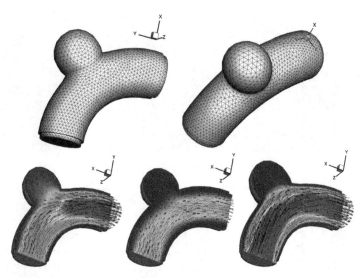

Fig. 12 Pulsatile fluid flow through an elastic tube with cavity. Flow field represented by velocity vectors and velocity magnitude at different times.

systems result from the finite element discretization by using the high order $Q_2 P_1$ FEM pair. The systems are solved monolithically via a discrete Newton iteration and special Krylov-multigrid approaches.

While we restricted our studies to the simplified case of Newtonian fluids and small deformations, the used numerical components allow the system to be coupled with additional models of chemical and electric activation of the active response of the biological material as well as power law models used to describe the shear thinning property of blood. Further extension to viscoelastic models and coupling with mixture based models for soft tissues together with chemical and electric processes allow to perform more realistic simulations for real applications.

We applied the presented numerical techniques to FSI benchmarking settings ('channel flow around cylinder with attached elastic beam', see [30]) which allow the validation and also evaluation of different numerical solution approaches for fluid-structure interaction problems. Moreover, we examined prototypically the influence of endovascular stent implantation on aneurysm hemodynamics. The aim was, first of all, to study the influence of the elasticity of the walls on the flow behavior inside the aneurysm. Moreover, different geometrical configurations of implanted stent structures have been analyzed in 2D. These 2D results are far from providing quantitative results for such a complex multiphysics configuration, but they allow a qualitative analysis w.r.t. both considered components, namely the elastic behavior of the structural parts and the multiscale flow behavior due to the geometrical details of the stents. We believe that such basic studies are helpful for the development of future 'Virtual Flow Laboratories' which individually assist to design personal medical tools.

Acknowledgements The authors want to express their gratitude to the German Research Association (DFG), funding the project as part of FOR493 and TRR30, the Jindřich Nečas Center for Mathematical Modeling, project LC06052 financed by MSMT, and the Higher Education Commission (HEC) of Pakistan for their financial support of the study. The present material is also based upon work kindly supported by the Homburger Forschungsförderungsprogramm (HOMFOR) 2008.

References

1. P. R. Amestoy, I. S. Duff, and J. Y. L'Excellent. Multifrontal parallel distributed symmetric and unsymmetric solvers. *Computer Methods in Applied Mechanics and Engineering*, 184 (2-4):501 – 520, 2000.
2. S. Appanaboyina, F. Mut, R. Löhner, E. Scrivano, C. Miranda, P. Lylyk, C. Putman, and J. Cebral. Computational modelling of blood flow in side arterial branches after stenting of cerebral aneurysm. *International Journal of Computational Fluid Dynamics*, 22:669–676, 2008.
3. D. N. Arnold, D. Boffi, and R. S. Falk. Approximation by quadrilateral finite element. *Math. Comput.*, 71:909–922, 2002.
4. R. Barrett, M. Berry, T. F. Chan, J. Demmel, J. Donato, J. Dongarra, V. Eijkhout, R. Pozo, C. Romine, and H. Van der Vorst. *Templates for the solution of linear systems: Building blocks for iterative methods*. SIAM, Philadelphia, PA, second edition, 1994.
5. D. Boffi and L. Gastaldi. On the quadrilateral Q_2P_1 element for the Stokes problem. *Int. J. Numer. Meth. Fluids.*, 39:1001–1011, 2002.
6. R. Bramley and X. Wang. *SPLIB: A library of iterative methods for sparse linear systems*. Department of Computer Science, Indiana University, Bloomington, IN, 1997. http://www.cs.indiana.edu/ftp/bramley/splib.tar.gz.
7. H. Damanik, J. Hron, A. Ouazzi, and S. Turek. A monolithic FEM approach for non-isothermal incompressible viscous flows. *Journal of Computational Physics*, 228:3869–3881, 2009.
8. T. A. Davis and I. S. Duff. A combined unifrontal/multifrontal method for unsymmetric sparse matrices. *ACM Trans. Math. Software*, 25(1):1–19, 1999.
9. M. A. Fernandez, J-F. Gerbeau, and V. Martin. Numerical simulation of blood flows through a porous interface. *ESAIM: Mathematical Modelling and Numerical Analysis*, 42:961–990, 2008.
10. M. Fortin. Old and new finite elements for incompressible flows. *International Journal for Numerical Methods in Fluids*, 1(4), 1981.
11. Y. C. Fung. *Biomechanics: Mechanical Properties of Living Tissues*. Springer-Verlag, New York, 1993.
12. R. Glowinski. Finite element method for incompressible viscous flow. In P. G Ciarlet and J. L. Lions, editors, *Handbook of Numerical Analysis*, Volume IX,, pages 3–1176. North-Holland, Amsterdam, 2003.
13. P. M. Gresho. On the theory of semi-implicit projection methods for viscous incompressible flow and its implementation via a finite element method that also introduces a nearly consistent mass matrix, part 1: Theory. *Int. J. Numer. Meth. Fluids.*, 11:587–620, 1990.
14. G. A. Holzapfel. *A continuum approach for engineering*. John Wiley and Sons, Chichester, UK, 2000.
15. G. A. Holzapfel. Determination of meterial models for arterial walls from uni-axial extension tests and histological structure. *International Journal for Numerical Methods in Fluids*, 238(2):290–302, 2006.
16. J. Hron, A. Ouazzi, and S. Turek. A computational comparison of two FEM solvers for nonlinear incompressible flow. In E. Bänsch, editor, *Challenges in Scientific Computing*, LNCSE:53, pages 87–109. Springer, 2002.

17. J. Hron and S. Turek. A monolithic FEM/multigrid solver for ALE formulation of fluid structure interaction with application in biomechanics. In H.-J. Bungartz and M. Schäfer, editors, *Fluid-Structure Interaction: Modelling, Simulation, Optimisation*, LNCSE:53. Springer, 2006.
18. J. C. Lagarias, J. A. Reeds, M. H. Wright, and P. E. Wright. Convergence properties of the Nelder-Mead simplex method in low dimensions. *SIAM J. Optim.*, 9:112–147, 1998.
19. R. Löhner, J. Cebral, and S. Appanaboyina. Parabolic recovery of boundary gradients. *Communications in Numerical Methods in Engineering*, 24:1611–1615, 2008.
20. D. A. McDonald. Blood flow in arteries. In *second ed.* Edward Arnold, 1974.
21. J. A. Nelder and R. Mead. A simplex method for function minimization. *Computer Journal*, 7(4):308–313, 1965.
22. R. Rannacher and S. Turek. A simple nonconforming quadrilateral Stokes element. *Numer. Methods Partial Differential Equations.*, 8:97–111, 1992.
23. M. Razzaq. *Numerical techniques for solving fluid-structure interaction problems with applications to bio-engineering*. PhD Thesis, TU Dortmund, to appear, 2010.
24. M. Razzaq, J. Hron, and S. Turek. Numerical simulation of laminar incompressible fluid-structure interaction for elastic material with point constraints. In R. Rannacher and A. Sequeira, editors, *Advances in Mathematical Fluid Mechancis-Dedicated to Giovanni paolo Galdi on the Occasion of his 60th Birthday*. Springer, in print, 2009.
25. T. E. Tezduyar, S. Sathe, T. Cragin, B. Nanna, B.S. Conklin, J. Pausewang, and M. Schwaab. Modeling of fluid structure interactions with the space time finite elements: Arterial fluid mechanics. *International Journal for Numerical Methods in Fluids*, 54:901–922, 2007.
26. T. E. Tezduyar, S. Sathe, M. Schwaab, and B.S. Conklin. Arterial fluid mechanics modeling with the stabilized space time fluid structure interaction technique. *International Journal for Numerical Methods in Fluids*, 57:601–629, 2008.
27. R. Torri, M. Oshima, T. Kobayashi, K. Takagi, and T.E. Tezduyar. Influence of wall elasticity in patient-specific hemodynamic simulations. *Computers and Fluids*, 36:160–168, 2007.
28. R. Torri, M. Oshima, T. Kobayashi, K. Takagi, and T.E. Tezduyar. Numerical investigation of the effect of hypertensive blood pressure on cerebral aneurysm dependence of the effect on the aneurysm shape. *International Journal for Numerical Methods in Fluids*, 54:995–1009, 2007.
29. S. Turek. *Efficient Solvers for Incompressible Flow Problems: An Algorithmic and Computational Approach*. Springer-Verlag, 1999.
30. S. Turek and J. Hron. Proposal for numerical benchmarking of fluid-structure interaction between an elastic object and laminar incompressible flow. In H.-J. Bungartz and M. Schäfer, editors, *Fluid-Structure Interaction: Modelling, Simulation, Optimisation*, LNCSE:53. Springer, 2006.
31. S. Turek, L. Rivkind, J. Hron, and R. Glowinski. Numerical study of a modified time-steeping theta-scheme for incompressible flow simulations. *Journal of Scientific Computing*, 28:533–547, 2006.
32. S. Turek and M. Schäfer. Benchmark computations of laminar flow around cylinder. In E.H. Hirschel, editor, *Flow Simulation with High-Performance Computers II*, volume 52 of *Notes on Numerical Fluid Mechanics*. Vieweg, 1996. co. F. Durst, E. Krause, R. Rannacher.
33. S. Turek and R. Schmachtel. Fully coupled and operator-splitting approaches for natural convection flows in enclosures. *International Journal for Numerical Methods in Fluids*, 40:1109–1119, 2002.
34. A. Valencia, D. Ladermann, R. Rivera, E. Bravo, and M. Galvez. Blood flow dynamics and fluid–structure interaction in patient-specific bifurcating cerebral aneurysms. *International Journal for Numerical Methods in Fluids*, 58:1081–1100, 2008.
35. S. P. Vanka. Implicit multigrid solutions of Navier-Stokes equations in primitive variables. *J. of Comp. Phys.*, 65:138–158, 1985.

Numerical Simulation of Fluid–Structure Interaction Using Eddy–Resolving Schemes

M. Münsch and M. Breuer

Abstract Eddy–resolving schemes such as large–eddy simulation (LES) or detached-eddy simulation (DES) have become popular due to their favorable capabilities of predicting complex turbulent flows. That is especially true for instantaneous flow processes involving large–scale flow structures such as separation, reattachment and vortex shedding. Flow phenomena of such kind are very often encountered when the flow around or through a device enforces the structures to be deformed or displaced, i.e. for fluid–structure interaction (FSI). The present study deals with several aspects which have to be taken into account when LES is married to FSI. That comprises the coupling scheme, the handling of moving or deformable grids and the question how their quality requirements can be achieved. A coupling scheme leading to strong coupling among flow and structure, but also maintaining the advantageous properties of explicit time–marching schemes used for LES, was set up and analyzed. Thus a new and favorable coupling procedure for FSI within the LES context was developed. This and other issues of the numerical methods applied such as the measures to maintain the grid quality are discussed in detail. Results of validation test cases of FSI in the context of rigid body motions (e.g. an elastically mounted cylinder or a swiveling flat plate) as well as benchmark results with flexible structures computed with a finite–element code are presented.

M. Münsch
Institute of Fluid Mechanics, Friedrich-Alexander University Erlangen-Nuremberg, Cauerstraße 4, D–91058 Erlangen, Germany
e-mail: mmuensch@lstm.uni-erlangen.de

M. Breuer
Department of Fluid Mechanics, Institute of Mechanics, Helmut-Schmidt-University Hamburg, Holstenhofweg 85, D–22043 Hamburg, Germany
e-mail: breuer@hsu-hh.de

H.-J. Bungartz et al. (eds.), *Fluid Structure Interaction II*, Lecture Notes in Computational Science and Engineering 73, DOI 10.1007/978-3-642-14206-2_9,

1 Introduction

The interaction between fluids and structures is a topic of interest in many fields such as mechanical engineering (e.g. airfoils), civil engineering (e.g. towers) or medicine technique (e.g. artificial heart valves). Beside experimental investigations also numerical simulations have become an important and valuable tool for solving this kind of problems. For numerical investigations two basic approaches can be used. The monolithic approach is based on a unique numerical formulation of the whole FSI problem (e.g. a finite–element formulation) which can be solved by one sole code. By using a partitioned approach the entire FSI problem is divided into a fluid and a structure domain. For both fields separate numerical formulations are set up. The solution of the subproblems is done separately by the usage of highly specialized codes for each subtask. However, the coupling between both domains has to be done via an additional coupling interface. Here aspects such as code-to-code communication or grid-to-grid interpolation, e.g., of loads and displacements, have to be considered.

In general the disparity of the scales is also characteristic for FSI problems. The largest scales can be in the range of several hundred meters, e.g. suspension bridges, whereas the smallest scales of turbulent flows, i.e. the Kolmogorov scale l_k, are in a range of 0.1 millimeters. In terms of flow simulation among others three categories of methodologies can be chosen for turbulent flows: DNS, RANS and LES. The direct numerical simulation (DNS) is the most accurate one for turbulence computation, see e.g., [1]. The Navier–Stokes equations are solved without averaging or approximations for all motions in the flow field. For that purpose a high number of grid points and time steps are required which leads to a costly and time–consuming simulation restricted to low or moderate Reynolds numbers. A second group of methods are the so-called Reynolds–Averaged Navier–Stokes (RANS) equations which arise from averaging the Navier–Stokes equations in time. In contrast to the DNS approach the RANS equations have to be closed by a statistical turbulence model. The resulting flow field is usually time–averaged. Previous studies investigated laminar and also turbulent FSI applications using this RANS approach. In [13, 14] a partitioned fully implicit FSI scheme was applied to couple a three–dimensional finite–volume based multi–block flow solver for incompressible fluids [9] with a finite–element code for the structural subproblem. Efficient coupling was performed for large time step sizes used for implicit time–stepping schemes within RANS simulations. Depending on the flow problem RANS simulations are not always suitable for the prediction of the flow field. Large–scale flow structures such as vortex shedding or instantaneous separation and reattachment require more advanced methodologies such as large–eddy simulations (LES) [1–3] leading to the third category of methodologies for the simulation of turbulent flows. In contrast to RANS the LES technique is based on a spatial filtering of the Navier–Stokes equations and not on temporal averaging. Thus by computing the large scales directly and modeling solely the small scales an unsteady field of physical values is obtained and the prediction of complex turbulent flow fields such as the flow past bluff bodies is enabled. Therefore, LES is also a preferential method in the context of FSI applications when large–scale flow structures are expected. The task of the project was to marry FSI and LES by considering all relevant aspects playing a significant role for the (new) combination of these advanced techniques.

2   Governing Equations for the Fluid Flow

For the present study the inhouse code FASTEST-3D [9, 10] was used and extended for the FSI–LES application intended. The code is based on a finite–volume scheme which is used to discretize the filtered Navier–Stokes equations for an incompressible fluid. The discretization is done on a curvilinear, blockstructured body–fitted grid with colocated variable arrangement by applying standard schemes. Linear interpolation of the flow variables to the cell faces and a midpoint rule approximation for the integrals is used to obtain a second-order accurate central scheme. Within a FSI application the fluid forces acting on the structure lead to the displacement or deformation of the structure. Thus the computational domain is no longer fixed but changes in time, which has to be taken into account. Besides other numerical techniques, the most popular one is the so-called Arbitrary Lagrangian–Eulerian (ALE) formulation. Here the conservation equations for mass, momentum (and energy) are re-formulated for a temporally varying domain, i.e., control volumes (CV) with time-dependent volumes $V(t)$ and surfaces $S(t)$. The governing equations in ALE formulation read:

Mass Conservation:

$$\frac{d}{dt} \int_{V(t)} \rho \, dV + \int_{S(t)} \rho(u_j - u_{g,j}) \cdot n_j \, dS = 0. \tag{1}$$

Momentum Conservation:

$$\frac{d}{dt} \int_{V(t)} \rho u_i \, dV + \int_{S(t)} \rho u_i (u_j - u_{g,j}) \cdot n_j \, dS = \int_{S(t)} \tau_{ij} \cdot n_j \, dS - \int_{S(t)} p \cdot n_i \, dS. \tag{2}$$

The density is denoted by ρ, the pressure by p and the three Cartesian components of the velocity vector by u_i. The molecular momentum transport tensor is indicated by τ_{ij} whereas n_j describes the unit normal vector pointing outwards. In case of LES, an additional tensor denoted τ_{ij}^{SGS} has to be taken into account in the surface integral on the right-hand side of Eq. (2) describing the influence of the non–resolved subgrid–scales (SGS) on the resolved flow field. Since the grid is deformable, the grid velocity with which the surface of a CV is moving is taken into account via $u_{g,j}$. Here, the volume integrals now describe local changes in a CV of variable shape and thus the additional mass and momentum fluxes due to $u_{g,j}$. Since the system of equations given by (1) and (2) is not closed anymore, the unknown grid velocity $u_{g,j}$ has to be determined. To consider the conservation principle and to avoid the loss of mass and momentum the so-called *space conservation law (SCL)* [7, 8] is applied to compute the unknown grid velocities $u_{g,j}$.

Space Conservation Law:

$$\frac{d}{dt} \int_{V(t)} dV - \int_{S(t)} u_{g,j} \cdot n_j \, dS = 0. \tag{3}$$

This extra conservation law assures that within a change of the position or the shape of a CV no space is lost. In discretized form the SCL is expressed by the swept volumes of the corresponding cell faces. Inserting Eq. (3) into (1) the mass conservation equation for a fixed grid is obtained. Therefore, the original pressure–correction scheme applied for the solution of the incompressible Navier–Stokes equations on fixed grids has not to be changed concerning the mass conservation equation in the context of moving grids. The additional grid fluxes in the momentum equation are consistently determined by applying the SCL. Here, an incompressible fluid with temperature–independent fluid properties is considered. Thus the conservation equation for the energy is not taken into account.

In principle, the LES concept leads to a closure problem similar to that obtained by RANS. However, the non–resolvable small scales in a LES are much less problem–dependent than the large–scale motion so that the subgrid–scale turbulence can be represented by relatively simple models, e.g., zero–equation eddy–viscosity models. Like other eddy–viscosity models the well-known and most often used Smagorinsky model [21] is based on Boussinesq's approximation which describes the stress tensor τ_{ij}^{SGS} as the product of the strain rate tensor $\overline{S}_{ij}$ and an eddy viscosity v_T,

$$\tau_{ij}^{SGS,a} = \tau_{ij}^{SGS} - \delta_{ij}\tau_{kk}^{SGS}/3 = -2\,v_T\,\overline{S}_{ij} \text{ with } \overline{S}_{ij} = \frac{1}{2}\left(\frac{\partial \overline{u}_i}{\partial x_j} + \frac{\partial \overline{u}_j}{\partial x_i}\right), \tag{4}$$

where $\tau_{ij}^{SGS,a}$ is the anisotropic (traceless) part of the stress tensor τ_{ij}^{SGS} and δ_{ij} is the Kronecker delta. The trace of the stress tensor is added to the pressure resulting in the new pressure $P = \overline{p} + \tau_{kk}^{SGS}/3$. The eddy viscosity v_T itself is a function of the strain rate tensor $\overline{S}_{ij}$ and the subgrid length l, as

$$v_T = l^2 \, |\overline{S}_{ij}| \quad \text{with} \quad l = C_s \, \Delta \left[1 - \exp\left(\frac{-y^+}{A^+}\right)^3\right]^{0.5},$$

$$\Delta = (\Delta x \cdot \Delta y \cdot \Delta z)^{1/3}, \quad y^+ = \frac{y\,u_\tau}{v}, \quad u_\tau = \sqrt{\frac{\tau_w}{\rho}} \quad \text{and} \quad A^+ = 25. \tag{5}$$

C_s is the well-known Smagorinsky constant which has to be prescribed as a fixed value or can be determined as a function of time and space by the dynamic procedure originally proposed by Germano et al. [12] and later improved by several authors, e.g. Lilly [20]. In the first case, a Van Driest damping function is required [see Eq. (5)] in order to take the reduction of the subgrid length l near solid walls into account. In the present investigations, both the fixed parameter version of the Smagorinsky model was applied using the well established standard constant $C_s = 0.1$ but also the dynamic version. The latter applies a test filter of filter width 2Δ taking the 27 neighboring control volumes into account.

The different structural models applied will be described in Sect. 6 in conjunction with the corresponding test cases.

3 Coupling Schemes for Partitioned FSI

For the coupling between the fluid solver and the structure solver in general two different methodologies can be applied in the context of a partitioned approach. Within the so-called loose coupling approach the fluid and the structure subproblem are only solved once per time step leading to severe stability problems in most applications. For the strong coupling approach, fluid and structure solutions are repeated in a staggered manner. Here, a coupling scheme was finally developed meeting the requirements for an efficient large–eddy simulation on the one hand and a stable coupling algorithm on the other hand. This approach is characterized by a strong coupling between fluid and structure and explicit time marching for the fluid subproblem. For comparison also a coupling scheme for implicit time marching used in previous studies [13, 14] is presented.

3.1 Strong Coupling with Explicit Time Marching

For LES small time steps are required to resolve the turbulent flow field in time which leads to a preferred usage of explicit time-marching schemes. Here, for example, a predictor–corrector scheme of second–order accuracy forms the kernel of the fluid solver. In the predictor step an explicit three substep low–storage Runge–Kutta scheme advances the momentum equation in time leading to a prediction of the velocities u^*, v^* and w^*. These predicted velocities do not satisfy mass conservation. Thus, in the following corrector step the mass conservation equation has to be fulfilled by solving a Poisson equation for the pressure–correction based on a solver relying on an incomplete LU decomposition. The pressure-correction algorithm is repeated until a predefined convergence criterion ε for the mass conservation is reached leading to the final velocities u, v, w and the corresponding pressure field p of the new time step. For that purpose, typically 5 to 10 pressure–correction iterations are required until the mass conservation equation is numerically satisfied, e.g. $\Delta \dot{m} < \varepsilon = \mathcal{O}(10^{-8})$. For FSI computations this algorithm was extended towards a code coupling scheme [4–6] which is shown in Fig. 1. Receiving the corresponding velocity and pressure field from the pressure–correction step, the resulting forces, i.e. pressure and shear stresses, or the associated moments are computed. These forces or moments are transferred to an equation of motion if rigid body motion is

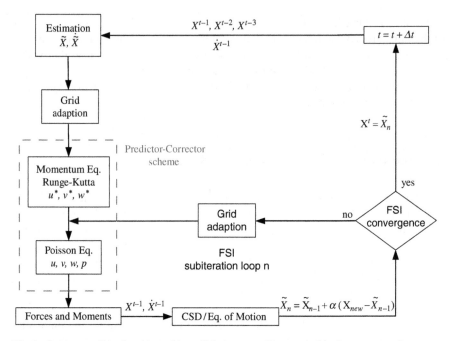

Fig. 1 Strong coupling algorithm with explicit time marching as used in the present study.

considered. For more complex structures such as shells or membranes the forces are transferred to an appropriate computational structure dynamics code (CSD), here a finite–element solver (Carat [24]). The response of the structure, i.e. the displacements X_{new}, are transferred to the fluid solver. In the case of a loose coupling approach the displacements are considered to be the physically correct displacements of the actual time step. Thus the computation passes on to the successive time step ($t = t + \Delta t$). This loose coupling between fluid and structure is in general only stable for low density ratios of the fluid to the structure, i.e., a low so-called added mass effect. For cases in which the added mass effect plays a dominant role, the loose coupling scheme, especially in connection with incompressible flows, tends to fail.

To overcome this difficulty, a so-called FSI subiteration loop is introduced in the scheme preferred [4–6]. If the FSI convergence criterion is not fulfilled, i.e., the dynamic equilibrium between fluid and structure is not achieved, the computational grid of the fluid domain is re-adapted. The grid adaptation is based on an underrelaxation of the structural response X_{new} by taking an underrelaxation factor α and the displacement of the previous subiteration loop (n-1) into account. Subsequent to the grid adaptation solely the corrector step of the predictor–corrector scheme is performed again and a new velocity and pressure field is obtained. Afterwards new loads for the structure solver are generated leading to an update of the corresponding displacements. The dynamic equilibrium between fluid and structure is numerically obtained if the change of the resulting displacements within the subiteration cycle reaches a convergence criterion. Then the subiteration process is stopped and the computation for the next time step is started. The new time step begins with an estimation of the displacement $\tilde{X}$ and velocity $\dot{\tilde{X}}$ of the structure which serves for convergence acceleration within the whole coupling process. Here, a second–order extrapolation for the displacements is applied by taking the displacement values of three former time steps indicated by the superscripts $t-1$, $t-2$, and $t-3$ into account:

$$\tilde{X}^t = 3X^{t-1} - 3X^{t-2} + X^{t-3} . \tag{6}$$

For consistency, a first–order extrapolation is applied on the structural velocity $\dot{\tilde{X}}$:

$$\dot{\tilde{X}}^t = 2\dot{X}^{t-1} - \dot{X}^{t-2} . \tag{7}$$

According to these estimated boundary values, the entire computational grid has to be adapted as it is done in each FSI subiteration loop. Presently, this grid adjustment is performed based on a transfinite interpolation (see also Sect. 4). Once the grid is adapted, the predictor–corrector scheme is applied and the first cycle of the FSI subiteration loop is entered.

In contrast to this strong coupling scheme relying on explicit time marching, a strong coupling scheme based on implicit time stepping as used in [13, 14] is depicted in Fig. 2. Since the flow is solved with a fully implicit scheme, an

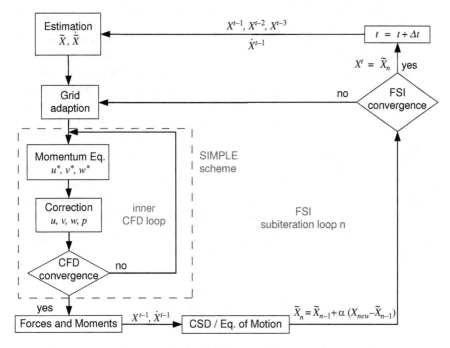

Fig. 2 Strong coupling algorithm with implicit time marching shown for comparison.

inner CFD loop becomes necessary within the applied SIMPLE scheme [15]. Both schemes significantly differ with respect to the momentum equations within the FSI–subiteration loop. For the implicit scheme the momentum equations are solved repeatedly in each subiteration sweep. In contrast they are only solved once per time step for the explicit case which strongly reduces the computational effort. Furthermore, the number of FSI–subiterations required to reach the convergence criterion is typically at least one order of magnitude smaller for the explicit scheme than for the implicit variant. In conclusion, instabilities due to the added mass effect known from loose coupling schemes are avoided by the newly developed coupling scheme and the explicit character of the time–stepping scheme beneficial for LES is still maintained.

3.2 Data Interpolation and Transfer

In the elementary case of a rigid body motion, here described via an equation of motion embedded within the fluid solver, the forces and/or moments are computed based on the pressure and the shear stresses on the surface. These pressures and shear stresses are located in the face center nodes of the control volumes which stick to the fluid–structure interface. Depending on the direction of the forces or moments they are applied as source terms in the corresponding equations of state for driving the displacement. The predicted displacements are directly transferred to the corresponding grid nodes.

When dealing with more complex structures, a more sophisticated structure solver, i.e. a finite–element solver, has to be considered. Due to different discretization techniques applied for the subtasks (finite volumes vs. finite elements) also different types of grids and different grid resolutions are used leading to non–matching surfaces meshes. Consequently, a grid to grid data interpolation and transfer becomes necessary. The fluid solver FASTEST-3D is based on a cell–centered variable arrangement for the pressure and the velocities, whereas the grid coordinates and displacements are cell–vertex bound. Therefore, in an initial step, the fluid loads F_C^i predicted based on the pressure and shear forces at the cell face nodes (CFN, center of the cell) are conservatively interpolated to the face vertices, i.e., the grid node (GN), see Fig. 3. That step is necessary since the displacements are required at GN. The whole interpolation procedure is as follows:

1. Conservative interpolation of the fluid loads located in the cell centers to the cell vertices of the fluid domain (CFD→CFD):

$$\sum_{j=1}^{nj} \mathbf{F}_{GN,CFD}^j = \sum_{i=1}^{ni} \mathbf{F}_{C,CFD}^i \qquad (8)$$

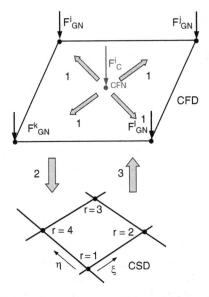

Fig. 3 Interpolation of loads and displacements between the fluid domain (CFD) and the structure domain (CSD).

2. Conservative interpolation of the fluid loads from the cell vertices of the fluid domain to the grid nodes of the structure domain (CFD→CSD):

$$\sum_{k=1}^{nk} \mathbf{F}_{CSD}^k = \sum_{j=1}^{nj} \mathbf{F}_{GN,CFD}^j \tag{9}$$

3. Bilinear interpolation of the displacements from the structure domain $\mathbf{u}_{r,CSD}$ to the cell vertices of the fluid domain (CSD→CFD):

$$\Delta \mathbf{x}_{GN,CFD}^j = \sum_{r=1}^{4} (N_r^e(\xi_j, \eta_j) \mathbf{u}_{r,CSD}) \tag{10}$$

Here, the resulting displacements in the fluid domain are denoted by $\Delta \mathbf{x}_{GN,CFD}^j$ and are obtained by the displacements of the structure domain $\mathbf{u}_{r,CSD}$ weighted by the shape functions N_r^e of the CSD element e. The interpolation steps 2 and 3 are done via the coupling interface CoMA [11] developed by the Chair of Structural Analysis of the Technical University Munich. This interface is based on the Message–Passing–Interface (MPI) and thus runs in parallel to the fluid and structure solver. The communication in-between the codes is performed via standard MPI commands. Since MPI is also used for code parallelization in FASTEST-3D, effective coupling with respect to high–performance computing is enabled.

4 Grid Smoothing

Within the FSI algorithm the computational grid of the fluid domain is adjusted to
the actual position of the structure. This is done during each subiteration cycle or at
least once in a time step. To maintain grid quality, i.e., orthogonality and smoothness, and according to these properties the accuracy of the numerical prediction
for time–dependent and moving grids, is a further challenge for FSI algorithms,
especially for turbulent flows using the LES approach (see Sect. 5). Beside using
algebraic approaches based on transfinite interpolations also developments on elliptic grid smoothing [22, 25] have been carried out within the project. The objective
is to maintain some designated grid quality, in order to resolve boundary layers
for example with a minimum computational effort. Here, the approach chosen is
based on a composite mapping which is performed separately on every block of
the flow field for the reason of parallelization. The composite mapping consists of
a nonlinear transfinite algebraic transformation and an elliptic transformation. The
algebraic transformation maps the computational space C one-to-one onto a param-
eter space P, with $P \subset \mathbf{R}^3$. The computational space C, with $C \subset \mathbf{R}^3$, as well as
the parameter space P are introduced as a Cartesian unit cube with $\xi = (\xi, \eta, \zeta)$
and $\mathbf{s} = (s, t, u)$, respectively. By applying the algebraic transformation the grid
points in the inner region of P are set based on the prescribed boundary grid point
distribution on ∂C. Due to the one-to-one mapping relation between the boundary
∂C and the boundary ∂D of the physical space D, the grid point distribution in P
is directly related to the prescribed boundary point distribution in D. The physical
space $D \subset \mathbf{R}^3$, a block of a block-structured grid for example, is spanned via Cartesian coordinates $\mathbf{x} = (x, y, z)$. Once discretized, the nonlinear system of equations
for calculating the position of the points $\mathbf{P_i} = (s_P, t_P, u_P)_i$ in space P is solved by
the Newton approach. The resulting grid distribution defines 18 control functions
(for the 3D case) of a Poisson system which has to be solved to obtain the final grid
point distribution in D. This differentiable one-to-one mapping of the parameter
space P onto the physical space D is defined by the elliptical transformation. The
resulting nonlinear equation system is linearized by the use of Picard iterations and
then solved via the strongly implicit procedure, i.e., the so-called SIP-solver already
available in FASTEST-3D. The grid obtained in D is basically non–orthogonal at
the boundaries.

In [22] an extended approach is suggested. In a first step the basic algorithm
described above is applied to generate an initial grid. In a following step the Laplace
equations $\Delta s = 0$, $\Delta t = 0$ and $\Delta u = 0$ are solved. Here, Neumann boundary conditions, i.e., $\partial s / \partial n = 0$ for example, are applied for the coaxial edges of P and
boundary conditions of Dirichlet type, i.e., $s = 1$ for example, for the remaining
edges. In a final step, again the Poisson system with control functions based on
cubic Hermite interpolations are solved leading to the desired orthogonality at the
boundaries. Here, the computational effort is very high and even higher in the context of FSI applications, where these steps have to be repeated several times during
one time step. Therefore, investigations on two alternative approaches have been
started to obtain a satisfying grid on the one hand with less computational effort

on the other hand. The first approach follows the idea of readjusting the source terms, in the vicinity of walls for example, to modify the grid with respect to grid point distances or angles between the grid lines. Here, the method of Hilgenstock [18] is applied and added to the basic algorithm described above. In theory the computational effort is reduced in contrast to the extended algorithm described in [22]. But special care has to be taken since grid folding may occur and convergence may restricted due to severe manipulations of the source terms. The second approach starts from an initially more or less perfect grid in respect to orthogonality and refinement, from which the corresponding source terms are computed. Then these source terms are adjusted, in the course of an FSI computations, to adapt the grid to the moving boundaries and to maintain grid quality. Here, only principle investigations have been performed up to now.

Beside the grid quality close to the walls also the grid quality close to the block-to-block connectivities is of interest. The basic approach described in [22] will lead to a smooth grid in the inner region of the corresponding block. Since the grid point on the boundary ∂D are not moved, the grid quality of the block-to-block intersection may be reduced. The extended approach in [22] leads to an orthogonal grid at the boundaries and thus to a smooth block-to-block intersection of grid lines. However, the computational effort is very high and total orthogonality may not be necessary.

To overcome this drawback, in the present study the source terms in parameter space P, which are influencing the grid quality also at the boundaries, are computed with the help of overlapping grids at the block-to-block connectivities. Due to this procedure a smooth but non–orthogonal grid is generated at the connectivities. Since the computation of these source terms can be incorporated into the basic approach of [22], the total computational effort is of the same order than the basic algorithm which is strongly advantageous for FSI–LES applications.

5 Large–Eddy Simulation on Moving Grids

For a finite–volume scheme as applied in the present investigation, the grid and thus the size of the control volumes varies in time. In the context of LES relying on an implicit filtering approach as used in many practically relevant LES predictions, the grid movement means that the filter width defined by Eq. (5) now additionally varies in time, i.e. $\Delta = f(\mathbf{x}, t)$. That leads to additional commutation errors depending on $d\Delta/dt$ and $d\Delta/dx$ [19]. In order to investigate the influence of the temporally varying filter width on the quality of the results, predictions of a plane channel flow at $Re_\tau = 590$ on oscillating grids (domain: $2\pi\delta \times 2\delta \times \pi\delta$ where δ denotes the channel half–width; grid: $128 \times 128 \times 128$ CVs) were carried out applying the classical Smagorinsky model [21] with Van Driest damping near solid walls as well as the dynamic model by Germano et al. [12] and Lilly [20].

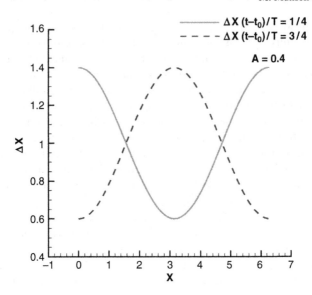

Fig. 4 Distribution of the grid spacing Δx for $(t - t_0) = \frac{1}{4}T$ and $(t - t_0) = \frac{3}{4}T$ over the channel length x.

5.1 Grid Movement

The sinusoidal grid movement was defined by

$$x_{move}/\delta = A \cdot \sin(2\pi(t - t_0)/T) \cdot \sin(x_0/\delta). \qquad (11)$$

Here, the new coordinate of the grid point is denoted with x_{move} whereas x_0 is the reference coordinate of the fixed grid computation. The grid movement and with it the effective grid spacing Δx is controlled via the amplitude A and the period T of the sinusoidal deformations. In Fig. 4 the resulting grid spacing Δx along the channel length x is plotted for two different instants in time within a period of motion. For an amplitude of $A = 0.4$ the grid is compressed to 60% of the initial grid spacing in the middle of the channel at $(t - t_0) = \frac{1}{4}T$, see also Fig. 5(a). At $(t - t_0) = \frac{3}{4}T$ the grid is stretched to 140% of the initial grid spacing in the middle of the channel. The time step Δt was equal to 0.01 s for all computations.

A variety of cases was considered (see Table 1). In one subset of cases, the amplitude A of the internal grid deformation was varied from 0.1 to 0.85 and the period T of the deformations was set to 1 flow–through time (T_{FT}). In a second subset, computations for different cycle periods T at a constant amplitude were performed.

5.2 Results

The results were compared to LES predictions on a fixed grid (reference) by computing the root-mean-square deviations (rmsd) of the time- and space–averaged stresses

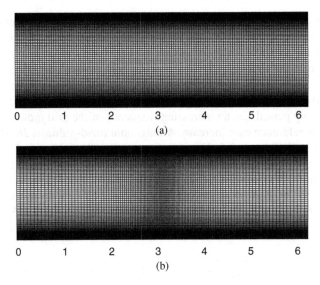

Fig. 5 Reference grid (a) and deformed grid (b) at $(t - t_0) = \frac{3}{4}T$.

Table 1 Overview of moving grid computations performed showing amplitudes **A** and cycle periods **T** with the resulting root-mean-square deviations (rmsd) of $\overline{u'u'}$, $\overline{v'v'}$, $\overline{w'w'}$, $\overline{u'v'}$ and $\overline{U}$ in [%].

		Smagorinsky model: variation of amplitude **A**			
A	$\mathbf{T/T_{FT}}$	rmsd$(\overline{u'u'})$	rmsd$(\overline{v'v'})$	rmsd$(\overline{w'w'})$	rmsd$(\overline{u'v'})$	rmsd$(\overline{U})$
0.10	1	2.63	0.81	0.85	8.93	0.16
0.25	1	2.98	1.11	1.16	5.22	0.21
0.50	1	2.55	2.68	2.36	9.36	0.48
0.65	1	2.63	3.71	2.53	5.74	0.67
0.75	1	3.34	5.13	3.63	13.73	0.81
0.85	1	4.19	6.35	3.92	12.06	1.06

		Dynamic model of Germano et al.: variation of amplitude **A**			
A	$\mathbf{T/T_{FT}}$	rmsd$(\overline{u'u'})$	rmsd$(\overline{v'v'})$	rmsd$(\overline{w'w'})$	rmsd$(\overline{u'v'})$	rmsd$(\overline{U})$
0.25	1	1.70	4.37	2.46	33.69	0.41
0.50	1	2.71	6.79	4.28	20.02	0.68

		Smagorinsky model: variation of cycle period **T**			
A	$\mathbf{T/T_{TF}}$	rmsd$(\overline{u'u'})$	rmsd$(\overline{v'v'})$	rmsd$(\overline{w'w'})$	rmsd$(\overline{u'v'})$	rmsd$(\overline{U})$
0.50	1	2.55	2.68	2.36	9.36	0.48
0.50	0.2	4.98	10.58	5.70	12.80	1.40
0.50	0.1	10.15	26.57	17.17	18.54	3.04

$\overline{u'u'}$, $\overline{v'v'}$, $\overline{w'w'}$, $\overline{u'v'}$ and of the mean flow velocity $\overline{U}$. As can be seen in Table 1 for a fixed period T an increasing amplitude A leads to increasing deviations of the investigated flow properties for the classical Smagorinsky model. Beside the shear stresses all other quantities reach their maximum deviations at an amplitude of $A = 0.85$. The rmsd for $\overline{u'v'}$ reaches a maximum for an amplitude of $A = 0.75$. The mean velocity component $\overline{U}$ shows in general the smallest deviation from the reference test case. These results are also shown in Fig. 6. The results of the amplitude variation for the dynamic model of Germano et al. yield the same trend. Comparing the results of both models, the Smagorinsky model shows less deviations from the reference case than the dynamic model. The second test case, i.e. the variation of the period T, shows a bigger impact on the rmsd–values than the first test case. For a decreasing period, i.e. an increasing frequency of the grid motion, the deviations from the reference case increase. A maximum rmsd–value of 26.57% for $\overline{v'v'}$ is obtained for $T/T_{FT} = 0.1$. Here, again the mean flow velocity $\overline{U}$ is the least sensitive variable with a deviation of 3.04%. Exemplarily, Fig. 7 depicts the results for the variation of the period T. In the first case, one cycle of the grid deformation is equivalent to one flow–through time. In that case ($T/T_{FT} = 1$) the deviations from the reference case are negligible. If the grid oscillates faster (i.e. 5 oscillations per flow–through, $T/T_{FT} = 0.2$) first deviations from the reference case appear.

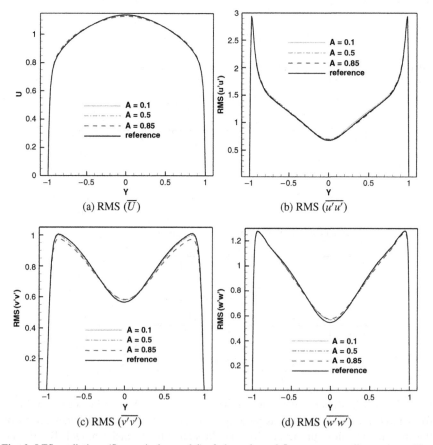

Fig. 6 LES predictions (Smagorinsky model) of plane channel flow on temporally varying grids, three different amplitudes A, grid cycle periods $T = T_{FT}$ (RMS values normalized by u_τ).

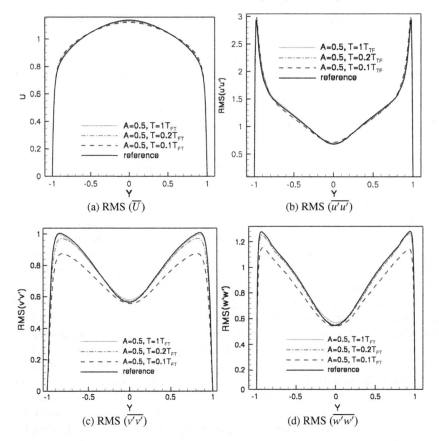

Fig. 7 LES prediction of plane channel flow on temporally varying grids, $A = 0.5$, three different grid cycle periods T (RMS values normalized by u_τ).

A further increase to 10 grid oscillations per flow–through leads to strong deviations. In that case the errors introduced by the grid movement are obvious. Although different effects such as time–varying local resolution and the commutation error itself cannot clearly be distinguished, in summary, these additional errors due to moving grids significantly affect the LES results. The impact on the mean flow properties is a minor one as long as the amplitudes and corresponding oscillations are not too high. Therefore, LES was already applied to fully coupled FSI computations within this study (see Sect. 6). The effect on the secondary moments is much stronger leading to severe restrictions on tolerable amplitudes and cycle periods. These results provide hints on what kind of grid motions within an FSI application using LES can be accepted.

6 FSI Test Cases

In this section, a subset of test cases is presented which have been computed to verify the FSI coupling algorithm developed. For the sake of simplicity only rigid body motion is considered. The test cases are dealing with laminar as well as turbulent flows. Laminar flows are considered here because of less computational effort to derive data sets for comparison, although this work focuses on turbulent flows.

6.1 Elastically Mounted Cylinder at Re = 200

The flow around an elastically supported circular cylinder was computed for different values of the reduced damping parameter Sg (defined below) according to the paper of Zhou et al. [26]. The Reynolds number based on the reference velocity $U_\infty = 0.514$ m/s and the cylinder diameter of $D = 6 \cdot 10^{-3}$ m was equal to 200 and thus in the laminar regime. Hence the subgrid-scale model was switched off (no LES). The configuration is depicted in Fig. 8(a).

For the prediction of the flow field a grid with 262,144 control volumes was used. The spanwise direction was resolved with 8 CVs since the flow is assumed to be two–dimensional. In spanwise direction symmetry boundary conditions are applied. At the inflow a constant velocity U_∞ was specified. A no–slip boundary condition was set for the cylinder surface. At the outflow a convective outflow condition was prescribed. The time-step size applied was $\Delta t = 1.5 \cdot 10^{-5}$ seconds. Assuming the cylinder to be fixed, i.e. the classical test case of the flow around a cylinder, the well–known vortex shedding phenomenon in the wake of the cylinder was obtained. The corresponding shedding frequency was computed to be $f_s = 16.7$ Hz leading to a Strouhal number of $St = f_s \cdot D / U_\infty \approx 0.2$. The result obtained by the fixed cylinder computation was taken as initial conditions for the fluid–structure interaction case, i.e., the released and thus oscillating cylinder. The response on the resulting fluid forces is described by equations of motions [Eq. (12)] which are solved numerical by a classical Runge–Kutta scheme. Thus the resulting displacements in x and y directions are controlled by the mass of the cylinder m, a damping coefficient d, a spring constant c and the related forces $F_x(t)$ and $F_y(t)$. The forces are determined by numerical integration of the resulting pressure field p and shear stresses τ acting on the cylinder surface:

$$
\begin{aligned}
m\,\ddot{x}(t) + d\,\dot{x}(t) + c\,x(t) &= F_x(t), \\
m\,\ddot{y}(t) + d\,\dot{y}(t) + c\,y(t) &= F_y(t).
\end{aligned}
\tag{12}
$$

In the work of Zhou et al. [26] a so-called reduced damping parameter $Sg = 8\pi^2 St^2 d^* M^*$ with the mass ratio M^* and the normalized damping parameter d^* was introduced. The reduced mass ratio is defined as $M^* = m^*/\rho D^2$ with the normalized cylinder mass $m^* = m/l_c$. The normalized damping parameter is given by $d^* = d/l_c$. Here, the fluid density is denoted by ρ and l_c describes the cylinder length. Additionally, the relation between the natural cylinder frequency f_n

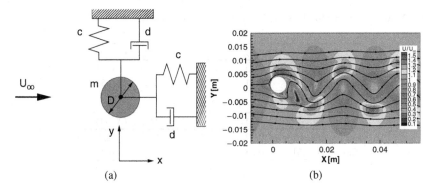

Fig. 8 (a) Spring–mass–damper model of the cylinder test case; (b) Flow field of the normalized velocity magnitude for $Sg = 1.0$ and $Re = 200$.

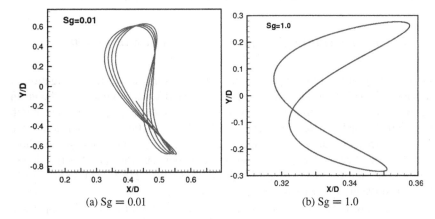

Fig. 9 Normalized displacements of the cylinder center for (a) $Sg = 0.01$ and (b) $Sg = 1.0$.

and the shedding frequency f_s given by $f_n = 1/(2\pi) \sqrt{c/m} = 1.3 f_s$ provides the corresponding values for the spring constant c and the cylinder mass m.

In Fig. 8(b) a snapshot of the flow field depicted by the normalized velocity magnitude is exemplarily shown for $Sg = 1.0$. Downstream of the cylinder the vortex shedding phenomenon occurring is visible. Furthermore, the recirculation area behind the cylinder is deflected due to the cylinder movement. In Fig. 9 the normalized traces of the cylinder are shown. As observed by Zhou et al. the X/D–Y/D plots describe an asymmetric 8. The point of intersection is located off-centered for both cases. In Table 2 the predicted results of the absolute normalized locations are presented in comparison with the results by Zhou et al. [26]. A decreasing reduced damping parameter Sg leads to an increase of the displacements. The agreement between the predicted and the reference Y/D values is better than for the X/D values. Overall the results show satisfactory correspondence with the reference values. The deviations maybe generally attributed to the different solution techniques,

Table 2 Absolute normalized displacements for $Sg = 0.01$ and $Sg = 1.0$ at $Re = 200$.

$Sg = 0.01$:

	$(X/D)_{min}$	$(X/D)_{max}$	$(Y/D)_{min}$	$(X/D)_{max}$
Present prediction	0.33	0.55	−0.64	0.61
Zhou et al. [26]	0.31	0.47	−0.62	0.56

$Sg = 1.0$:

	$(X/D)_{min}$	$(X/D)_{max}$	$(Y/D)_{min}$	$(X/D)_{max}$
Present prediction	0.32	0.34	−0.29	0.28
Zhou et al. [26]	0.28	0.39	−0.27	0.27

i.e. finite–volume method versus discrete vortex method, and/or to the unsteadiness of the flow leading to different solutions due to bifurcation effects. For this test case the maximum number of FSI subiterations n_{max} was very low, i.e., equal to 1.

6.2 Elastically Mounted Square Cylinder at $Re = 13,000$

Further testing of the FSI coupling scheme was performed by the investigation of the sub-critical flow around an elastically mounted square cylinder as shown in Fig. 10(a). With a free–stream velocity of $U_\infty = 10$ m/s and a cylinder edge length of $b = 0.02$ m the Reynolds number is equal to $Re = 13,000$. The extension of the cylinder in spanwise direction is equal to $4b$, whereas the dimension of the whole domain is specified by $40b \times 11b \times 4b$. The cylinder center is placed at $10b$ downstream of the inflow section. The grid applied consists of 2,228,224 CVs using 130 CVs to resolve the spanwise direction.

At the inflow a velocity block profile was specified with U_∞. For the upper and lower face of the domain slip boundary conditions were set. In spanwise direction a periodic boundary condition was used. A convective outflow condition was applied with $U_c = U_\infty$ at the outflow section. On the cylinder surface a no–slip boundary condition is applied. For the computation of the turbulent flow field a LES prediction using the Smagorinsky model was performed. The Smagorinsky constant was set to $C_S = 0.1$. In order to compare the coupling schemes presented in Figs. 1 and 2 the time-step size has been chosen such that the step size fits to the numerical characteristics of the considered coupling scheme. Thus, a time-step size of $\Delta t = 10^{-4}$ seconds was applied for the implicit time–marching algorithm leading to a CFL number of 20. A CFL number of 0.59 was obtained for the explicit time marching algorithm with a time step size of $\Delta t = 3 \cdot 10^{-6}$ which is more appropriate for the simulation of turbulent flows with LES than the implicit counterpart. The displacement of the structure are considered by equations of motions fitting to a spring–mass–damper system [Eq. (12)]. The mass of the cylinder is equal to $m = 0.005$ kg, the damping coefficient is set to $d = 0.1$ Ns/m and a spring constant of $c = 5$ N/m is applied. As before, the equations are solved by

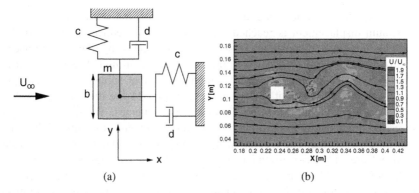

(a)   (b)

Fig. 10 (a) Spring–mass–damper model of the square cylinder test case; (b) Normalized velocity magnitude and streamlines of the oscillating cylinder test case at $Re = 13,000$.

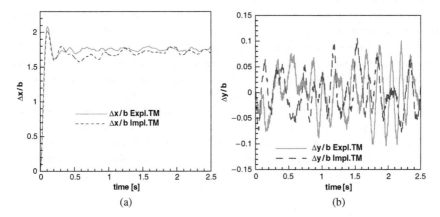

(a)   (b)

Fig. 11 Normalized displacements obtained by the strong coupling algorithm with either explicit or implicit time–marching schemes: (a) x–component and (b) y–component.

a classical Runge–Kutta scheme. As initial conditions the displacements, velocities and accelerations of the structure were set to zero.

A snapshot of the resulting normalized velocity magnitude $\sqrt{U_x^2 + U_y^2 + U_z^2}/$ U_∞ and the corresponding streamlines is plotted in Fig. 10(b). For this case the flow at the inlet is laminar. Flow separation occurs at the leading and partially at the trailing edge of the cylinder forming free shear layers in which the transition to turbulence takes place. Oscillations of the cylinder in x and y–direction are excited due to the phenomenon of alternating vortex shedding from the cylinder which is visible in the wake of the cylinder. In addition, a drag–induced movement of the cylinder in x–direction occurs. The resulting normalized displacements $\Delta x/b$ and $\Delta x/b$ are shown in Fig. 11.

Starting from the initial conditions $\Delta x/b = 0$ and $\Delta y/b = 0$ a transitional behavior is visible for the x–component for both FSI coupling schemes replying either on explicit or implicit time–marching. After a real time of about 0.1 seconds a maximum displacement is reached. The displacement reduces until the transitional time interval ends at about 0.5 seconds. In the following, the x–components vary around $\Delta x/b = 1.6$. The y–component oscillates around $\Delta y/b = 0$ caused by the vortex shedding cycle. The maximum number of subiterations was equal to $n_{max} = 3$ for the explicit time marching. In Fig. 11 the displacement values are not perfectly matching for both time–stepping schemes. The displacement in x–direction predicted by the explicit time–marching scheme is slightly higher than for the implicit scheme. The y-components do not correspond in their chronological sequence. Here, only a correspondence in the displacement magnitudes can be seen. The differences can be attributed to different starting time steps when the cylinder is released, used for both LES computations. Furthermore, the non–deterministic nature of turbulence has to be taken into account. Thus only statistical values such as mean values or standard deviations can be directly compared but not the time histories of instantaneous processes.

For further verification of the results a Fast–Fourier–Transformation was applied on the displacement data sets by neglecting the phase of transition, i.e., data sets for real time values greater than 0.8 seconds. Whereas no characteristic peak is visible in the frequency spectra of the x–displacements, such peaks show up for the y–displacements. For the data predicted by the implicit time–marching a peak was found for a frequency of $f = 69$ Hz which equals the shedding frequency of the vortices. The explicit time–marching scheme delivers a slightly shifted characteristic frequency of $f = 71$ Hz. Summing up, the results obtained by the different coupling strategies relying on different time–marching schemes show good agreement. Thus, the functionality of the developed strong coupling scheme using an explicit time–marching method within a LES prediction is confirmed.

6.3 Swiveling Plate in the Turbulent Regime

The fluid–structure interaction of a pivot–mounted flat plate within a turbulent channel flow [6] was considered for further testing of the developed FSI scheme. For that purpose also experimental investigations have been performed at LSTM Erlangen [17]. In Fig. 12 the setup of the computational domain is plotted. A flat plate of dimensions $177mm \times 64mm \times 2mm$ with a chord length $c = 64$ mm is mounted in a channel with rectangular cross-section.

The turbulent flow is simulated by LES using the Smagorinsky model with a Smagorinsky constant of $C_S = 0.1$ and Van Driest damping near solid walls. At the inflow a block velocity profile is set. Here, two inflow velocities have been considered with $U_\infty = 1.07$ m/s and $U_\infty = 0.6$ m/s leading to Reynolds numbers based on the chord length c of $Re = 68,000$ and $38,000$. At the outlet a convective outflow condition is set with an outflowing velocity $U_c = U_\infty$. At the upper and lower channel wall as well as for the flat plate no–slip conditions have been applied. In spanwise direction two cases have been considered. On the one hand only a section

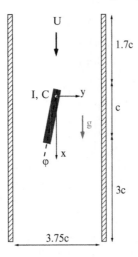

Fig. 12 Setup of the domain for the swiveling flat plate inside a channel.

Table 3 Grid resolutions and spanwise extensions used to resolve the turbulent flow field.

Grid	Number of CVs	Spanwise Resolution	Spanwise Extension [mm]
1	434,176	4	64
2	9,409,536	64	64
3	31,757,184	177	177

of one chord length of the full spanwise extension of the plate used in the experimental setup is taken into account. On the other hand, i.e. for one computation, the spanwise direction is set as in the experimental setup. For both cases periodic boundary conditions have been applied. Thus, the influence of the side walls is neglected. This is at least partially justified due to the fact that in the experimental setup a small gap between the plate ends and the spanwise channel walls is installed. The simulations have been performed on three different grids (see Table 3). The preliminary grid 1 has a total number of 434,176 CVs resolving the spanwise direction (64 mm) with only 4 CVs. Grid 2 has a total number of 9,409,536 CVs. Here, the spanwise direction (64 mm) is resolved with 64 CVs. Grid 3 has a total number of 31,757,184 CVs with 177 CVs in spanwise direction taking the experimental extension in spanwise extension (177 mm) into account.

For the non–deflected plate a wall–normal resolution given by $y^+ = 3$ for $Re = 68,000$ and by $y^+ = 0.7$ for $Re = 38,000$ is achieved for the wall–nearest control volume center to resolve the flow at the trailing edge of the plate. The time–step size was $\Delta t = 10^{-5}$ s for all cases considered. The response of the plate on the outer moment $M_z(t)$ imposed by pressure and shear forces is described by a spring–mass–model leading to an ordinary differential equation for the plate angle φ. This equation is solved by a classical Runge–Kutta scheme (RK) on the one hand and, for verification reasons, with an implicit Euler scheme (IE) on the other hand.

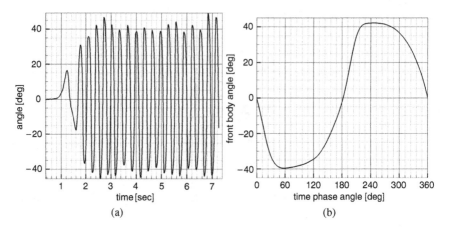

Fig. 13 Computed angular displacements of the plate: (a) as a function of time, (b) for one period as a function of the time–phase angle for $Re = 68,000$ computed on grid 1 and solving the equation of motion with the Runge–Kutta scheme.

$$I \ddot{\varphi}(t) + C \sin(\varphi(t)) = M_z(t) \tag{13}$$

Here, I describes the moment of inertia of the plate and the axle and C is equal to the resulting moment due to gravity and buoyancy forces. The latter results from density difference between the fluid and the structure. The outer moment $M_z(t)$ was linearly scaled for the cases with reduced spanwise extension for fitting the experimental setup.

In Fig. 13(a) the temporal development of the plate angle is shown. For initialization purposes of the turbulent flow field the plate is kept in fixed position, i.e. $\varphi = 0°$, for about 0.36 s real time. Then the plate is released and the coupling between fluid and structure is performed. After a period of transition, which takes about 2 s, amplitudes of $\varphi_{min} = -45°$ to $\varphi_{max} = -45°$ are reached. In Fig. 13(b) the instantaneous time–phase resolved plate angle for an exemplarily chosen period is shown. In Table 4 the computed results for the different Reynolds numbers, computational grids and solvers used for the equation of motion are shown. Obviously, the swiveling frequency obtained increases with the Reynolds number. Furthermore, the absolute magnitude of the angular displacements reduces by refining the grid in spanwise direction. For grid 3 only preliminary results, i.e., not averaged results, are shown but the tendency is confirmed. The choice of an implicit Euler or a classical Runge–Kutta scheme for solving the equation of motion barely does not effect the angular amplitudes nor the frequencies. Performing the FSI subiterations in combination with this classical Runge–Kutta scheme, an averaged number of $n = 5$ subiterations was required to obtain dynamic equilibrium between fluid and structure. When taking the experimental results into consideration, a mismatch between the computed and the measured results becomes obvious. The angular displacements are measured to be in a range of $\varphi = \pm 22°$ and $\varphi = \pm 27°$, respectively. The effect of increasing angular amplitudes with increasing Reynolds number was

Table 4 Overview of the simulations performed concerning to different grid sizes, Reynolds numbers Re and solvers for the equation of motion with the corresponding findings in comparison with the experimental results.

Grid 1: 434,176 CVs with 4 CVs in spanwise direction

U_∞[m/s]	Re	Structure	φ_{max}[°]	φ_{min}[°]	Frequency[Hz]
0.6	38,000	RK	+49	−49	1.94
1.07	68,000	RK	+49	−45	3.14
1.07	68,000	IE	+46	−47	3.45

Grid 2: 9,409,536 CVs with 64 CVs in spanwise direction

U_∞[m/s]	Re	Structure	φ_{max}[°]	φ_{min}[°]	Frequency[Hz]
0.6	38,000	RK	+44	−44	1.95
1.07	68,000	RK	+44	−45	3.23

Grid 3: 31,757,184 CVs with 177 CVs in spanwise direction

U_∞[m/s]	Re	Structure	φ_{max}[°]	φ_{min}[°]	Frequency[Hz]
1.07	68,000	RK	>+40	<−40	>3.0

Experiment:

U_∞[m/s]	Re	φ_{max}[°]	φ_{min}[°]	Frequency[Hz]
0.6	38,000	+22	−22	1.5
1.07	68,000	+27	−27	2.49

not detected in the computations. The behavior of the frequency as a function of the Reynolds number is predicted in the right way, but the absolute frequencies computed are higher than the experimental ones. So far, only the frequency ratio $[Freq(Re = 38,000)/Freq(Re = 68,000)]_{EXP} = 0.602$ of the measured data show good agreement with the ratio $[Freq(Re = 38,000)/Freq(Re = 68,000)]_{CFD} = 0.604$ obtained from the numerical investigations (grid 2).

In Fig. 14 the computed instantaneous velocity field and the phase–averaged velocity field obtained by measurements are shown for different time–phase angles. The time–phase angle is defined as $tpa = t/T \cdot 360°$ with T denoting the swiveling period and t describing the real time passed within the period. Being aware of comparing averaged with instantaneous data sets, nevertheless basic differences in the results can be detected. The vortex generated on the suction side of the inclined plate is of larger scale in the computations than for the experimental case for $tpa \approx 300°$. This vortex travels downstream and with it a zone of low pressure. When passing the center of rotation the low pressure region together with the high pressure on the pressure side of the plate reduces the rotational speed of the plate. Finally, the direction of rotation is changed and the plate tends towards the next angular amplitude passing the point of origin at $tpa = 360°$. Here, the vortex is going to leave the plate region for the experimental case (Fig. 14(d)). However, in the simulation the vortex still maintains on the former suction side, close to the trailing edge, affecting the movement of the plate. Obviously for this kind of test case, there is a strong coupling between the vortices generated by the angular displacements on the one hand and the effect of the vortices on the angular displacement within the next instant of

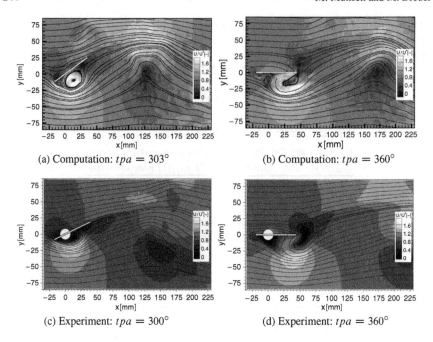

(a) Computation: $tpa = 303°$ (b) Computation: $tpa = 360°$

(c) Experiment: $tpa = 300°$ (d) Experiment: $tpa = 360°$

Fig. 14 Computed (instantaneous) and experimentally measured (phase–averaged) velocity fields and plate deflections at two different time instants of the swiveling motion at $Re = 68,000$.

time on the other hand. In spite of extensive numerical investigations taking different resolutions, different numerical algorithms for the structure and a variety of other issues not presented here into account, up to now, the reason for the over-prediction of angular displacements and frequencies has not been found. Beside further numerical improvements such as grid quality or pressure–correction on curved grids, also the starting procedure for recomputing the experimental cases has to be considered. Whereas the fluid and the movement of the plate is accelerated over a certain time interval in the experimental case, the fluid is already initialized with the desired flow velocities and the plate is released not until the flow computation is settled for the numerical predictions. Furthermore, the influence of damping effects in the bearings of the swiveling plate was investigated but could not fully excluded up to now. Finally, the small gap between the plate and the channel walls mentioned above for the experimental setup could not be replicated in the simulations and might be a reason for the deviations observed.

7 Benchmark Test Cases

Within the activities of the DFG research group FOR 493 several benchmark test cases have been developed (see [23]). Here, pure CFD as well as pure CSD test cases, and numerical FSI test cases within the laminar flow regime are available [23] to be solved with the fluid–structure interaction algorithms developed. Furthermore, also experimental test cases for laminar as well as turbulent flows were investigated and are available for benchmark purposes, see [16, 17] and the corresponding contribution in this issue. First verifications of the codes developed have been performed on the pure fluid dynamic applications (named CFD1–CFD3 in [23]) or structure mechanics benchmarks (named CSM1–CSM3, see also the paper in this issue). Since good agreement with the proposed reference data sets was obtained, fully coupled FSI computations on numerical as well as experimental benchmark test cases have been carried out in the present project.

7.1 Numerical Benchmark FSI3

For testing the coupling algorithm developed (see Sect. 3.1) as well as the data transfer and interpolation strategies (see Sect. 3.2), the benchmark test case named FSI3 in [23] and depicted in Fig. 15 was considered. In a channel of length $L = 2.5$ m and a height of $H = 0.41$ m a cylinder is mounted. The cylinder position is slightly off–centered, with the cylinder center located at 0.2 m downstream of the inflow section and with a distance of 0.2 m from the lower lateral channel wall. The cylinder has a diameter of $D = 0.1$ m. With a mean inflow velocity of $U_\infty = 2$ m/s and a kinematic viscosity of $v^f = 0.001$ m^2/s a Reynolds number of $Re = 200$ is obtained. The density of the incompressible and Newtonian fluid is set to $\rho^f = 1000$ kg/m^3. In the wake of the cylinder, a flexible beam is attached to the cylinder. The beam is of length $l = 0.35$ m and has a thickness of $h = 0.02$ m. The cylinder is assumed to be fixed and rigid, whereas the material of the beam is of St. Venant–Kirchhoff type characterized by a Poisson ratio of $v^s = 0.4$, a Young modulus of $E = 5.6 \cdot 10^6$ kg/(m s^2) and a density of $\rho^s = 1000$ kg/m^3.

For the flow predictions two different grids with either only 27,040 or 108,160 CVs in a 2D plane were used, respectively. Since the code is three–dimensional, in total 108,160 CVs or 865,280 CVs were used in 3D. Since for low Reynolds numbers the stability of explicit time–marching schemes is dominated by viscous effects, either the computational grid has to be coarse or the time–step size has to be

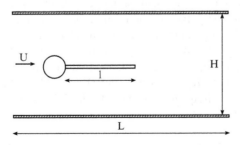

Fig. 15 Setup of the computational domain for the FSI 1–3 test case.

rigorously reduced or a combination of both has to be applied to stay in the stable limits[1]. Therefore, relatively coarse grids in combination with time–step sizes of $\Delta t_f = 9 \cdot 10^{-5}$ s (coarse grid) and $\Delta t_f = 2 \cdot 10^{-5}$ s (medium grid) were chosen to predict the flow field. At the inflow a parabolic velocity profile was set with:

$$u_{in}(y) = 1.5 U_\infty \frac{y(H-y)}{(\frac{H}{2})^2}. \qquad (14)$$

At the lateral boundaries as well as for the structure, i.e. the fixed cylinder and the beam, no–slip boundary conditions were applied. In spanwise direction slip conditions were chosen. At the channel outlet a convective outflow condition with $U_c = U_\infty$ was specified. The response of the structure was computed with the finite–element code Carat [24] provided by the Chair of Structural Analysis of TU Munich. Here, the beam was modeled using in total 30 four–noded shell elements. The time–step size of the solver for the structure was adapted to the time–step size of the fluid solver, i.e., $\Delta t_s = \Delta t_f$. For the code coupling the coupling tool CoMA [11] was used as described in Sect. 3.2. Within the coupled FSI prediction a constant underrelaxation factor of $\alpha = 0.5$ was considered for transferring the computed displacements from the structure domain to the fluid domain. For this case, a first–order extrapolation for the estimation of the structural displacements at the beginning of each time step was performed (see Sect. 3.1).

In the present configuration, vortex shedding occurs at the cylinder. These vortices travel downstream and start to interact with the flexible structure which leads to an oscillation of the beam. The resulting displacements are depicted in Fig. 16(a) and 16(c) for point A located at the trailing edge on the center line of the beam (medium grid). After some transition phase the amplitude of the oscillation reaches constant values for the x– and y–components (see Fig. 16(b) and 16(d)). The mean values and the amplitude of the displacements in both direction are given in Table 5 for both grids. It is obvious that the displacements in y–direction are about one order of magnitude larger than those in x–direction. Furthermore, an improvement of the results of the refined grid is visible with respect to the reference data. The frequencies of the displacements were determined to $f_x = 10.12$ Hz and $f_y = 5.05$ Hz on the medium grid, respectively. The actual reference value [23] is specified with $f_{x,ref} = 10.93$ Hz for the x–component and $f_{y,ref} = 5.46$ Hz for the y–component.

Due to the off–centered location, the structure is affected by lift forces right from the beginning of the computation. The resulting lift and drag forces are depicted in Fig. 17. For the lift minimum and maximum values of $L_{min} = -172.5$ N and $L_{max} = 204.9$ N are predicted on the medium grid. The DFT delivers a frequency of $f_L = 5.05$ Hz, which is equal to the oscillation frequency of Δy. For the drag force corresponding results of $D_{min} = 428$ N and $D_{max} = 507$ N have been obtained. Similar to the predicted frequency value of Δy the DFT leads to $f_D = 10.12$ Hz.

---

[1] Since the FSI coupling scheme is mainly intended for high–Re flows within the LES context, this restriction solely applies to the present test case.

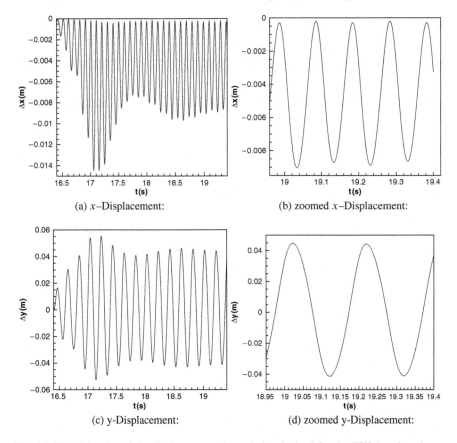

(a) x–Displacement:

(b) zoomed x–Displacement:

(c) y-Displacement:

(d) zoomed y-Displacement:

Fig. 16 Time histories of the displacements Δx and Δy obtained for the FSI3 benchmark test case on the medium grid.

The actual reference values are $L_{min,ref} = -151.4$ N and $L_{max,ref} = 156.4$ N for the lift force with a frequency of $f_{L,ref} = 5.46$ Hz and $D_{min,ref} = 432.7$ N and $D_{max,ref} = 488.2$ N for the drag with a reference frequency of $f_{D,ref} = 10.93$ Hz. The deviations between the predicted results and the reference values are obvious. They in general decrease with increasing grid resolution. Taking the restrictions on the grid resolution into account, the predicted results are within the expected range. These restrictions on grid resolution and time–step size are, by far, less severe when applying the algorithm to turbulent flows connected with higher flow velocities (see Sect. 7.2). Nevertheless, by performing this test case the proper behavior of the whole partitioned FSI setup was proven.

Table 5 Results of numerical benchmark FSI3 on two different grid levels; forces are given for a reference width of 1 m (values in parenthesis are the corresponding frequencies in Hz).

	Δx [$\times 10^{-3}$ m]	Δy [$\times 10^{-3}$ m]	Drag [N]	Lift [N]
Coarse	-5.18 ± 5.04 [10.14]	1.12 ± 45.1 [4.99]	477.0 ± 48.0 [10.14]	7.0 ± 223.0 [4.99]
Medium	-4.54 ± 4.34 [10.12]	1.50 ± 42.5 [5.05]	467.5 ± 39.5 [10.12]	16.2 ± 188.7 [5.05]
Ref. [23]	-2.88 ± 2.72 [10.93]	1.47 ± 34.9 [5.46]	460.5 ± 27.74 [10.93]	2.50 ± 153.91 [5.46]

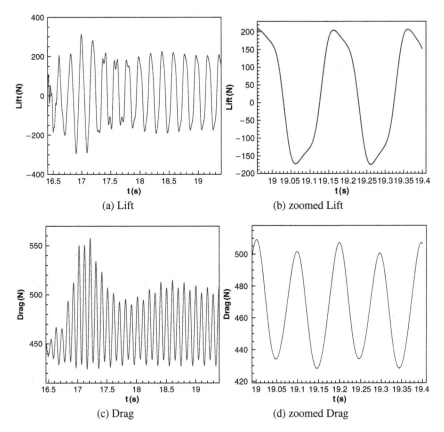

(a) Lift   (b) zoomed Lift

(c) Drag   (d) zoomed Drag

Fig. 17 Time histories of lift and drag obtained for the FSI3 benchmark test case on the medium grid.

7.2 Turbulent Experimental Benchmark

First steps have been taken to compute a FSI case which was defined to serve as a benchmark test case supported by experimental investigations on FSI in laminar and turbulent flow regimes [16]. On the centerline of a channel a rigid cylindrical front body with diameter $D = 22$ mm is mounted which is joined to a flexible steel membrane of 0.04 mm thickness. A rigid rectangular rear mass is attached to the membrane and forms the trailing edge of the whole structure. The rear mass has a

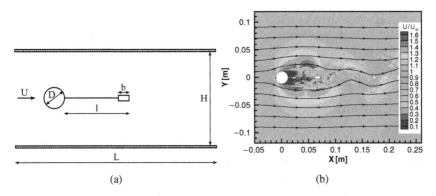

Fig. 18 (a) Setup of the computational domain; (b) Normalized velocity magnitude and streamlines of the fixed structure at $Re = 14,885$.

thickness of 4 mm and a length of $b = 10$ mm. In total, the length of the membrane and the rear mass is specified with $l = 60$ mm. The monovariant structure is free to rotate around an axle located in the center point of the cylindrical front body. The densities are specified with 2828 kg/m^3 for the aluminum front body, 7855 kg/m^3 for the stainless steel membrane and 7800 kg/m^3 for the stainless steel rear mass. In [16] the Young modulus of the membrane is given with 200 kN/mm^2. The channel, i.e. the fluid domain in the simulation, has a length of $L = 15.4D$ and a height of $H = 10.9D$. The center of the cylindrical front body is located at $2.5D$ downstream of the inflow section. The computational domain is depicted in Fig. 18(a).

Water with a density of $\rho = 997$ kg/m^3 and a dynamic viscosity of $\eta = 1.002 \cdot 10^{-3}$ kg/ms was chosen. With an inflow velocity of $U_\infty = 0.68$ m/s a Reynolds number of $Re = 14,885$ is obtained by taking the cylinder diameter D as reference length. The fluid domain was resolved by $6,503,760$ control volumes considering a resolution of 121 control volumes in spanwise direction. The wall–nearest control volume center is located at $y^+ = 3$ at the trailing edge of the structure, i.e. at the rear mass. The prediction of the turbulent flow field was done via LES applying the Smagorinsky model ($C_s = 0.1$) with Van-Driest damping. At the inflow section a velocity block profile with U_∞ was applied. An convective outflow boundary condition was used with a convective velocity of $U_c = U_\infty$. The channel walls as well as the structure have been set to no–slip boundary conditions. In a first step, the structure is assigned to be fixed. The resulting normalized velocity magnitude is depicted in Fig. 18(b). In a next step, the membrane and the rear mass are released and allowed to move freely. The computation of the structure response is done as in the FSI3 test case via the finite–element code Carat [24]. The coupling was again done via the coupling tool CoMA [11]. For a first check of the code coupling, a reduced grid resolution of $812,970$ control volumes was applied for the fluid domain to compute the whole FSI problem on an eight–core local machine. As consequence spurious pressure oscillations appeared in the region of the flexible membrane and thus no valid FSI data set was obtained yet. In a prospective step the grid size will be refined again and the whole structure will be released, i.e. also the cylindrical front body. As a result of this preliminary test, the whole setup was tested to be ready for high–performance computing now, which will become necessary for solving this problem properly.

8 Conclusions

The computation of fluid–structure interaction in turbulent flows requires advanced techniques such as LES. To resolve the turbulent flow properly in time and to account for the spectrum of spatial scales in the flow field, which also might have an influence on the response of the structure, for LES small time–step sizes are inevitable. Thus, explicit time–marching algorithms are favored. Within this basic framework a coupling scheme was developed which conserves the character of this explicit time–marching schemes on the one hand and serves for a strong coupling between fluid and structure on the other hand. This is achieved by directly connecting the computation of the structure subproblem with the corrector step of the fluid subproblem leading to a dynamic equilibrium between fluid and structure.

Furthermore, to step towards FSI applications replying on LES, the influence of deforming grids on the quality of LES predictions in the context of the ALE approach have been investigated. For this purpose, grid movements with specific amplitudes and periods have been forced to a plane channel flow at $Re_\tau = 590$. This results provide hints on what kind of grid motions are tolerable within an FSI application using LES.

Since the grid quality, which should be maintained during the movement or deformation of the structure, is found to be an important issue, a lot of effort was spent on grid smoothing techniques. Here the dilemma is twofold. On the one hand orthogonality and smoothness of the grid are known to be of special interest in the LES context and therefore it is worth to spend CPU–time for this issue. On the other hand adjustments of the grids might be necessary within each subiteration. Thus the procedure for each smoothing step is not tolerated to be highly CPU–time consuming. Besides methods relying on transfinite interpolations also elliptic grid smoothing techniques based on composite mapping were taken into consideration. The composite mapping consists of a nonlinear transfinite algebraic transformation and an elliptic transformation. Special emphasis had to be put on maintaining the grid quality near boundaries and block connectivities, everything under the constraint of low computational costs per iteration or time step.

Results of the new coupling scheme developed have been presented for various cases, e.g., the flow around cylindrical structures for laminar ($Re = 200$) and sub-critical ($Re = 13,000$) flows. Here, the response of the structure, i.e. the cylinders, was determined by solving equations of motion. For the laminar test case the results are found to be in good agreement with [26]. The sub-critical case was designed to compare the results obtained by the explicit scheme developed with the result delivered by a classical scheme based on implicit time-marching [13,14]. Here, satisfying results have been obtained for the displacement magnitudes and the frequencies of the cylinder oscillations. Aiming at the verification with experimental results, the response of a loosely mounted flat plate confined to turbulent flows at Reynolds numbers of $Re = 38,000$ and $68,000$ was computed. Here, different grid resolutions, extensions of the plate in spanwise direction and solution algorithms for solving the structure subproblem were investigated. The presented results deviate from the experimental findings. Consequently, further investigations considering the influence of the channel length, the influence of the gaps between plate and channel walls and the proper numerical realization of the experimental startup procedure have to be performed.

The coupling algorithm developed was extended towards code-to-code communication with external codes, since proper computation of the structure domain requires more convenient approaches, such as finite-element solvers. Here, coupling with the finite-element solver Carat [24] by the use of the coupling tool CoMA [11] was enabled leading to the computation of benchmark test cases [23]. For the numerical benchmark case of the laminar flow ($Re = 200$) around a cylinder with an attached flexible beam, satisfactory results have been achieved. Here, restrictions on the grid resolution and the time step have to be taken into account to apply explicit time–marching on low Reynolds number flows. Nevertheless, the proper functionality of the developed partitioned fluid-structure approach was shown. Thus, these developments were finally applied to compute an experimental benchmark test case in turbulent flow. Due to the complexity of the structure and a huge number of CVs in the fluid domain this extensive simulation is still in progress. In summary, significant steps have been done to marry LES with FSI. Reasonable results have been obtained for numerical test cases. The computational reproduction of experimentally obtained results is still challenging in connection with FSI. Here, further investigations are necessary to investigate the influence of parameters, which are usually simplified or not considered within a CFD approach up to now, i.e. the experimental startup processes or small gaps or structural tolerances.

Acknowledgements The project is financially supported by the *Deutsche Forschungsgemeinschaft* within the research group *'Fluid–Struktur–Wechselwirkung: Modellierung, Simulation, Optimierung' (FOR 493)* under contract number BR 1847/6. We gratefully acknowledge the cooperation with the Chair of Structural Analysis of the Technical University Munich providing the codes Carat [24] and CoMA [11] including intensive support, especially Dipl.–Ing. Th. Gallinger and Dr.–Ing. R. Wüchner.

References

1. Breuer, M.: Direkte Numerische Simulation und Large–Eddy Simulation turbulenter Strömungen auf Hochleistungsrechnern. Habilitationsschrift, Universität Erlangen–Nürnberg, Berichte aus der Strömungstechnik, ISBN 3–8265–9958–6, Shaker Verlag, Aachen, (2002).
2. Breuer, M.: A Challenging Test Case for Large–Eddy Simulation: High Reynolds Number Circular Cylinder Flow. Int. J. Heat Fluid Flow **21**, 648–654 (2000)
3. Breuer, M.: Large–Eddy Simulation of the Sub-Critical Flow Past a Circular Cylinder: Numerical and Modeling Aspects. Int. J. Numer. Meth. Fluids **28**, 1281–1302 (1998)

4. Breuer, M., Münsch, M.: LES Meets FSI – Important Numerical and Modeling Aspects, Proc. of the Seventh Int. ERCOFTAC Workshop on DNS and LES: DLES-7, Trieste, Italy, Sept. 8–10, 2008, ERCOFTAC Series, vol. 13, Direct and Large–Eddy Simulation VII, eds. Armenio, V., Geurts, B., Fröhlich, J. Springer, Netherland (2010)

5. Breuer, M., Münsch, M.: Fluid–Structure Interaction Using LES – A Partitioned Coupled Predictor–Corrector Scheme. Proc. in Applied Mathematics and Mechanics, PAMM, **8**, 10515–10516 (2008)

6. Breuer, M., Münsch, M.: FSI of the Turbulent Flow around a Swiveling Flat Plate Using Large-Eddy Simulation, In: Proc. of the Int. Workshop on Fluid-Structure Interaction (2008), eds. Hartmann, S., Meister, A., Schäfer, M., Turek, S., 31–42, Kassel University Press, ISBN 978-3-89958-666-4, Kassel (2009)

7. Demirdžić, I. and Perić, M.: Finite–Volume Method for Prediction of Fluid Flows in Arbitrarily Shaped Domains with Moving Boundaries. Int. J. Numer. Meth. Fluids **10**, 771–790 (1990)

8. Demirdžić, I. and Perić, M.: Space Conservation Law in Finite–Volume Calculations of Fluid Flow. Int. J. Numer. Meth. Fluids **8**, 1037-1050 (1988)

9. Durst, F., Schäfer, M.: A Parallel Block–Structured Multigrid Method for the Prediction of Incompressible Flows. Int. J. Numer. Meth. Fluids **22**, 549–565 (1996)

10. Durst, F., Schäfer, M., Wechsler, K.: Efficient Simulation of Incompressible Viscous Flows on Parallel Computers. In: Hirschel, E.H. (ed.) Flow Simulation with High–Performance Computers II, pp. 87–101. Vieweg, Braunschweig (1996)

11. Gallinger, T., Kupzok, A., Israel, U., Bletzinger, K.-U., Wüchner, R.: A Computational Environment for Membrane–Wind Interaction. Int. Workshop on Fluid–Structure Interaction: Theory, Numerics and Applications, Herrsching am Ammersee, Germany, Sept. 29 – Oct. 1, (2008)

12. Germano, M., Piomelli, U., Moin, P., Cabot, W.H.: A Dynamic Subgrid Scale Eddy Viscosity Model. Phys. Fluids A **3**, 1760–1765 (1991)

13. Glück, M., Breuer, M., Durst, F., Halfmann, A., Rank, E.: Computation of Wind–Induced Vibrations of Flexible Shells and Membranous Structures. J. Fluids Structures **17**, 739–765 (2003)

14. Glück, M., Breuer, M., Durst, F., Halfmann, A., Rank, E.: Computation of Fluid–Structure Interaction on Lightweight Structures. Int. J. of Wind Eng. Indus. Aerodyn. **89**, 1351–1368 (2001)

15. Patankar, S.V., Spalding, D.B.: A Calculation Procedure for Heat, Mass and Momentum Transfer in Three–Dimensional Parabolic Flows, Int. J. Heat Mass Transfer, **15**, 1787–1806 (1972)

16. Pereira Gomes, J., Lienhart, H.: Experimental Study on a Fluid–Structure Interaction Reference Test Case. In: Bungartz, H.-J., Schäfer, M. (eds.) Fluid–Structure Interaction, Lecture Notes Comput. Sci. & Eng., LNCSE **53**, pp. 356–370, Springer, Heidelberg (2006)

17. Pereira Gomes, J., Münsch, M., Breuer, M., Lienhart, H.: Flow–induced Oscillation of a Flat Plate — A Fluid–Structure Interaction Study Using Experiment and LES, 16. DGLR–Fach–Symposium der STAB, RWTH Aachen, Germany, Nov. 3–5, 2008, In: Notes on Numerical Fluid Mechanics and Multidisciplinary Design, in press, Springer, Berlin

18. Hilgenstock, A.: A Method for the Elliptic Generation of Three-Dimensional Grids with Full Boundary Control. Deutsche Forschungsanstalt für Luft– und Raumfahrt (DFLVR), DFVLR-IB **A 09**, 221–87 (1989)

19. Leonard, S., Terracol, M., Sagaut, P.: Commutation Error in LES with Time–Dependent Filter Width. Computers & Fluids **36**, 513–519 (2007)

20. Lilly, D.K.: A Proposed Modification of the Germano Subgrid Scale Closure Method. Phys. Fluids A **4**, 633–635 (1992)

21. Smagorinsky, J.: General Circulation Experiments with the Primitive Equations. I: The Basic Experiment. Month. Weath. Rev. **91**, 99–165 (1963)

22. Spekreijse, S.P.: Elliptic Grid Generation Based on Laplace Equations and Algebraic Transformations. J. of Comput. Physics **118**, 38–61 (1995)

23. Turek, S., Hron, J.: Proposal for Numerical Benchmarking of Fluid–Structure Interaction between an Elastic Object and Laminar Incompressible Flow. In: Bungartz, H.-J., Schäfer, M. (eds.) Fluid–Structure Interaction, Lecture Notes Comput. Sci. & Eng., LNCSE **53**, pp. 371–385, Springer, Heidelberg (2006)

24. Wüchner, R., Kupzok, A., Bletzinger, K.-U.: A Framework for Stabilized Partitioned Analysis of Thin Membrane-Wind Interaction. Int. J. Numer. Meth. Fluids **54**, 945–963 (2007)

25. Yigit, S., Schäfer, M., Heck, M.: Numerical Investigation of Structural Behavior During Fluid Excited Vibrations. REMN **16**, 491–519 (2007)

26. Zhou, C.Y., So, R.M.C., Lam, K.: Vortex-Induced Vibrations of an Elastic Circular Cylinder. J. Fluids & Structures **13**, 165–189 (1999)

Partitioned Simulation of Fluid-Structure Interaction on Cartesian Grids

H.-J. Bungartz, J. Benk, B. Gatzhammer, M. Mehl, and T. Neckel

Abstract This contribution describes recent developments and enhancements of the coupling tool preCICE and the flow solver Peano used for our partitioned simualtions of fluid-structure interaction scenarios. Peano brings together hardware efficiency and numerical efficiency exploiting advantages of tree-structured adaptive Cartesian computational grids that, in particular, allow for a very memory-efficient implementation of parallel adaptive multilevel solvers – an efficiency which is crucial facing the large computational requirements of multi-physics applications and the recent trend in computer architectures towards multi- and many-core systems. preCICE is the successor of our coupling tool FSI✳ce and offers a solver-independent implementation of coupling strategies and data mapping functionalities for general multi-physics problems. The underlying client-server-like concept maintains the full flexibilty of the partitioned approach with respect to exchangeability of solvers. The data mapping relies on fast spacepartitioning tree algorithms for the detection of geometric neighbourhood relations between components of non-matching grids.

1 Introduction

The importance of multi-physics simulations becomes evident looking at applications such as aerodynamic design optimisation, noise control, biomedical flow and structure simulations, and many others. Within the last years, computing power has reached a state that allows to compute more and more realistic and detailed models for such problems – provided that efficient simulation codes are available as stability and accuracy issues involved still are very challenging for standard approaches. Efficiency here refers to numerical methods as well as hardware-efficient

H.-J. Bungartz, J. Benk, B. Gatzhammer, M. Mehl, and T. Neckel
Department of Computer Science, Technische Universität München, Boltzmannstr. 3, 85748 Garching, Germany
e-mail: {bungartz,gatzhamm,mehl,neckel}@in.tum.de

H.-J. Bungartz et al. (eds.), *Fluid Structure Interaction II*, Lecture Notes in Computational Science and Engineering 73, DOI 10.1007/978-3-642-14206-2_10,

implementations and efficient software engineering. The latter is crucial to be able to flexibly and easily establish, extend, or maintain simulation environments according to new requirements such as improved models.

In terms of the software engineering aspect, partitioned approaches for the simulation of multi-physics scenarios offer a great potential: They reuse existing, established, tested, and trusted codes for single-physics problems and couple them to a multi-physics simulation environment via suitable software tools and numerical coupling strategies. In comparison to monolithic approaches tackling the whole system in one code as one large set of equations, the partitioned approach thus allows for a fast prototyping and for exchanging solvers according to new developments or modified scenarios. This advantage, of course, does not come for free but raises the new challenge of developing numerical coupling strategies that yield stable transient coupled simulations. The coupling of incompressible fluid flow to (elastic) structures is one example where this holds in particular. Another numerical aspect that has to be kept in mind – in particular with respect to physical conservation laws – is the mapping of data between the in general non-matching solver grids. But also the technical realisation of the code coupling, which may seem to be a minor issue, is non-trivial if the whole flexibility of the partitioned approach is to be maintained. This means, for example, that the independence and black-box property of the involved codes must not be destroyed by data mapping routines that directly connect the grid information of two solvers. Also the implementation of the coupling strategy and control in one (or even both) of the solvers would contradict the idea of a partitioned approach. Thus, developing a new, separate coupling component is favourable.

This paper focusses on the simulation of fluid-structure interactions as a special case of surface-coupled multi-physics applications. We developed the coupling tool **preCICE** (precise **C**ode **I**nteraction **C**oupling **E**nvironment) for such kinds of coupled simulations which aims at maintaining the full flexibility of the partitioned approach and provides a variety of coupling and data mapping functionalities as well as clearly defined interfaces for the implementation of additional variants. From the point of view of computing time and memory requirements, the simulation of the fluid side is the most costly part of fluid-structure interaction simulations. Our flow solver, implemented in the partial differential equation framework Peano, achieves very low memory requirements and a high data access efficiency – which is essential with memory access becoming a more and more severe bottleneck of any data-intensive simulation – on the basis of octree-like adaptively refined Cartesian grids. In addition to its memory efficiency, the Peano flow solver offers a state-of-the-art numerical efficiency: dynamical grid adaptivity, multigrid solvers, load balanced domain decomposition, and accurate force computations at the coupling surface. These features are not trivial to combine with hardware efficiency.

In Sect. 2, we describe the coupling environment preCICE. Section 3 presents the essential basics of the PDE framework Peano and details of the implementation of the Navier-Stokes solver. Numerical results achieved with these two codes are shown in Sect. 4.

2 The Coupling Environment preCICE

The development of the coupling environment **preCICE** (Precise Code Interaction Coupling Environment) has been guided by three main principles: preserving the inherent flexibility of the partitioned approach for multi-physics simulations, implementing hardware- and numerically efficient methods, and providing a reasonable user comfort. Of course, preCICE is not the first tool capable of coupling different simulation codes. MpCCI [2] is probably the best known tool offering an MPI-like programming interface and a rich data mapping and communication functionality, but no implementation of coupling numerics. In addition, particular scientific solutions for specific fields of applications exist such as The Model Coupling Toolkit MCT [36], which is widespread in climate simulation, or Uintha [47], a component-oriented realisation of a coupling unit. Although Uintha is restricted to solvers based on Cartesian grids, the successful integration of a fluid-structure interaction model [47] shows the general potential of component-based approaches. These approaches establish the coupling tool as a separate software component with functionalities going far beyond pure data mapping and communication, including the coupling control and, for some coupling strategies, also the solution of interface equations. The Component Template Library [45] is a computer science oriented approach to realise distributed component-based simulations and provides data communication for a wide variety of computer architectures and programming languages.

In contrast to these existing tools, preCICE aims at bringing together the component-based approach with an efficient data mapping technology, a general applicability for a wide range of solvers and solver grids, and easy plug-in mechanisms and exchangeability of solvers.

2.1 FSI✻ce – the First-Generation Coupling Tool

preCICE is an improved and generalised version of FSI✻ce (Fluid-Structure Interaction Coupling Environment). Developed and validated in [12] and [14], the main goal of FSI✻ce is to separate coupling functionalities from the solvers to be coupled and to pertain the flexibility of exchanging coupled solvers readily and independently from each other. The implementation of the data mapping is left to the solver codes, in order to allow for full customization. However, support is provided by spacetree data structures accelerating neighbour search and projection tasks occurring in data mapping. Coupling with FSI✻ce is realised in a client-server approach as depicted in Fig. 1. FSI✻ce provides a coupling supervisor, an extra program instance acting as client in the coupled simulation. The supervisor steers the coupled simulation by sending simulation requests to the solvers which are seen as servers and have to deliver a corresponding result back to the supervisor. The coupling supervisor holds an explicit representation of the coupling interface in form of a triangulated mesh. Solvers map their data to and from this coupling mesh only. Together with the client-server approach, this allows to separate the internals of coupled solvers from each other, such that an independent exchange of a solver can be performed.

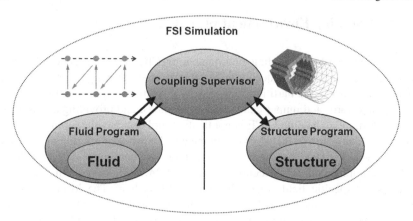

Fig. 1 Client-server approach realised in the coupling tool FSI✳ce. The solvers are solely interacting with the coupling supervisor. The coupling partner is, hence, logically decoupled from the
solver and can thus be exchanged independently.

2.2 Coupling Approach of preCICE

preCICE provides its functionality in form of a modular programming library with
defined interfaces for all main coupling tasks (cf. Sects. 2.3–2.6). Multi-physics scenarios to be simulated are specified by xml files. This makes preCICE applicable to a
wide range of surface-coupled multi-physics – not only fluid-structure interaction –
simulation scenarios. As FSI✳ce, preCICE uses an explicit representation of the
coupling surface by a triangulated mesh. In addition, preCICE provides a polygon
variant for 2D scenarios. In contrast to FSI✳ce, the data mapping is implemented
as a black-box functionality in preCICE. Except for data mapping, the geometric
information of the coupling surface can also be used as a geometry interface for
grid generation and modification of flow and structural solvers as it is done for the
Peano CFD solver (cf. Sect. 3).

An important difference between FSI✳ce and preCICE is related to the physical
deployment of the coupling components. In contrast to the client-server approach
of FSI✳ce, preCICE realises a peer-to-peer communication layout. The coupling
supervisor dedicated to the whole coupled simulation is replaced by local realisation of a coupling supervisor for every solver involved. (see Fig. 2 (a)). This local
control instance has all necessary information to steer the associated solver and to
exchange data with other solvers. The peer-to-peer communication layout avoids
duplicated communication of coupling data as in FSI✳ce and allows for a unified
implementation of the coupling control logic. In contrast to MpCCI, which also
works with a peer-to-peer communication concept, preCICE still includes both the
convergence control and the implementation of the coupling strategy and, thus, acts
as an independent numerical compenent. Despite of the peer-to-peer communication

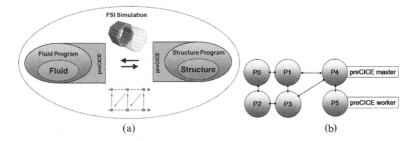

Fig. 2 (a) Peer-to-peer concept of preCICE with local realizations of coupling supervisors and direct communication between coupled solvers. (b) Parallel coupling concept of preCICE. P0–P3 are solver processes, while P4 and P5 are processes running preCICE. P4 acts as master, similar to the coupling supervisor of FSI✳ce, while P5 is a worker taking up jobs from the master.

layout, the actual deployment of resources to computing processes is flexible. When running a coupled simulation with parallelised solvers, e.g., the coupling supervisor can be realised in separate processes as depicted in Fig. 2 (b). Via the proxy software design pattern, this behaviour is hidden from a solver using preCICE.

2.3 Data Mapping

preCICE supports the exchange of data on the coupling surface of the involved solvers. This can be a simple copying of data in case of matching grids or projections and interpolations if grids are non-matching. An overview and comparison of data mapping methods can be found in [22]. There, the most important properties of a data mapping are defined to be (i) global conservation of energy over the interface, (ii) global conservation of loads over the interface, (iii) accuracy, (iv) conservation of the order of the coupled solvers, and (v) computational efficiency. For fluid-structure interaction, displacements u and forces f have to be mapped between the discrete coupling surface representation of the fluid Γ_F and of the structure Γ_S. The mappings can be written as

$$u_F = \mathbf{H_u} u_S, \quad f_S = \mathbf{H_f} f_F, \tag{1}$$

where the matrix $\mathbf{H_u}$ denotes the mapping from the vector u_S of discrete displacement values at Γ_S to the vector u_F of discrete displacement values at Γ_F, and $\mathbf{H_f}$ determines the mapping of discrete forces f_F on the fluid surface to forces f_S on the structure side. Global conservation of energy $W_F = f_F^T u_F = f_S^T u_S = W_S$, of total forces, and of rigid body motion is achieved if

$$\mathbf{H_f} = \mathbf{H_u}^T, \tag{2}$$

$$\sum_i (\mathbf{H_f})_{ij} = 1 \quad \forall j, \text{ and} \tag{3}$$

$$\sum_j (\mathbf{H_u})_{ij} = 1 \quad \forall i. \tag{4}$$

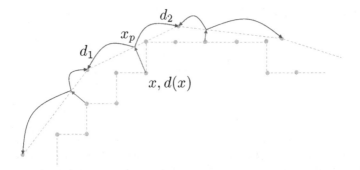

Fig. 3 Mapping of data from a Cartesian grid to a preCICE coupling mesh in 2D, with step 1, finding the closest projection point, and step 2, interpolating the data of the mapping point.

As preCICE uses its own, central explicit representation of the coupling surface, it only has to provide data mapping from this coupling mesh to arbitrary solver grids and vice versa. These mappings consist of two steps: first, given a grid point x from the solver grid, neighbouring elements of the coupling mesh have to be determined. Second, the data at vertices of the determined elements on the coupling mesh are mapped to data $d(x)$ at point x of the solver grid (or vice versa, respectively).

Currently, a mapping with linear accuracy is implemented (see Fig. 3): for each grid point x from the solver grid, a closest projection point x_p on the preCICE coupling mesh has to be found. The barycentric coordinates α, β, and γ of x_p are used to linearly interpolate the value $d(x)$ at x from the vertex values d_1, d_2, and d_3 of the coupling mesh triangle including the projection point x_p:

$$d(x) = \alpha d_1 + \beta d_2 + \gamma d_3, \quad \alpha + \beta + \gamma = 1. \tag{5}$$

α, β, and γ denote the barycentric coordinates of x_p in the respective triangle.[1] For the mapping from the solver grid to the coupling mesh, we use the transpose of this mapping such that condition (2) is fulfilled. In addition, also the fulfillment of (3) and (4) can be shown.

2.4 Spacetree Acceleration for Geometry Queries

The data mapping described in the previous section involves determining neighbouring elements of the preCICE coupling mesh for solver grid points[2]. preCICE uses spacetrees to speed up such geometrical queries. The basic idea of a spacetree

---

[1] In case x_p is a vertex $\gamma = \beta = 0$ and in case x_p lies on an edge $\gamma = 0$, e.g.

[2] A similar situation appears for other geometry queries used to set up the Cartesian solver grid of our flow solver in Peano.

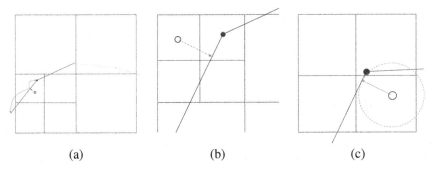

(a)   (b)   (c)

Fig. 4 (a) Part of a quadtree refined during a query for the closest projection point on the preCICE coupling mesh. The query point is drawn as a circle, and the projection and interpolation steps of the mapping are indicated by arrows. (b) Invalid situation where the spacetree cell with the mapping point is not intersected by the coupling mesh due to a too deep spacetree refinement. (c) Special case where the closest projection point found in the spacetree cell is actually not the closest on the coupling mesh. The circle indicates the distance to the projection point.

is to recursively divide the space under consideration into smaller elements and to use information inheritance to restrict expensive operations to local areas of interest. The spacetree variants employed in preCICE are quad- and octrees as well as binary kd-trees with different subdivision rules. We limit the discussion here to the case of quadtrees in 2D.

In the beginning of a simulation, the quadtree consists of the root cell only, covering the whole computational domain. Depending on the location of the incoming geometrical (projection) queries (cf. Fig. 4(a)), arbitrary, dynamical, and recursive refinements are performed. A leaf cell of the quadtree is then split into four equally sized children and the elements of the coupling mesh contained in the father cell are distributed to the newly created children cells according to their location. An element being partly contained in several quadtree cells is owned by all of these cells. For the computation of a projection point, the quadtree is traversed in a top-down manner. The tree structure allows to exclude cells from the search already at a very high level such that small cells have to be examined only in a small area. This leads to a reduction of the computational complexity from $O(NM)$ in the case of N solver grid points and M coupling mesh elements to $O(N \log M)$.

The underlying locality principle works only if the solver grid points and the coupling mesh are close to each other when compared to the cell size of the quadtree. Fig. 4(b) shows an example, where the coupling mesh elements and the solver grid point are not in the same quadtree cell, which is an invalid situation and has to be avoided.

A special case appearing when searching for closest projection points is illustrated in Fig. 4(c): If the distance to the projection point is larger than the one to the sides of the spacetree cell, there might be a closer projection point in a neighbouring cell. This situation is circumvented by comparing the closest distance found to the distance to the sides of the cell, and to redo the projection point search within all cells of the parent quadtree-node, if required. This can lead to an uprising cascade within the hierarchical structure of the quadtree. However, according to our experiences (cf. Sect. 4.3), such cases occur only in $O(1)$ cases and are negligible for the overall complexity.

2.5 Coupling Schemes

The partitioned coupling approach separates the system of equations describing the whole multi-physics scenario into subsystems describing only single fields. In fluid-structure interaction, we denote the overall equation system by $M(f, u) = 0$ and divide it into the fluid subsystem $F(u) = f$ and structure subsytem $S(f) = u$, where u are the structural displacements and f are the fluid forces. The goal of coupling schemes is to find a solution of the original system $M(u, f) = 0$ by solving the subsystems $F(u) = f$ and $S(f) = u$ only.

Explicit coupling schemes solve the systems $F(u) = f$ and $S(f) = u$ once per discrete simulation time step, either synchronised or staggered[3]. The updated coupling surface values u and f are exchanged and used for the next time step. While these schemes show very good computational efficiency compared to solving the full system $M(u, f) = 0$, they become instable for many FSI scenarios, in particular for incompressible fluids. This is due to the added-mass-effect described in [46] and [24], which is predominant in incompressible flow FSI. Implicit coupling schemes set up a fixed-point iteration for each timestep. The iteration can be written as

$$u_{n+1}^{i+1} = S_n(F_n(u_{n+1}^i)),   \tag{6}$$

where n denotes the timestep and i the iteration count. The solution of $M(u, f) = 0$ is retrieved when convergence is achieved. However, these schemes are instable for the same problems as the explicit schemes. The remedy is to apply an underrelaxation $0 < \omega \leq 1$ in modifying the structural displacements:

$$\tilde{u}^{i+1} = (1 - \omega)u^i + \omega u^{i+1}.   \tag{7}$$

Since convergence can be achieved only with very small relaxation factors ω resulting in a slow convergence, strategies for accelerating the convergence speed have to be sought after. One such technique is the Aitken acceleration for vector sequences, which is described in [35] and applied to FSI in [21]. The Aitken technique is an adaptive underrelaxation with variable ω. [57] applies Newton's method to the coupling surface values and uses reduced order models for fluid and structure solver to compute the required Jacobians in a black box manner.

These coupling schemes can be enhanced by subcycling in either of the involved solvers. This allows non-coupled intermediate timesteps and, thus, decouples the

---

[3] Note that this may involve also implicit time integration within a field (intrafield).

time scales of individual fields. In fluid-structure interaction, the fluid solver usually has much more restrictive stability requirements on the timestep length than the structure solver. Subcycling is, hence, typically applied to the fluid solver.

Since preCICE provides its functionality in form of a black box library, only coupling schemes fulfilling this requirement can be implemented. This has the benefit that all implemented coupling schemes can be used for any solver combination. preCICE currently implements staggered explicit and implicit coupling schemes with constant or Aitken based relaxation. Implicit coupling iterations can be performed with a constant number of iterations per timestep, with an absolute convergence limit, or with a relative convergence limit. A solver is allowed to do arbitrary subcycling, but has to obey the synchronised global timestep length given by preCICE. All the implemented coupling schemes are decoupled from the solvers and can be configured by providing xml-files to specific multi-physics scenarios such as FSI. The real potential of preCICE is in the implementation of more sophisticated coupling schemes involving more complicated numerics such as solvers for interface equations. The realisation of such schemes will be the focus of future work in preCICE.

2.6 Communication

A rather technical task in the coupling of partitioned multi-physics applications is the exchange of coupling data (forces and displacements for FSI) and control information between the coupled solvers. Typical communication means are file-based, use the Message Passing Interface (MPI) [42], or socket communication libraries. preCICE currently offers communication based on MPI. To meet the requirements of different types of solver codes as good as possible, solvers can be coupled in three different ways, which can be configured by providing xml-files.

The first way of coupling two solvers with preCICE is rather untypical for partitioned multi-physics simulations. It requires to compile the solver codes into one executable and to start the executable with several processes, at least as many as coupled solvers. The startup code has to separate the processes into the respective solver codes, the rest of the communication between the solver processes is setup by preCICE automatically.

The second approach works with separate solver codes which are run with one MPI call. MPI then puts the processes of all solvers into the global communication space and preCICE automatically separates the processes into the right groups and setting up the connections between them. This approach does not work for parallel solvers using the global communication space. Such solvers need to be modified to set up a separate communication space for its processes in that case, which might be tedious in some cases and can be impossible for commercial codes.

A third approach uses MPI 2.0 features and works similar to socket-based communciation. The solvers can be started individually, and a connection is achieved automatically by preCICE through MPI ports. This approach works fine with parallelized solvers, but requires MPI 2.0 features which are not commonly installed on supercomputers.

2.7 Application Programming Interface

The functionality described in Sects. 2.3 - 2.6 is provided to a solver code by the application programming interface (API) of preCICE. The main interface is written in C++. A C wrapper is also available to make preCICE work with solvers written in C code. A simple application example for the preCICE API can be as follows:

```
precice::CouplingInterface cplInterface ( solverName );
cplInterface.configure ( xmlConfigurationFileName );
double limitDt = cplInterface.initialize ();
while ( cplInterface.isCoupledSimulationOngoing() ) {
  limitDt = cplInterface.advance ( computedDt );
}
cplInterface.finalize ();
```

An object of class `CouplingInterface` has to be created and used to access any coupling functionality. The coupled simulation is configured from an xml file, where the coupling data, coupling mesh, spacetrees, involved solvers, and coupling schemes are specified. The call of `initialize()` creates all required data structures and initiates the contact to the other coupled solver(s). If a solver is not starting the coupling simulation, coupling data is received at that point already. A limit for the length of the first timestep is returned back. This limit is a maximum limit, i.e. a solver is always allowed to do a smaller timestep, but will be eventually synchronized to a global timestep length by the limits set by preCICE. The while loop represents the main timestepping loop of the solver. Here, the control of the overall length of the coupled simulation is given to preCICE by using the `isCoupledSimulationOngoing()` method. The computed data is exchanged when calling `advance()`. The solver makes the computed timestep length available to preCICE, which is necessary to steer the simulation in time. Finally, the communication and data structures are terminated in the call of `finalize()`.

In order to apply implicit coupling schemes, additional routines need to be used, since a caching of the solver state becomes necessary within the coupling iterations. The API functions to be used have the following syntax:

```
cplInterface.requireAction ( actionName );
cplInterface.fulfilledAction ( actionName );
```

The first function makes queries of preCICE known to a solver. The second function tells preCICE that the required action has been fulfilled by a solver and prevents the invalid use of coupling functionalities. For implicit coupling, two actions are required by a solver. The first action is to store the state of a solver, the second to load the state of the solver.

The examples above do not show how a solver can acccess the coupling data. This can be done by using the following methods

```
int forcesID = cplInterface.getDataID (nameForces);
cplInterface.readData (forcesID, pointCoords, force);
cplInterface.writeData (forcesID, pointCoords, force);
```

In order to access data, an identifier has to be retrieved from preCICE. Then, a simple call to `readData()` or `writeData()` transfers coupling data from or to the coupling mesh of preCICE. The actually applied mapping is hidden from the solver code.

As can be seen from these examples, the preCICE API can be integrated in solver codes with minimal effort and, in particular, requires neither the (re-)implementation of data mapping nor of the coupling strategy[4].

3 The Cartesian Flow Solver in Peano

Our flow solver has been implemented in the PDE framework Peano. The idea of Peano is to exploit the potential of octree-like adaptively refined Cartesian grids for hardware and numerically efficient implementations of PDE solvers. Cartesian grids have undergone a revival within the last years due to their simplicity and structuredness [26, 40, 51, 54, 58]. In some sense, Peano is the successor of our flow solver F3F (see for example [14] for a description). In contrast to F3F, Peano allows for adaptively refined grids as well as dynamical grid refinement, for the generic implementation of solvers for further PDEs and varying discretisations, and for multilevel algorithms. We describe the underlying algorithmical concept of Peano in a very short manner. For more detailed descriptions, refer to [13, 18, 19, 38, 44, 59].

3.1 Octree-Like Cartesian Grids

Octree-like grids are constructed in a (locally) recursive way: in each (local) refinement step, each grid cell can be subdivided into a fixed and constant number of child cells. This way, they correspond to a tree of grid cells, the so-called spacetree. Octrees represent three-dimensional grids that divide each grid cell into eight equal child cells. Peano uses slightly different grids that emerge from a partitioning into three parts per coordinate direction and refinement step. We call the corresponding trees ($k = 3$)-spacetrees. Fig. 5 shows examples of a two- and a three-dimensional Peano grid together with the associated trees.

In Peano, data are assigned to grid vertices and grid cells. It is up to the application to decide which data such as degrees of freedom and grid refinement

---

[4] This holds as long as standard methods are sufficient. Otherwise, the preCICE interfaces can be used to implement user-defined methods.

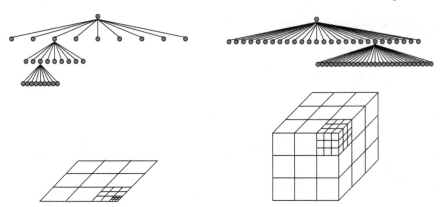

Fig. 5 Examples for ($k = 3$)-spactree grids together with the corresponding spacetrees. Left: two-dimensional grid with maximal refinement depth three; right: three-dimensional grid with maximal refinement depth two.

information are stored at these points. Hanging nodes, that is vertices at boundaries between different refinement depths that do not have the regular number of neighbouring cells, are not allowed to hold degrees of freedom.

The discretisation and evaluation of all operators involved in a simulation is realised in a strictly cell-oriented way. That is, the algorithm processes the grid cell by cell and computes the part of the respective operator that can be calculated solely with vertex and midpoint data of the current cell. For adaptively refined grids, this offers the great advantage that these operator cell parts are evaluated in an equal manner for all cells no matter whether they are located at a boundary between different refinement depths or not. For finite element discretisations, the cell part of the operator is given by the integral of the discrete weak form over the current cell [10]. During a grid traversal, the cell-parts of the operators are accumulated to the complete operator values. Hanging nodes get values of the respective unknown function by interpolation before the cell-wise operator evaluation starts and deliver restricted values of their cell-part operator values to father nodes after the operator evaluation is completed.

3.2   Data and Parallelisation Algorithms

As a basis for the data access algorithms in Peano, we use the space-filling Peano curve [50]. Space-filling curves have been established as a simple and efficient tool for traversal and domain decomposition of various types of adaptively refined grids (see for example [11, 18, 26, 27, 43, 48, 49, 53, 59, 60] and citations therein). In fact, not the space-filling curve itself but discrete iterates of the curve define a (processing) order of the grid cells. Figure 6 displays this for a two-dimensional ($k = 3$)-spacetree grid traversed along the Peano curve.

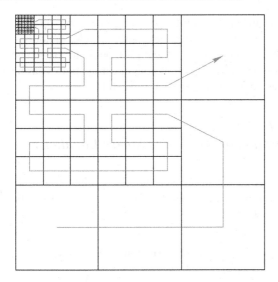

Fig. 6 Two-dimensional adaptive ($k = 3$)-spacetree grid with the corresponding iterate of the Peano curve.

The locality properties of the Peano curve ensure a quasi-optimal time locality of data access. This leads to improved cache-efficiency in various codes described in literature (see for example [4, 61]). The PDE framework Peano uses the Peano curve not only as a processing order for the grid cells but also as a tool supporting the construction of highly efficient data structures and data access algorithms. For that purpose, further properties of the curve – that could not be shown to hold for any Hilbert curve in 3D, e.g. – are decisive: The projection property (projections to lower-dimensional manifolds are lower-dimensional Peano curves) and the palindrome property (projections to lower-dimensional manifolds from different sides in space result in the same Peano curve but with changing direction). These two properties allow to base the whole data management on a few streams and stacks. Such data-structures inherently enforce a spatially highly local data access, which further improves the cache-hitrates of the code.

For parallelisation, Peano uses a parallel and balanced tree-based domain partitioning [18, 59] – again exploiting the locality properties of the Peano curve to achieve both load-balancing and quasi-minimal communication costs (compare also [11, 53, 61]).

As the second focus in the implementation of the Peano framework was on numerical efficiency, it offers state-of-the art numerical methods: The hierarchical tree structure of the grid inherently includes multilevel data that are traversed by a recursive version of the Peano curve in a top-down-bottom-up order. On this basis, Peano provides geometric multigrid solvers in an efficient and natural way. In addition, this traversal and the stack concept allow for an easy implementation of dynamical grid adaptivity without affecting the data access efficiency of the code [39].

To fully exploit all benefits of the spacetree grid and the sophisticated data administration based on the Peano curve, the framework idea is not based on an iterator concept as used for example in [5, 6] and [37], but relies on an event concept. That is, the user can trigger calls of his functions at certain points during a grid traversal, for example when entering or leaving a grid cell [59].

3.3 Discretisation of the Navier-Stokes Equations

Various approaches exist and are successfully applied which allows to compute numerical solutions for incompressible flows represented by the incompressible Navier-Stokes equations

$$\frac{\partial \mathbf{u}}{\partial t} - \nu \Delta \mathbf{u} + (\mathbf{u} \cdot \nabla)\mathbf{u} + \nabla p = 0 \,, \tag{8}$$

$$\nabla \cdot \mathbf{u} = 0 \,. \tag{9}$$

In Peano, we are just starting to investigate space-time methods (see [41], e.g.) discretising both space and time simultaneously in a single mesh via four-dimensional adaptive Cartesian grids available in the framework. So far, Peano's flow solver uses the more common separated discretisation in space and time with a d-dimensional computational grid (regular or adaptive, with $d = 2$ or $d = 3$) for the domain and an additional 1D "mesh" for time integration where only the necessary active time levels are actually stored.

3.3.1 Space Discretisation

To discretise the Navier-Stokes equations in space, two different approaches exist in Peano's flow solver: a low-order finite element method (FEM) and the interpolated differential operators (IDO). The $Q_1 Q_0$ elements of the FEM approach with constant pressure and d-linear velocities (see Fig. 7(a)) allow for a straigtforward and efficient cell-wise evaluation of the operators and fit well with local and dynamical adaptivity. Various boundary conditions are supported. For 2D, a divergence-free variant of ansatz functions is available (for details see [7, 14, 44]). The concept of IDO represents a higher-order discretisation scheme for parabolic, hyperbolic, and elliptic PDEs originally proposed in [3] and further developed in [31–34], e.g. In spite of a moderate support of the discretisation stencil of IDO, a wide range of wave-numbers up to high values can numerically be resolved on a given grid resulting in advantageous stability properties. The basic idea of this scheme is to use the governing equations in point-wise and in integrated form. Hence, a coupled system of point-values, line-integrated data, and cell-integrated values is solved. From the technical point of view, the successful realisation of IDO in Peano represents the

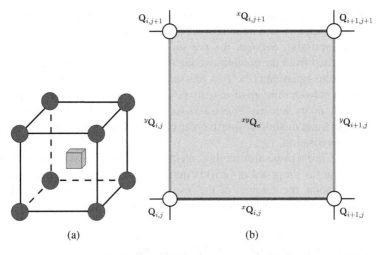

Fig. 7 Storage locations for the low order FEM approach in three dimensions (a): Trilinear veloc-ities (dark-blue spheres) and constant pressure (light-blue box) stored on vertices and the center of the cell, respectively. 2D IDO discretisation scheme (b) with vertex data, face- and cell-centered values to be stored.

proof of concept that larger stencils with additional degrees of freedom on faces (see Fig. 7(b)) may be evaluated efficiently via interlayed iterations over the grid.

Both discretisation types for the Navier-Stokes equations (8) and (9) in space yield the following semi-discrete equations:

$$A\frac{d\mathbf{u}_h}{dt} + D\mathbf{u}_h + C(\mathbf{u}_h)\mathbf{u}_h - M^T p_h = 0 , \qquad (10)$$

$$M\mathbf{u}_h = 0 . \qquad (11)$$

The operators $D, C(\mathbf{u}_h)$, M^T, and M correspond to diffusion, convection, gra-dient, and divergence. A represents the mass matrix. The concrete form of these operators depends, of course, on the type of discretisation (FEM or IDO, respec-tively).

3.3.2 Time Discretisation

The continuity and momentum equations in their semi-discrete form (10), (11) rep-resent a system of ordinary differential equations (ODE). Various ODE integration methods exist which differ in their order (accuracy), their computational complexity, and their storage demands (see [23, 28, 29, 52]). For efficiency reasons, spatial and temporal error contributions have to be balanced. Therefore, the order of the ODE integration method in use depends on the order of the spatial discretisation scheme and vice versa.

The Peano flow solver implements an explicit time discretisation following Chorin's projection method (see [20]) decoupling the solution of the velocity and pressure fields in two separate steps, similar to a predictor-corrector scheme. First, an intermediate velocity is computed via (10) that does not yet respect the mass-conservation constraint. Second, the pressure is computed from a Poisson-type equation generated from the continuity equation (11) applied to the right hand side of the momentum equation (10). This pressure is used to project the intermediate velocity onto a mass-conservative velocity field according to (10). For the computation of the preliminary velocity, Peano uses forward Euler for FEM and the classical explicit Runge-Kutta method of fourth order (in a low-storage variant, cf. [8, 9]) for IDO space discretisation.

Besides the Chorin projection method, an implicit Θ method with automatic time step size control (as proposed in [25]) is implemented depending on the scenario under consideration, the solution of the resulting non-linear system in each time step may pay off as it allows larger time steps avoiding stability restrictions.

3.4 Memory Consumption and Performance

Peano in general and its flow solver in particular possess low memory requirements due to the efficient storage format of the grid data in the spacetree. The vast majority of data is stemming from the degrees of freedom (DoF) such as velocity or pressure data of the simulation. The automatic generation of the corresponding data classes via the in-house tool DaStGen (see [16, 17]) allows for an additional packing of data: While C++ stores specific data types in technically larger blocks of memory than necessary (one byte for boolean values, e.g., depending on the computer architecture), DaStGen can compress several of such variables in one block and save memory.

Of course, different space or time integration schemes for the flow solver such as the ones described above require a different number of local data in vertices or cells. Peano's flow solver minimises the amount of data by choosing the concrete data in use at runtime. Table 1 shows a survey on the memory consumption for a single cell and vertex in two and three dimensions in case of the FEM space discretisation and different time discretisation methods. We emphasise the low amount of memory limited by 200 bytes compared to typical requirements of several kilobytes in comparable simulation packages using unstructured grids.

Table 1 Memory requirement for FEM discretisation with three different time integration schemes on adaptive grids in Peano: forward Euler (FE), classical Runge-Kutta (RK), and adaptive Θ method. The values indicated in bytes include, besides necessary attributes for grid management and flow simulation, all variables used in the program.

dimension	cell in bytes		vertex in bytes		
	default	Θ adap.	FE	RK	Θ adap.
2D	40	48	92	124	144
3D	44	52	124	172	200

Concerning the performance of Peano and its flow solver, different benchmark comparisons have been evaluated (see [44]). To summarise, a comparison of the adaptive grid implementation and a standard regular grid implementation with lexicographic grid traversal resulted in runtime overheads of 30–80% only (decreasing with increasing problem size) for two-dimensional simulations, dropping to less than 3% in 3D. Comparing such an adaptive setup with a tuned, specific-purpose 2D IDO code on merely regular grids showed an overhead of a factor of about 1.7–7.9 (again decreasing with increasing problem size) which is in the order of what can be expected. Thus, Peano's flow solver is able to efficiently simulate adaptive scenarios in two and three dimensions. Note that the grid adaptivity pattern — even steep adaptivity near geometric boundaries — does not influence the runtime per degree of freedom or grid cell.

3.5 Moving Geometry

Peano's spacetree representation of the Cartesian grid is combined with a marker-and-cell approach to define obstacles in the fluid. If such obstacles move in the domain, a clever handling of subtrees allows to create new and delete old cells as well as to switch markers (inner to outer or vice versa) by local computations only.

In addition to these grid updates, three additional numerical features have been implemented in the flow solver: First, a mass-conservative interpolation is done for velocities on newly created cells. Second, velocity data are restricted to parent nodes when deleting cells/vertices. Third, an additional projection step similar to the Chorin projection method may be carried out to generate a fully mass-conservative flow field after velocity changes at boundaries (see [13] for details).

Figure 8 shows two snapshots of the computational grid in a scenario with an obstacle moving in a channel[5].

3.6 Consistent Forces

Accurate forces at the surface between fluid and structure are crucial for the computation of FSI simulations. Therefore, we adapted the elegant and efficient method of consistent forces proposed by Gresho and Sani in [25] to our requirements. We will restrict all formulations to the case of a finite element space discretisation in the following. Note that there are analogue formulations for finite volume discretisations (see [14, 15]). The basic idea is to use the residual of the momentum equations on the structure surface Γ_S which has to be in equilibrium with the sustaining forces F_s exerted by the structure on the fluid. This results in one minor drawback and two major advantages: It is not possible to separate the pressure and viscous parts of the

---

[5] This scenario has been used as a test case to compare the case of flow around a fixed obstacle with its identical counterpart, a moving obstacle in a non-moving liquid (see [56]).

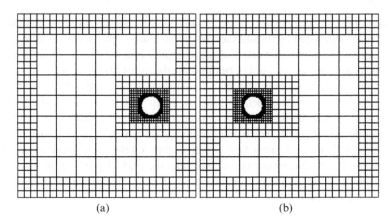

(a)   (b)

Fig. 8 2D adaptive Cartesian grid with moving, spherical obstacle at start (a) and end (b) position.

consistent forces. Typically, however, such a separation is not necessary for FSI simulations. Concerning the advantages, we get the desired force or stress values with
nearly no additional computation. All components in (15) have already been calculated by the flow solver for the boundary cells. Furthermore, all properties of the
discretisation concerning for example discretisation order or physical conservation
laws carry over to the force computation (see [25]).

To apply the method of consistent forces, the semi-discrete momentum equations (10) are enlarged to hold not only for the nodes $i = 1, \ldots, N$ inside the fluid
domain, but also for nodes $i = N + 1, \ldots, N_S$ on the boundary Γ_S. Virtual degrees
of freedom are introduced which possess ansatz functions $\tilde{f}_j$ and corresponding
coefficients $\tilde{F}_j$ on these boundary points[6]. The ansatz functions $\tilde{f}_j$ are of the same
type as used for the velocities (bilinear/ trilinear, e. g.) and represent the sustaining
forces F_s as follows:

$$F_s(x) = \sum_{j=N+1}^{N_S} \tilde{F}_j \tilde{f}_j(x) \quad \forall x \in \Gamma_S . \tag{12}$$

Using the equilibrium between sustaining forces and residual of the momentum
equations for a boundary node $i \in \{N + 1, \ldots, N_S\}$ in the weak form, we obtain

$$\int_{\Gamma_S} \tilde{f}_i(x) \cdot F_s(x) \, ds = \sum_{j=N+1}^{N_S} \tilde{F}_j \int_{\Gamma_S} \tilde{f}_i(x) \cdot \tilde{f}_j(x) \, ds,$$

$$= \left(A \frac{d\mathbf{u}_h}{dt} + D\mathbf{u}_h + C(\mathbf{u}_h)\mathbf{u}_h - M^T p_h \right)_i , \tag{13}$$

---

[6] For nodal basis functions, the coefficients $\tilde{F}_j$ denote the local forces on the boundary nodes j.

which shows the coupling of the force coefficients $\tilde{F}_j$ by a boundary mass matrix. To avoid the solution of the linear system of equations (13) in each time step, this mass matrix is lumped, resulting in

$$\int_{\Gamma_S} \tilde{f}_i(x) \cdot F_s(x) \, ds = \tilde{F}_i \int_{\Gamma_S} \tilde{f}_i(x) \, ds \, . \tag{14}$$

Hence, we obtain an explicit formula for the consistent local force coefficients

$$\tilde{F}_i = \frac{1}{\int_{\Gamma_S} \tilde{f}_i(x) \, ds} \left(A \frac{d\mathbf{u}_h}{dt} + D\mathbf{u}_h + C(\mathbf{u}_h)\mathbf{u}_h - M^T p_h \right)_i , i = N+1, \ldots, N_S.$$

$$\tag{15}$$

To determine the global force on a part of the boundary Γ_S, one has to integrate (12) using the coefficients $\tilde{F}_i$ calculated by (15). This integration of the ansatz functions can be realised analytically and is, thus, reduced to summing up weighted force coefficients $\tilde{F}_i$.

From the point of view of implementation, the consistent forces hardly require additional computations: The program only adds up contributions to the semi-discrete momentum Eq. (10) which have already been calculated before during the time-step.

4 Numerical Results/Coupled Simulations

4.1 Flow Simulations

Within the different benchmarks for fluid-structure interaction, a suite of three 2D scenarios are included which are intended to validate the flow solver separately (i.e. without any interaction). The setup consists of a channel with a fixed solid obstacle (a cylinder joint with a cantilever) near the inlet (see [30]). The three scenarios vary in the corresponding Reynolds number Re. The setups CFD1 and CFD2 at Re = 20 and Re = 100 exhibit a stationary behaviour whereas CFD3 at Re = 200 is fully time-dependent. As reference values, the total forces on the obstacle (and the frequency of the oscillations in the time-dependent case) are examined.

Peano uses the coupling mesh provided by the coupling library preCICE described in Sect. 2 as a basis of the generation of its adaptive Cartesian grid and also transmits the resulting force data to this mesh. To judge the influence of different grid adaptivity patterns and also the general potential of aptively refined grids to save degrees of freedom and, thereby, also computational costs, we used six different setups concerning the underlying, a priori refined grids with a maximum spacetree level of eight (see [44]). Fig. 9 and 10 show simulated pressure distributions for all

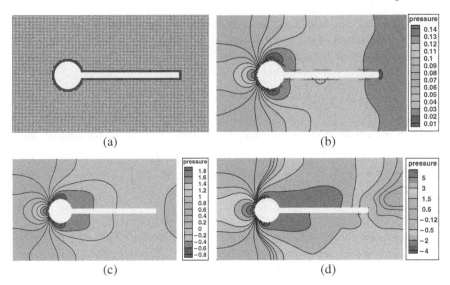

Fig. 9 Pressure disctributions in a part of the computational domain around the obstacle for the CFD1–CFD3 benchmarks. (a) Adaptive mesh in use with maximum and minimum refinement level eight and seven, respectively. (b) Steady-state solution of the CFD1 benchmark with Re = 20. (c) Steady-state solution of the CFD2 benchmark with Re = 100. (d) Solution of the CFD3 benchmark with Re = 200 at t = 8.12.

Fig. 10 Pressure distribution of the FSI benchmarks CFD3 (Re = 200, cf. [30]) at t = 8.12. The minimum and maximum tree level is 7 and 8.

three benchmark cases on adaptive grids with minimum and maximum refinement depth seven and eight and a purely geometry-driven refinement.

Table 2 contains the resulting drag and lift forces of the two steady-state scenarios CFD1 and CFD2. Up to the last line in this table, the grids used are refined only at the boundary of the geometry as shown in Fig. 9. The force coefficients converge towards the reference values. These results can be improved by using manually designed refinement boxes (see Fig. 11 and the last line of Table 2). A comparison of the third (regular grid with depth seven) and the last line of Table 2 shows the gain in accuracy of an adaptive grid compared to a regular grid with comparable number of cells. Dynamical grid adaptivity, which is currently being implemented in the Peano CFD solver will further reduce the number of cells required for a certain accuracy.

Table 2 Results for the steady-state benchmark scenarios CFD1 and CFD2 at Re $=$ 20 and Re $=$ 100 (cf. [30]). The force values in horizontal and vertical direction are denoted by f_x and f_y. The "box" grid setup corresponds to the setup in Fig. 11.

max. level	min. level	#cells	#vertices	#total DoF	#hang. nodes	Re = 20		Re = 100	
						f_x	f_y	f_x	f_y
6	6	4294	4096	12486	0	15.35	1.854	141.7	15.60
7	6	5096	4737	14570	340	15.58	1.243	149.3	11.44
7	7	38680	38083	114846	0	15.57	1.248	149.5	11.57
8	6	7636	6833	21302	1256	14.48	1.137	137.1	10.46
8	7	41220	40179	121578	916	14.48	1.146	137.3	10.81
8	6 (box)	39268	38344	115956	1744	14.46	1.143	137.2	10.74
ref. data	–	–	–	–		14.29	1.119	136.7	10.53

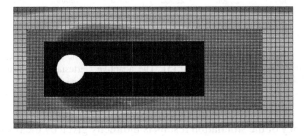

Fig. 11 Visualisation of the adaptive grid with levels 6 to 8 using manual refinement boxes for the FSI benchmark flow (cf. [30]). The colouring corresponds to the horizontal velocity values.

Table 3 Results of the time-dependent benchmark scenario CFD3 at Re $=$ 200 (cf. [30]) for different adaptive grids. The force values in horizontal and vertical direction, f_x and f_y, are indicated with mean value and oscillation amplitude, while the frequency is shown in the last column. The "box" grid setup corresponds to the grid shown in Fig. 11.

max. level	min. level	#cells	#total DoF	#hang. nodes	f_x	f_y	frequ.
6	6	4294	12486	0	–	–	–
7	6	5096	14570	340	–	–	–
7	7	38680	114846	0	494.8 ± 7.57	−12.97 ± 479.8	4.407
8	6	7636	21302	1256	433.0 ± 4.79	−23.33 ± 328.7	3.997
8	7	41220	121578	916	434.6 ± 5.14	−8.43 ± 381.7	4.332
8	6 (box)	39268	115956	1744	439.3 ± 6.25	−7.25 ± 447.6	4.421
ref. data	–	–	–		439.5 ± 5.62	−11.89 ± 437.8	4.396

For the time-dependent setup at Re $=$ 200 (CFD3), resulting forces are given in Table 3. The computations have been carried out using identical grid layouts as for the two steady-state setups CFD1 and CFD2. The first two runs with level combinations (6,6) and (7,6) are not resulting in a stable oscillation due to the very coarse mesh sizes. For finer resolutions, we observe good convergence towards the reference force data both in terms of mean values and amplitudes of the oscillations. The

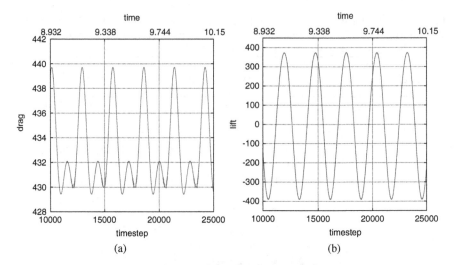

Fig. 12 Drag (a) and lift (b) force of the laminar flow around the FSI obstacle at Re = 200 (i.e. CFD3) in the time interval [8.932,10.15]. The adaptive Cartesian grid in use possesses a maximum and minimum level of refinement of eight and seven, respectively.

correct frequency of the simulation results is indirectly given by the resulting values of the Strouhal number. The time-dependent behaviour of the forces for refinement levels (8,7) is visualised in Fig. 12. Note that there are tiny oscillations in the drag force of the second minimum in each period. They show an amplitude of 0.0125 corresponding to 0.003% of the absolute drag values. Two different reasons might cause this behaviour (unknown from Peano's behaviour for the corresponding classical cylinder benchmark [55]): not accurate enough pressure solutions at those time intervals (in fact, the limiting number of 200 CG iterations is reached), or a too large time step size for the temporal discretisation error. These oscillations do not represent a major problem for the stability of the overall flow as they disappear quickly after the second minimum.

The fluid-structure interaction benchmark scenarios are merely two-dimensional. Peano and its flow solver support, of course, also 3D setups. A snapshot of streamlines of the well-known, laminar benchmark flow around a cylinder (see [55]) is shown in Fig. 13.

4.2 Structure Simulations

As for the fluid solver, [30] also describes a set of pure computational structural mechanics (CSM) benchmark scenarios to validate structural solvers. The scenarios consist of a 2D setup with the same geometry as placed into the fluid domain of the CFD tests, which is a fixed circle (cylinder) with a flexible attached cantilever. Only a gravitational force is applied to the structure, which results in a bending

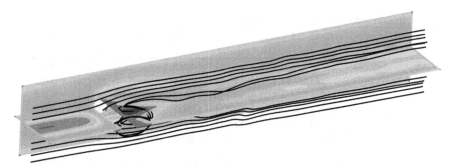

Fig. 13 Time-depending 3D flow field around a cylinder at Re = 100 (cf. [55]). Streamlines are plotted in black over the coloured horizontal velocity distribution. The adaptive grid possesses a refinement of level 6 in the total domain and of level 7 at the surface of the cylinder.

Table 4 Results and reference values of benchmark CSM1.

no elements	no dofs	$A_x[10^{-3}m]$	$A_y[10^{-3}m]$
183	840	−7.147	−65.89
723	3142	−7.177	−66.04
2928	12138	−7.184	−66.08
11712	47698	−7.186	−66.10
187392	752962	−7.187	−66.10
	ref. values	−7.187	−66.10

Table 5 Results and reference values of benchmark CSM2.

no elements	no dofs	$A_x[10^{-3}m]$	$A_y[10^{-3}m]$
69	348	−0.4659	−16.91
276	1246	−0.4682	−16.96
1104	4698	−0.4686	−16.97
4416	18226	−0.4688	−16.97
17664	71778	−0.4690	−16.97
	ref. values	−0.4690	−16.97

of the cantilever. The displacements A_x and A_y of point A, which is located in the center of the back edge of the cantilever, are provided as benchmark reference values. Scenarios CSM1 and CSM2 are evaluated stationary, while scenario CSM3 is without damping and results in a periodic oscillation.

We used the commercial simulation tool COMSOL Multiphysics [1] to compute the CSM scenarios. The results obtained for CSM1 and CSM2 are listed in Tables 4 and 5. They are computed with quadratic order finite elements and uniformly decreasing mesh size. CSM3 is computed with a backward-Euler method. The results are presented in form of a mean value, an oscillation amplitude, and the oscillation frequency, which are computed from the last oscillation period. Table 6 and Fig. 14 show the results of CSM3.

Table 6 Results and reference values of benchmark CSM3.

dt	no elements	no dofs	$A_{x,mean} \pm A_{x,ampl}$ $[10^{-3}m]$	$f_x[Hz]$	$A_{y,mean} \pm A_{y,ampl}$ $[10^{-3}m]$	$f_y[Hz]$
1e-3	183	840	-14.141 ± 14.139	1.0965	-64.359 ± 63.441	1.0965
1e-3	723	3142	-14.161 ± 14.159	1.0941	-64.374 ± 63.626	1.0929
1e-3	2928	12138	-14.180 ± 14.180	1.0953	-64.351 ± 63.649	1.0953
1e-3	11712	47698	-14.224 ± 14.224	1.0941	-64.388 ± 63.812	1.0941
1e-3	46848	189090	-14.155 ± 14.155	1.0952	-64.274 ± 63.626	1.0941
5e-4	46848	189090	-14.375 ± 14.375	1.0941	-64.503 ± 64.297	1.0941
		ref. values	-14.305 ± 14.305	1.0995	-63.607 ± 65.160	1.0995

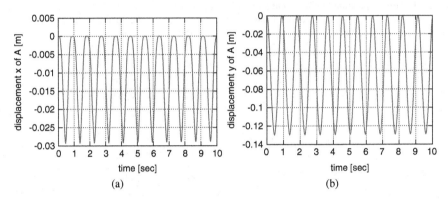

(a)   (b)

Fig. 14 Oscillating x- (a) and y-displacement (b) of point A, located at the back end of the cantilever, for benchmark scenario CSM3.

4.3 Data Mapping

In Sect. 2.4, we have described the spacetree algorithms used to answer geometrical queries occuring during data mapping between non-matching solver grids and the coupling mesh in preCICE. To analyse the performance of these algorithms, a simple 2D setup with a quadtree has been considered. We measured computational runtimes for finding nearest projection points onto a coupling mesh polygon. The coupling mesh approximates a circle with radius $r = 1.0$, the query points that are to be projected on the coupling mesh are located equispaced on a concentric circle with radius $r_{query} = r + d$, with $d > 0$. A quadtree with sidelengths $l = 2.0$ is covering both the circle geometry and the query points. To study the influence of different parameter settings, we varied the maximal cell size h_{quad} at the coupling surface, the distance of the query points to the coupling mesh d, and the discretisation width of the circle h_{geo}. We first studied the influence of different refinement levels h_{quad} on the runtime for different geometry discretisation widths h_{geo} and corresponding projection distances d. The choice of $h_{geo} = d$ mimics a convergence study where both, the solver grid and the coupling mesh, are refined repeatedly. The results in Fig. 15(a) illustrate the different effects occuring in spacetree accelerated projection

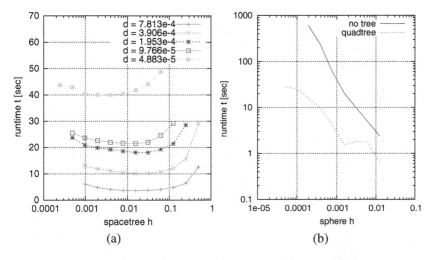

Fig. 15 (a) Runtimes t over different quadtree refinements h_{quad}. The different curves correspond to different values of d, which is set to equal h_{geo}. Optimal runtimes are achieved for $h_{quad} \approx 100\, h_{geo}$. (b) Comparison of the runtime of a quadtree with optimally chosen h_{quad} to a projection point computation without using a spacetree.

point search. The optimal choice of h_{quad} is limited from below by the special case described in Sect. 2.4, leading to repeated projection point searches. However, also if h_{quad} is chosen to large, the coupling mesh elements are not subpartitioned fine enough. For our test scenario $h_{quad} \approx 100 h_{geo}$ yields an optimal acceleration efficiency. Figure 15(b) compares the runtimes of a quadtree with optimal h_{quad} to that of using no spacetree.

4.4 Coupled Simulations

Corresponding to the three fluid benchmarks CFD1–CFD3, three benchmark cases for the full fluid-structure interaction scenario are proposed in [30]. The first scenario, FSI1 results in a stationary deflection of the cantilever, FSI2 and FSI3 in periodic oscillations. Accurate simulations for these scenarios are currently work in progress using the flow solver in Peano (cf. Sect. 3), the coupling tool preCICE (cf. Sect. 2), and the commercial structural solver in COMSOL [1]. Figure 16 shows snapshots of a simulation of the FSI2 benchmark at Re $= 100$.

In addition, various other scenarios such as particle transport in microscopic devices and deflection of a 'tower' have been simulated (see Fig. 17 and [13, 38] for some examples).

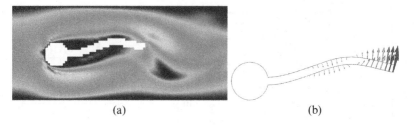

(a)   (b)

Fig. 16 (a) Snapshot of FSI2 benchmark Peano fluidsolver grid with velocity magnitudes. (b) Snapshot of FSI2 benchmark preCICE coupling mesh and channel geometry used to set up the Peano fluidsolver grid. Velocity vectors show the movement of the cantilever.

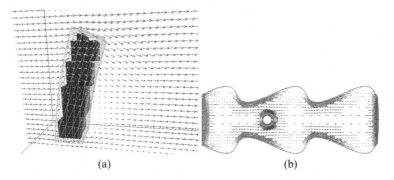

(a)   (b)

Fig. 17 (a) 3D coupled simulation with F3F and AdhoC, coupled by FSI❄ce. (b) Simulation of a drift ratchet with Peano and Comsol, coupled by preCICE.

5 Summary and Outlook

The results shown in this paper demonstrate the suitability of Cartesian grids for partitioned fluid-structure interaction simulations both in terms of achievable accuracy and in particular in terms of hardware performance of the respective codes. We proposed a Cartesian flow solver implemented in our PDE framework Peano with very low memory requirements and high cache hit-rates as a result of the structuredness of the chosen type of adaptive Cartesian grids. In spite of this hardware efficiency, the Peano CFD solver involves state-of-the-art numerical features such as highly local grid adaptivity, multigrid solvers, and accurate and efficient force computations via the method of consistent forces. Even large geometry or topology changes that frequently occur in fluid-structure interaction scenrios can be handled very naturally and cheaply based on a fixed grid (Eulerian) approach. The parallelisation of the solver is work in progress with some first promising results for simple examples. The same holds for dynamical, criterion-based grid adaptivity. The second tool described is the coupling tool preCICE that currently offers standard coupling functionality concerning data mapping, code-to-code communication, and coupling strategies. In contrast to other available coupling tools, preCICE combines mapping, communication, and coupling numerics in a central component. With this and the central coupling surface description, preCICE fully exploits the flexibility of the partitioned simulation approach that is allows for an easy and independent exchange of flow and structure solvers or coupling strategies.

In the near future, the basic functionality of both codes will be enhanced to further improve the efficiency in various aspects. The Cartesian geometry representation in Peano shall be improved by at least second order accurate methods such as cut-cell or immersed boundary methods. At the same time, completing the work on dynamical grid adaptivity and parallelisation with dynamical load balancing will allow for higher (and more focussed) grid resolutions. The solution of the involved systems of equations will be accelerated by improved matrix-free geometric multigrid solvers.

In the coupling tool preCICE, the most important step will be the implementation of further, non-standard coupling strategies such as reduced order methods, non-staggered coupling iterations, or multigrid methods. These methods strongly profit from or even are possible only by the component-based approach with the coupling tool as a separate numerical and not a mere data transfer component. Of course, also other, more specific mapping methods will be included as needed.

In the end, we will be able to provide both a flow solver fully exploiting the numerical and hardware-related advantages of Cartesian grids as well as a coupling tool offering maximum flexibility in terms of the choice of flow and structural solvers as well as in terms of (also non-standard highly sophisticated) coupling strategies.

Acknowledgements This work has partially been funded by the German Research Foundation (DFG) within the framework of the Research Group FOR493 – Fluid-Structure Interaction: Modelling, Simulation, and Optimisation. This funding is thankfully acknowledged.

References

1. COMSOL Multiphysics GmbH. COMSOL: Multiphysics Modeling and Simulation. http://www.comsol.de/.
2. R. Ahrem, P. Post, B. Steckel, and K. Wolf. MpCCI: A tool for coupling CFD with other disciplines. In *Proceedings of the 5th World Conference in Applied Fluid Dynamics, CFD - Efficiency and the Economic Benefit in Manufacturing*, 2001.
3. T. Aoki. Interpolated Differential Operator (IDO) scheme for solving partial differential equations. *Comput. Phys. Comm.*, 102:132–146, 1997.
4. M. Bader and Ch. Zenger. Cache oblivious matrix multiplication using an element ordering based on a Peano curve. *Linear Algebra and its Applications*, 417(2–3):301–313, 2006.
5. P. Bastian, M. Blatt, A. Dedner, C. Engwer, R. Klöfkorn, M. Ohlberger, and O. Sander. A Generic Grid Interface for Parallel and Adaptive Scientific Computing. Part I: Abstract Framework. *Computing*, 82(2–3):103–119, 2008.
6. P. Bastian, M. Blatt, A. Dedner, C. Engwer, R. Klöfkorn, M. Ohlberger, and O. Sander. A Generic Grid Interface for Parallel and Adaptive Scientific Computing. Part II: Implementation and Tests in DUNE. *Computing*, 82(2–3):121–138, 2008.
7. C. Blanke. Kontinuitätserhaltende Finite-Element-Diskretisierung der Navier-Stokes-Gleichungen. Diploma Thesis, Technische Universität München, 2004.

8. E. K. Blum. A modification of the Runge-Kutta fourth-order method. *Mathematics of Computation*, 1962.
9. M. E. Brachet, P. D. Mininni, D. L. Rosenberg, and A. Pouquet. High-order low-storage explicit Runge-Kutta schemes for equations with quadratic nonlinearities. Technical Report arXiv:0808.1883, CERN, Aug 2008.
10. D. Braess. *Finite Elements—Theory, Fast Solvers, and Applications in Solid Mechanics*. Cambridge University Press, 3rd edition, 2007.
11. V. Brázdová and D.R. Bowler. Automatic data distribution and load balancing with space-filling curves: implementation in conquest. *Journmal of Physics: Condensed Matter*, 20, 2008.
12. M. Brenk, H.-J. Bungartz, M. Mehl, R.-P. Mundani, A. Düster, and D. Scholz. Efficient interface treatment for fluid-structure interaction on cartesian grids. In *ECCOMAS COUPLED PROBLEMS 2005, Proc. of the Thematic Conf. on Computational Methods for Coupled Problems in Science and Engineering*. International Center for Numerical Methods in Engineering (CIMNE), 2005.
13. M. Brenk, H.-J. Bungartz, M. Mehl, I. L. Muntean, T. Neckel, and T. Weinzierl. Numerical simulation of particle transport in a drift ratchet. *SIAM Journal of Scientific Computing*, 30(6):2777–2798, 2008.
14. M. Brenk, H.-J. Bungartz, M. Mehl, and T. Neckel. Fluid-structure interaction on cartesian grids: Flow simulation and coupling environment. In H.-J. Bungartz and M. Schäfer, editors, *Fluid-Structure Interaction*, number 53 in Lecture Notes in Computational Science and Engineering, pages 233–269. Springer, 2006.
15. M. Brenk, H.-J. Bungartz, and T. Neckel. Cartesian discretisations for fluid-structure interaction – consistent forces. In P. Wesseling, E. Oñate, and J. Périaux, editors, *ECCOMAS CFD 2006, European Conference on Computational Fluid Dynamics*. TU Delft, 2006.
16. H.-J. Bungartz, W. Eckhardt, M. Mehl, and T. Weinzierl. Dastgen - A Data Structure Generator for Parallel C++ HPC Software. In M. Buback, G. D. van Albada, P. M. A. Sloot, and J. J. Dongarra, editors, *ICCS 2008 Proceedings*, Lecture Notes in Computer Science, Heidelberg, Berlin, 2008. Springer-Verlag.
17. H.-J. Bungartz, W. Eckhardt, T. Weinzierl, and C. Zenger. A precompiler to reduce the memory footprint of multiscale PDE solvers in C++. *Future Generation Computer Systems*, 2009. (accepted).
18. H.-J. Bungartz, M. Mehl, T. Neckel, and T. Weinzierl. The pde framework peano applied to computational fluid dynamics. *Computational Mechanics*, 2009. accepted.
19. H.-J. Bungartz, M. Mehl, and T. Weinzierl. A parallel adaptive Cartesian PDE solver using space–filling curves. In E. W. Nagel, V. W. Walter, and W. Lehner, editors, *Euro-Par 2006, Parallel Processing, 12th International Euro-Par Conference*, volume 4128 of *Lecture Notes in Computer Science*, pages 1064–1074, Berlin Heidelberg, 2006. Springer-Verlag.
20. A. J. Chorin. Numerical solution of the Navier-Stokes equations. *Math. Comp.*, 22:745–762, 1968.
21. W. A. Wall D. P. Mok and E. Ramm. Accelerated iterative substructuring schemes for instationary fluid-structure interaction. *Computational Fluid and Solid Mechanics*, pages 1325–1328, 2001.
22. A. de Boer, A. H. van Zuijlen, and H. Bijl. Review of coupling methods for non-matching meshes. *Comput. Methods Appl. Mech. Engrg.*, 196(8):1515–1525, 2007.
23. P. Deuflhard and F. Bornemann. *Numerische Mathematik II. Integration gewöhnlicher Differentialgleichungen*. de Gruyter, 1994.
24. Ch. Förster, W. A. Wall, and E. Ramm. Artifcial added mass instabilities in sequential staggered coupling of nonlinear structures and incompressible viscous flows. *Comput. Methods Appl. Mech. Engrg.*, 196(7):1278–1293, 2007.
25. P. M. Gresho and R. L. Sani. *Incompressible Flow and the Finite Element Method*. John Wiley & Sons, 1998.
26. M. Griebel and G. Zumbusch. Parallel multigrid in an adaptive PDE solver based on hashing and space-filling curves. *Parallel Computing*, 25(7):827–843, 1999.

27. F. Günther, A. Krahnke, M. Langlotz, M. Mehl, M. Pögl, and Ch. Zenger. On the parallelization of a cache-optimal iterative solver for pdes based on hierarchical data structures and space-filling curves. In *Recent Advances in Parallel Virtual Machine and Message Passing Interface: 11th European PVM/MPI Users Group Meeting Budapest, Hungary, September 19 - 22, 2004. Proceedings*, volume 3241 of *Lecture Notes in Computer Science*, Heidelberg, 2004. Springer-Verlag.

28. E. Hairer, S. P. Nørset, and G. Wanner. *Solving Ordinary Differential Equations I.Nonstiff Problems.*, volume 8 of *Springer Series in Comput. Mathematics*. Springer-Verlag, second edition, 1993.

29. E. Hairer, S. P. Nørset, and G. Wanner. *Solving Ordinary Differential Equations II.Stiff and Differential-Algebraic Problems.*, volume 14 of *Springer Series in Comput. Mathematics*. Springer-Verlag, second edition, 1996.

30. J. Hron and S. Turek. Proposal for numerical benchmarking of fluid-structure interaction between elastic object and laminar incompressible flow. In H.-J. Bungartz and M. Schäfer, editors, *Fluid-Structure Interaction*, number 53 in Lecture Notes in Computational Science and Engineering, pages 371–385. Springer-Verlag, 2006.

31. Y. Imai and T. Aoki. Accuracy study of the IDO scheme by Fourier analysis. *Journal of Computational Physics*, 217:453–472, 2006.

32. Y. Imai and T. Aoki. A higher-order implicit IDO scheme and its CFD application to local mesh refinement method. *Comput. Mech.*, 38:211–221, 2006.

33. Y. Imai and T. Aoki. Stable coupling between vector and scalar variables for the IDO scheme on collocated grids. *Journal of Computational Physics*, 215:81–97, 2006.

34. Y. Imai, T. Aoki, and K. Takizawa. Conservative form of interpolated differential operator scheme for compressible and incompressible fluid dynamics. *Journal of Computational Physics*, 227:2263–2285, 2008.

35. B. M. Irons and R. C. Tuck. A version of the aitken accelerator for computer iteration. *International Journal of Numerical Methods in Engineering*, 1:275–277, 1969.

36. J. Larson, R. Jacob, and E. Ong. The model coupling toolkit: A new fortran90 toolkit for building multiphysics parallel coupled models. *Int. J. High Perf. Comp. App.*, 19(3):277–292, 2005.

37. K. Long. Sundance: A rapid prototyping toolkit for parallel pde simulation and optimization. In M. Heinkenschloss L. T. Biegler, O. Ghattas and B. van Bloemen Waanders, editors, *Large-Scale PDE-Constrained Optimization, Lecture Notes in Computational Science and Engineering*, volume 30 of *Lecture Notes in Computational Science and Engineering*, pages 331–339. Springer-Verlag, 2009.

38. M. Mehl, M. Brenk, I.L. Muntean, T. Neckel, and T. Weinzierl. Benefits of structured cartesian gris for the simulation of fluid-structure interactions. In *Proceedings of the Third Asian-Pacific Congress on Computational Mechanics*, Kyoto, Japan, 2007.

39. M. Mehl, T. Weinzierl, and C. Zenger. A cache-oblivious self-adaptive full multigrid method. *Numerical Linear Algebra with Applications*, 13(2-3):275–291, 2006.

40. D. Meidner, R. Rannacher, and J. Vihharev. Goal-oriented error control of the iterative solution of finite element equations. *Journal of Numerical Mathematics*, 17:143–172, 2009.

41. D. Meidner and B. Vexler. Adaptive space-time finite element methods for parabolic optimization problems. *SIAM Journal on Control and Optimization*, 46(1):116–142, 2007.

42. Message Passing Interface Forum. MPI: A Message-Passing Interface Standard, Version 1.1. http://www.mpi-forum.org/docs/docs.html.

43. W. F. Mitchell. A refinement-tree based partitioning method for dynamic load balancing with adaptively refined grids. *Journal of Parallel and Distributed Computing*, 67(4):417–429, 2007.

44. T. Neckel. *The PDE Framework Peano: An Environment for Efficient Flow Simulations*. Verlag Dr. Hut, 2009.

45. R. Niekamp. Component template library (ctl). http://www.wire.tu-bs.de/forschung/projekte/ctl/e_ctl.html.

46. J. F. Gerbeau P. Caussin and f. Nobile. Added-mass effect in the design of partitioned algorithms for fluid-structure problems. *Computer Methods in Applied Mechanics and Engineering*, 194:4506–4527, 2005.

47. St. G. Parker and J. Guilkey und T. Harman. A component-based parallel infrastructure for the simulation of fluid-structure interaction. *Engineering with Computers*, 22(3–4):277–292, 2006.
48. A. Patra and J.T. Oden. Problem decomposition for adaptive hp finite element methods. *Computing Systems in Engineering*, 6(2):976–109, 1995.
49. S. Roberts, S. Klyanasundaram, M. Cardew-Hall, and W. Clarke. A key based parallel adaptive refinement technique for finite element methods. In *Proc. Computational Techniques and Applications: CTAC '97*, pages 577–584, Singapore, 1998.
50. H. Sagan. *Space-filling curves*. Springer-Verlag, New York, 1994.
51. R. S. Sampath, S. S. Adavani, H. Sundar, I. Lashuk, and G. Biros. Dendro: parallel algorithms for multigrid and amr methods on 2:1 balanced octrees. In *SC '08: Proceedings of the 2008 ACM/IEEE conference on Supercomputing*, pages 1–12, Piscataway, NJ, USA, 2008. IEEE Press.
52. J. Stoer and R. Bulirsch. *Numerische Mathematik 2*. Springer-Verlag, 4th edition, 2000.
53. H. Sundar, R.S. Sampath, and G. Biros. Bottom-up construction and 2:1 balance refinement of linear octrees in parallel. *SIAM J. Sci. Comput.*, 30(5):2675-2708, 2008.
54. T. Tu, D. R. O'Hallaron, and O. Ghattas. Scalable parallel octree meshing for terascale applications. In *SC '05: Proceedings of the 2005 ACM/IEEE conference on Supercomputing*, page 4, Washington, DC, USA, 2005. IEEE Computer Society.
55. S. Turek and M. Schäfer. Benchmark computations of laminar flow around a cylinder. In E. H. Hirschel, editor, *Flow Simulation with High-Performance Computers II*, number 52 in NNFM. Vieweg, 1996.
56. K. Unterweger. CFD simulations of moving geometries using Cartesian grids. Diploma Thesis, Institut für Informatik, Technische Universität München, 2009.
57. J. Vierendeels. Implicit coupling of partitioned fluid-structure interaction solvers using reduced-order models. In H.-J. Bungartz and M. Schäfer, editors, *Fluid-Structure Interaction*, number 53 in Lecture Notes in Computational Science and Engineering, pages 1–18. Springer-Verlag, 2006.
58. P. Gamnitzer W.A. Wall and A. Gerstenberger. Fluid-structure interaction approaches on fixed grids based on two different domain decomposition ideas. *International Journal of Computational Fluid Dynamics*, 22(6):411–427, 2008.
59. T. Weinzierl. *A Framework for Parallel PDE Solvers on Multiscale Adaptive Cartesian Grids*. Verlag Dr. Hut, 2009.
60. G. W. Zumbusch. On the quality of space-filling curve induced partitions. *Z. Angew. Math. Mech.*, 81:25–28, 2001.
61. G.W. Zumbusch. *Parallel Multilevel Methods*. Advances in Numerical Mathematics. Teubner Verlag, 2003.

An Explicit Model for Three-Dimensional Fluid-Structure Interaction using LBM and p-FEM

S. Geller, S. Kollmannsberger, M. El Bettah, M. Krafczyk, D. Scholz, A. Düster, and E. Rank

Abstract An explicit coupling model for the simulation of surface coupled fluid-structure interactions with large structural deflections is introduced. Specifically, the fluid modeled via the Lattice Boltzmann Method (LBM) is coupled to a high-order Finite Element discretization of the structure. The forces and velocities are discretely computed, exchanged and applied at the interface. The low compressibility of the Lattice Boltzmann Method allows for an explicit coupling algorithm. The proposed explicit coupling model turnes out to be accurate, very efficient and stable even for nearly incompressible flows. It was implemented in three software components: VIRTUALFLUIDS (fluid), ADHOC (structure) and FSI✸ce (a communication library). The validity of the approach is demonstrated in two dimensions by means of comparing numerical results to measurements of an experiment. This experiment involves a flag-like structure submerged in the laminar flow field of an incompressible fluid where the structure exhibits large, geometrically non-linear, self excited, periodic motions. The methodology is then extended to three dimensions. Its performance is first demonstrated via the computation of a falling sphere in a pipe. The close correspondence of the results obtained by application of the numerical scheme compared to a semi-analytic solution is demonstrated. The proposed explicit coupling model is then extended to a plate in a cross flow. We verify the results

S. Geller and M. Krafczyk
Institute for computational modeling in civil engineering, TU Braunschweig, Muehlenpfordtstraße 4-5, 38106 Braunschweig, Germany
e-mail: geller@irmb.tu-bs.de

S. Kollmannsberger, D. Scholz, and E. Rank
Computation in Engineering, TU München, Arcisstraße 21, 80333 München, Germany
e-mail: kollmannsberger@bv.tum.de

A. Düster
Numerische Strukturanalyse mit Anwendungen in der Schiffstechnik (M-10),
Technische Universität Hamburg-Harburg, Schwarzenbergstraße 95 C, D-21073 Hamburg
e-mail: alexander.duester@tu-harburg.de

M.E. Bettah
Department of Ocean Engineering, University of Rhode Island, Narragansett, RI 02882, USA
e-mail: melbettah@oce.uri.edu

H.-J. Bungartz et al. (eds.), *Fluid Structure Interaction II*, Lecture Notes
in Computational Science and Engineering 73, DOI 10.1007/978-3-642-14206-2_11,
© Springer-Verlag Berlin Heidelberg 2010

by comparing them to results obtained by application of the commercial ALE-Finite Volume—*h*-FEM Fluid-Structure interaction solver ANSYS MULTIPHYSICS. Additional examples demonstrate the applicability of the proposed methodology to problems of (arbitrarily) large deformations and of large scale.

1 Introduction

Examples for Fluid-Structure interaction (FSI) problems are found in various disciplines. In civil engineering bridges, sky scrapers, membrane structures, off shore platforms and many other constructions are exposed to loads that require FSI investigations. In mechanical engineering, valves, pumps, airfoils, ship propellers and many other devices have to be dimensioned considering FSI. The design of stirrers, extruders and injection systems in process engineering or artificial heart valves and blood vessels in medicine require the consideration of the bidirectional interaction between fluid and structure as well. The realistic transient simulation of such processes is rather challenging with respect to both, the numerical techniques and the required computer capacities. In addition to the effort resulting from the combination of structural and fluid mechanical computations, the interaction brings in a variety of complex numerical and mechanical additional phenomena which need special attention as well. This comprises questions of the adequate consideration of the dynamic and nonlinear interaction itself as well as the capabilities of partitioned solution approaches. Additionally, the suitability of methods which have established themselves in the individual disciplines may now suffer drawbacks when applied in the interaction. The robustness and efficiency of the methods, and the adequate description of complex changing geometries must be addressed as well.

In general, two approaches to solve fluid-structure interaction (FSI) problems exist: The monolithic approach [38,39,82] discretizes the two separate domains with a similar discretization scheme and solves the resulting, coupled system of equations within one solver. The compatibility conditions at the interface are treated inherently within this system of equations. By contrast, the partitioned approach [53] uses separate solvers for the fluid and the structural system. Strong coupling methods [48,83] as well as loose coupling methods [19, 20, 54, 61, 62, 62, 63, 63] exist. In the partitioned solution, the solvers need to communicate physical properties of their mutual boundary to fulfill the interface conditions. Each domain may utilize any type of discretization considered efficient for its field. For the fluid, mainly two approaches are used in the context of partitioned FSI for large deformations. The fluid is usually described either on an arbitrarily moving grid (arbitrary Lagrangian Eulerian (ALE) formulation) or on a fixed Cartesian grid. At first sight the primary advantage of the ALE method is its capability to handle boundary layers more naturally, though this is not true in all cases. By contrast, large displacements impose virtually no extra effort when coupling to fluid solvers based on fixed Cartesian grids but do cause difficulties for methods based on an ALE approach.

The first LB algorithm for an interaction problem between a fluid and rigid obstacles has been developed by Ladd [49, 50] for the simulation of particulate suspensions. A special treatment of boundary conditions and the activation/deactivation of fluid nodes have been developed in this work. Krafczyk et al. [47] extended these ideas to perform two-dimensional FSI of an artificial heart-valve with moving leaflets. The flow in the channel was driven by transient boundary conditions representing a physiologically relevant regime. Previous contributions to the subject matter include [27, 71], where the LB fluid-solver VIRTUALFLUIDS based on unstructured quadtree type grids was coupled to the high order finite element (p-FEM) structural dynamics solver ADHOC by an eigenvalue approach to predict bidirectional fluid-structure interaction for geometrically linear problems in two dimensions.

The methodology applied can be classified as a discrete forcing approach of an Embedded Domain Method. Here, the Fluid-Structure interface is represented as a sharp interface and the boundary conditions are directly imposed on the structure and on the fluid at this discrete interface. As such, this contribution is a continuation of the work presented in [27, 45, 71]. In [45], this methodology was laid out in detail, benchmarked and verified against the two-dimensional, numerical examples proposed in [80, 81]. By contrast, the present paper shows that the method can reproduce the experimental results described in [32] where a detailed comparison of the results obtained in comparison to the application of other methods is presented as well.

Additionally, the methods applied in two dimensions are now extended to three dimensions and verified against a semi-analytic solution for a falling sphere in a cylinder and the three-dimensional benchmark suggested in [1, 2].

An immersed boundary method lattice Boltzmann approach was presented in [21, 34, 35] for solving fluid-particles interaction problems. The method was extended by a multi-block strategy allowing for mesh refinement [76]. Other approaches for computing Fluid-Structure interactions with the Lattice Boltzmann Method have been developed recently as well. In [73] a distributed-Lagrange-multiplier/fictitious-domain method (DLM/FD) is introduced to deal with the fluid/elastic–solid interactions. In [72, 74] this approach is extended to three dimensions. The fluid motion is solved by LBM and the deformation of the solid body is solved by a low order finite element method.

The contribution at hand is structured as follows: Sect. 2 recapitulates the Lattice Boltzmann Method followed by a description of the necessary transformation of the physical quantities into the lattice Boltzmann system in Sect. 3. A brief description of the p-Version of the Finite Element Method applied to structural dynamics is given in Sect. 4. The coupling approach itself is described in Sect. 5 and the boundary conditions are laid out in Sect. 6. Section 7 presents the validation of the methodology via comparison with the experimental setup given in [32]. It consists of a rigid cylinder in the cross flow of a Newtonian, incompressible fluid with a flexible tail attached to its downstream side. The method is then extended in Sect. 8 to three dimensions and applied to an example of a rigid body motion, namely a falling sphere in a cylinder. Section 9 then applies the methodology to a three-dimensional plate in a cross flow which exhibits geometrically non-linear deformations induced by the fluid. Section 10 then demonstrates the applicability of the developed methods to problems of very large deflections in FSI. The full scale computation of a bridge in Fluid-Structure interaction is also laid out in Sect. 10 to discuss an application to problems of large scale before conclusions are drawn in Sect. 11.

2   The Lattice Boltzmann Method for single-phase flow on non-uniform grids

A smoothed tree-type grid (quadtrees and octrees) is used. Typical quadtree meshes are shown in Fig. 1. In the interest of algorithmic simplicity, we employ smoothed trees i.e. the neighboring cells can only differ by at most one grid level. In Fig. 1, a non-smoothed quadtree, a smoothed quadtree and a smoothed quadtree with a minimum width of 3 cells per grid level are shown. Regions around arbitrarily shaped objects can be refined locally. In essence, the LBM is a discretization of the Boltzmann equation in space *and* time on these grids. The basic, physical notion of the method is to let fluid-mass fractions[1] collide on the nodes formed by this lattice and propagate them only among the links formed between the nodes of this lattice. To introduce the method formally, the following abbreviations are used: $\mathbf{x}$ represents a 3D vector in space and f is a b-dimensional vector, where b is the number of microscopic velocities along the links formed by the grid. The $d3q19$ model [64] with the following microscopic velocities is considered:

$$\{\mathbf{e}_i, i = 0, \ldots, 18\} = \begin{Bmatrix} 0 & c & -c & 0 & 0 & 0 & 0 & c & -c & c & -c & c & -c & c & -c & 0 & 0 & 0 & 0 \\ 0 & 0 & 0 & c & -c & 0 & 0 & c & -c & -c & c & 0 & 0 & 0 & 0 & c & -c & c & -c \\ 0 & 0 & 0 & 0 & 0 & c & -c & 0 & 0 & 0 & 0 & c & -c & -c & c & c & -c & -c & c \end{Bmatrix},$$

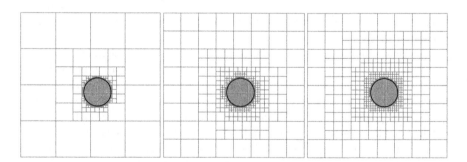

Fig. 1 Quadtree, smoothed quadtree and quadtree with minimum width of 3 cells.

---

[1] i.e. particles

where $c = \frac{\Delta x_l}{\Delta t_l}$ is a constant velocity determining the speed of sound c_s and $c_s^2 = c^2/3$. The lattice Boltzmann equation using the multi-relaxation time model [10,52] for the grid level l is then given by

$$f_{i,l}(t + \Delta t_l, \mathbf{x} + \mathbf{e}_i \Delta t_l) = f_{i,l}(t, \mathbf{x}) + \Omega_{i,l}, \quad i = 0, \ldots, b - 1, \tag{1}$$

where Δt_l is the time step and $\Omega_{i,l}$ the collision operator. The collision operator on grid level l is given by:

$$\Omega_l = \mathsf{M}^{-1} \mathsf{S}_l \left[(\mathsf{M} f) - m^{eq} \right]. \tag{2}$$

M represents the transformation matrix, which is composed of 19 orthogonal basis vectors $\{ \Phi_i, i = 0, \ldots, b - 1 \}$. Its definition can be found in the appendix. They are orthogonal with respect to the inner product $\langle \Phi_i, \Phi_j w \rangle$ (in contrast to [11], where $\langle \Phi_i, \Phi_j \rangle = 0$, if $i \neq j$). The vector w constitutes the weights $\{ w_i, i = 0 \ldots, b - 1 \}$:

$$w = (\frac{1}{3}, \frac{1}{18}, \frac{1}{18}, \frac{1}{18}, \frac{1}{18}, \frac{1}{18}, \frac{1}{18}, \frac{1}{36}, \frac{1}{36}, \frac{1}{36}, \frac{1}{36}, \frac{1}{36}, \frac{1}{36}, \frac{1}{36}, \frac{1}{36}, \frac{1}{36}, \frac{1}{36}, \frac{1}{36}, \frac{1}{36}).$$

The moments $m = \mathsf{M} f$ are labeled as

$$m = (\rho, e, \epsilon, j_x, q_x, j_y, q_y, j_z, q_z, 3p_{xx}, 3\pi_{xx}, p_{ww}, \pi_{ww}, p_{xy}, p_{yz}, p_{xz}, m_x, m_y, m_z).$$

m^{eq} is the vector symbolizing the equilibrium moments given in Eqs. (4) and $\mathsf{S}_l = \{ s_{l,i,i}, i = 0, \ldots, b - 1 \}$ is the diagonal collision matrix. The nonzero collision parameters $s_{l,i,i}$ (the eigenvalues of the collision matrix $\mathsf{M}^{-1} \mathsf{S}_l \mathsf{M}$) are:

$$s_{l,1,1} = s_{l,a},$$
$$s_{l,2,2} = s_{l,b},$$
$$s_{l,4,4} = s_{l,6,6} = s_{l,8,8} = s_{l,c},$$
$$s_{l,10,10} = s_{l,12,12} = s_{l,d},$$
$$s_{l,9,9} = s_{l,11,11} = s_{l,13,13} = s_{l,14,14} = s_{l,15,15} = -\frac{\Delta t_l}{\tau_l} = s_{l,\omega},$$
$$s_{l,16,16} = s_{l,17,17} = s_{l,18,18} = s_{l,e}.$$

The relaxation time τ_l is defined as:

$$\tau_l = 3 \frac{\nu}{c^2} + \frac{1}{2} \Delta t_l, \tag{3}$$

where ν is the kinematic viscosity. The parameters s_a, s_b, s_c, s_d and s_e can be chosen in the range $[-2, 0]$ and tuned to improve stability [52]. The optimal values depend on the specific characteristics of the system (considering geometry, initial and boundary conditions) and can not be computed in advance. Some reasonable values for these parameters are given in [11]. We choose $s_a = s_b = s_c = s_d = s_e = \max\{s_{l,\omega}, -1.0\}$. The nonzero equilibrium distribution functions $\{m_i^{eq}, i = 0, \ldots, 18\}$ are given by:

$$m_0^{eq} = \rho, \quad m_3^{eq} = \rho_0 \, u_x, \quad m_5^{eq} = \rho_0 \, u_y, \quad m_7^{eq} = \rho_0 \, u_z, \tag{4a}$$

$$m_1^{eq} = e^{eq} = \rho_0 \, (u_x^2 + u_y^2 + u_z^2), \tag{4b}$$

$$m_9^{eq} = 3 p_{xx}^{eq} = \rho_0 \, (2\, u_x^2 - u_y^2 - u_z^2), \tag{4c}$$

$$m_{11}^{eq} = p_{zz}^{eq} = \rho_0 \, (u_y^2 - u_z^2), \tag{4d}$$

$$m_{13}^{eq} = p_{xy}^{eq} = \rho_0 \, u_x \, u_y, \tag{4e}$$

$$m_{14}^{eq} = p_{yz}^{eq} = \rho_0 \, u_y \, u_z, \tag{4f}$$

$$m_{15}^{eq} = p_{xz}^{eq} = \rho_0 \, u_x \, u_z, \tag{4g}$$

where ρ_0 is a constant density and ρ a density variation.

The macroscopic quantities (density and momentum) and the stress tensor are given by (omitting the index l):

$$\rho = \sum_i f_i, \tag{5a}$$

$$\rho_0 \mathbf{u} = \sum_i \mathbf{e}_i f_i, \tag{5b}$$

$$S_{xx} = -(1 - \frac{\Delta t}{2\tau})(\frac{1}{3} e + p_{xx} - \rho_0 u_x^2), \tag{5c}$$

$$S_{yy} = -(1 - \frac{\Delta t}{2\tau})(\frac{1}{3} e - \frac{1}{2} p_{xx} + \frac{1}{2} p_{ww} - \rho_0 u_y^2), \tag{5d}$$

$$S_{zz} = -(1 - \frac{\Delta t}{2\tau})(\frac{1}{3} e - \frac{1}{2} p_{xx} - \frac{1}{2} p_{ww} - \rho_0 u_z^2), \tag{5e}$$

$$S_{xy} = -(1 - \frac{\Delta t}{2\tau})(p_{xy} - \rho_0 u_x u_y), \tag{5f}$$

$$S_{yz} = -(1 - \frac{\Delta t}{2\tau})(p_{yz} - \rho_0 u_y u_z), \tag{5g}$$

$$S_{xz} = -(1 - \frac{\Delta t}{2\tau})(p_{xz} - \rho_0 u_x u_z). \tag{5h}$$

Using the Chapman-Enskog expansion [6, 42], which in essence is a gradient expansion around the local equilibrium state, it can be shown that the LB Method:

- is a second order scheme both in space and time for the compressible Navier-Stokes equations in the low Mach number limit using the advective scaling
- is a scheme of first order in time and second order in space for the *incompressible* Navier-Stokes equations using the diffusive scaling

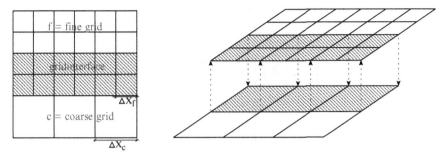

Fig. 2 Grid-interface and scale directions.

In order to model spatial refinement where one level of refinement is represented by one grid level, nested time-stepping is used in this paper. If the speed of sound, the kinematic viscosity, the Mach and the Reynolds number are assumed to be equal on all grids, then two time steps on the finer grid have to be performed during one time step on the coarser level [22]. An approach where the Mach number is lowered on the finer grid levels to ensure a faster convergence to the *in*compressible Navier-Stokes equations can be found in [66]. This approach requires to perform four time steps on the finer grid during one time step on the coarser mesh.

A typical grid interface between two grid levels is shown in Fig. 2. The overlap of the interface is due to the fact that missing distributions on one grid level have to be computed from the adjacent grid level. Ensuring the continuity of pressure, the velocity, and also of their derivatives, the non-equilibrium parts (f_i^{neq} [9, 22, 85] or alternatively m_i^{neq}) have to be rescaled. Performing the rescaling after the propagation step one obtains:

$$m_{i,l-1}^{neq} = \frac{s_{l,i,i}}{s_{l-1,i,i}} \frac{\Delta t_{l-1}}{\Delta t_l} m_{i,l}^{neq}. \tag{6}$$

The relaxation parameters $s_{l,a} = s_{l,b} = s_{l,c} = s_{l,d} = s_{l,e}$ can in principle be arbitrarily chosen in the range $[-2, 0]$, but we used the same scaling as for $s_{l,\omega}$.

$$\frac{s_{l,a}}{s_{l+1,a}} = \frac{s_{l,b}}{s_{l+1,b}} = \frac{s_{l,c}}{s_{l+1,c}} = \frac{s_{l,d}}{s_{l+1,d}} = \frac{s_{l,e}}{s_{l+1,e}} = \frac{s_{l,\omega}}{s_{l+1,\omega}}. \tag{7}$$

Different space and time interpolations have to be used because of the different mesh spacings Δx_l and time steps Δt_l. A cubic interpolation in space is used for the 'hanging' nodes. Fig. 3 shows the interpolation for the 'hanging' node P using the nodes P1, P2, P3 and P4. Linear interpolation is used in time. Details of the algorithm can be found in [8, 85]. The necessary boundary conditions are described in Sect. 6.

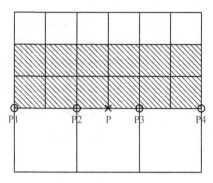

Fig. 3 Hanging node P and nodes P1-P4 required for interpolation.

3 Transformation of physical quantities

A reasonable set of dimensionless quantities which uniquely define an FSI problem are given e.g. by

- the Aerolastic number $Ae = \dfrac{E_s}{\rho_f U_{0f}^2}$,
- the Reynolds number $Re = \dfrac{U_{0f} H}{\nu_f}$,
- the density ratio $\beta = \dfrac{\rho_s}{\rho_f}$,
- and the Poisson ratio of the elastic structure ν_s.

Here E_s is the Young's modulus, ρ_s and ρ_f the density of structure and fluid respectively, U_{0f} a reference velocity, H a reference length and ν_f the kinematic viscosity.

The fluid solver rescales the system in such a way, that numerical errors are reduced (avoiding very small and/or large numerical values). Since this rescaling is inherent to most LB implementations, the procedure for rescaling two different systems labeled as *LB* and *real* is presented below. The scaling of the forces and the time from the fluid solver system (LB) to the real system is done by setting the dimensionless drag coefficient equal in both systems. The result is:

$$\mathbf{F}_{real} = \mathbf{F}_{LB} \cdot \frac{H_{real}^2 \cdot \rho_{real} \cdot u_{real}^2}{H_{LB}^2 \cdot \rho_{LB} \cdot u_{LB}^2}, \tag{8}$$

with reference height H, reference density ρ and reference velocity u. By multiplying the drag formula with the square of the Reynolds number we obtain the following force scaling formula:

$$\mathbf{F}_{real} = \mathbf{F}_{LB} \cdot \frac{\rho_{real} \cdot v_{real}^2}{\rho_{LB} \cdot v_{LB}^2}. \tag{9}$$

This force transformation is necessary, if no reference velocity is given. By equating the Reynolds number in both systems and using $u = \frac{H}{\Delta T}$ we obtain the following time scaling:

$$\Delta T_{real} = \Delta T_{LB} \cdot \frac{\nu_{LB} \cdot H_{real}^2}{\nu_{real} \cdot H_{LB}^2}. \tag{10}$$

Computing the displacements as well as the velocity of the structure the wall velocity (boundary condition) for the LB system is

$$\mathbf{u}_{LB} = \mathbf{u}_{real} \cdot \frac{\nu_{LB} \cdot H_{real}}{\nu_{real} \cdot H_{LB}}. \tag{11}$$

Further dimensionless quantities are the weight coefficients:

- $\gamma = \frac{g \cdot H^3}{\nu^2}$,
- $\delta = \frac{g \cdot H}{u^2}$,

The formula for rescaling the weights due to both systems is:

$$g_{LB} = g_{real} \cdot \frac{u_{LB}^2 \cdot H_{real}}{u_{real}^2 \cdot H_{LB}}. \tag{12}$$

If there is no reference velocity the scaling formula is:

$$g_{LB} = g_{real} \cdot \frac{\nu_{LB}^2 \cdot H_{real}^3}{\nu_{real}^2 \cdot H_{LB}^3}. \tag{13}$$

4 Structural Solver

The 3D LB solver VIRTUALFLUIDS was coupled with two structural models implemented in PE [41] and ADHOC [12]. PE is a rigid multi-body physics engine. It is a framework designed for both virtual reality scenarios with real time requirements as well as large-scale computer simulations with millions of interacting rigid bodies. In contrast to the physics engines for computer games, PE primarily aims at realistic simulation environments comprising rigid multi-body dynamics, as they arise for example in material sciences or fluid mechanics. For instance, for some years, rigid multi-body dynamics has been used in combination with lattice Boltzmann flow simulations to simulate the behavior of rigid bodies in a flow [41]. This coupled simulation system is used to simulate fluidization processes with a high number of buoyant rigid bodies or sedimentation processes with large numbers of floating bodies.

For modeling the interaction with fully three-dimensional thin- and thick walled structures exhibiting transient deformations and deflections, a high-order Finite Element Method (p-FEM) is employed. This has been implemented in ADHOC [12]. Our implementation of high order finite elements in three dimensions utilizes a hexahedral element formulation, using the hierarchic shape functions introduced in [77].

These three-dimensional shape functions can be classified into four groups: the nodal or vertex modes, the edge modes, the face modes, and the internal modes. The nodal or vertex modes are defined by the standard trilinear shape functions, well known from the isoparametric eight-noded brick element. The edge and face modes are non-zero on the edges and faces which they are associated to and vanish on all other edges and faces, whereas the internal modes are purely local being zero on all faces and edges of the hexahedral element.

Three different types of trial spaces can be defined: the *trunk space* $S_{ts}^{p_\xi, p_\eta, p_\zeta}$ (Ω_{st}^h), the *tensor product space* $S_{ps}^{p_\xi, p_\eta, p_\zeta}(\Omega_{st}^h)$ and an *anisotropic tensor product space* $S^{p,p,q}(\Omega_{st}^h)$. A detailed description of these trial spaces can be found in [13, 14, 77, 78]. The polynomial degree for the trial spaces $S_{ts}^{p_\xi, p_\eta, p_\zeta}(\Omega_{st}^h)$ and $S_{ps}^{p_\xi, p_\eta, p_\zeta}(\Omega_{st}^h)$ can be varied separately in each local direction. It is possible to construct discretizations where the polynomial degree for the in-plane and thickness direction of thin-walled structures can be treated differently. High-order solid elements can therefore provide a fully three-dimensional solution also including arbitrary three-dimensional stress states, and, nevertheless can cope with high aspect ratios of thin-walled structures. Moreover, they are less prone to locking effects, in contrast to classical low order elements. Thus, only one element type for thin- as well as thick-walled structures is sufficient, without making transition elements between thin-walled and massive parts of the structure necessary. A detailed investigation of the advantages of high order solid elements for thin-walled (nonlinear) continua can be found in [13, 15–17, 36, 37, 65, 71, 77, 78].

Time integration is performed using classical second order accurate finite difference methods such as the generalized-α method [7] or the method of Newmark [59]. Arising nonlinearities are treated by a Newton-Raphson procedure.

For Fluid-Structure Interaction problems, one advantage of three-dimensional models is especially obvious, i.e. the inherently correct representation of the *wet surface* being in contact with the fluid. Unlike for dimensionally reduced beam, plate and shell models, there is no necessity to reconstruct this skin surface from the middle surface and certain kinematic assumptions. Curved geometries are described by means of the blending function method. This method also works well for three-dimensionally curved surfaces. A detailed description is given in [78]. Moreover, by using an adaptive selection of the polynomial degree in the three directions, a very cost-efficient discretization can be found while maintaining control of the approximation and modeling errors. See [17, 56] for elastostatic problems and [68–70] for elastodynamic problems as further references.

5 Description of the Coupling approach

The solution of the two-field problem of Fluid-Structure interaction is achieved by a partitioned approach solving each field separately. The coupling is performed by exchange of information at the boundary. The coupling framework is described below.

5.1   Previous work

In the course of this work we investigated several coupling algorithms which were especially devised to cope with difficulties we faced at the beginning of the research work. These are mentioned here for completeness. However, only the most efficient one will be described in more detail in Sect. 5.2.

A first version used an eigenvalue approach for the structural integration in time [27, 43, 71] which by design could only account for geometrically linear structural deformations. The coupling algorithm was of explicit nature since the boundary information was only exchanged once at each time step.

A second, completely different and more stable implicit version which could also take geometrically nonlinear structural displacements into account is described in [28]. This concept proved to be quite involved from a computational point of view and was therefore not further investigated.

The most efficient algorithm turned out to be the version described next.

5.2   Coupling algorithm

Setup process

At the boundary, two completely different discretizations are present. The structural boundary is described by coarse and curved p-elements while the fluid boundary is described by relatively small cubes and a locally refined octree type grid. We therefore introduce an interface mesh to couple both discretizations. The interface mesh is a moving surface mesh consisting of flat triangles. On each node the values for the velocity, the load vector and other physical quantities required for the exchange are stored. The mesh is constructed as follows. The nodes are defined by the Gaussian integration points on the surface of the p-Elements or as a set of equidistant points situated on the surface of the structure. A triangulation is then carried out to obtain the interface mesh. The interface mesh can be adapted by the p-FEM solver as well as by the LB solver. A Gaussian interface mesh and the coupling is shown in Fig. 4.

The framework (for details see [4]) is based on a client-server concept and realized through MPI [23]. A master process manages the exchange of data between the structural and fluid code.

The values to be transferred (i.e. the tractions and displacements) are only allowed to vary linearly between the nodes of the interface mesh.

As part of the setup, the structural solver provides the supervisor process with the interface mesh in its initial configuration. This mesh is handed to the fluid solver.

Core algorithm

The coupling coupling algorithm is depicted in Fig. 5. It incorporates nested time stepping. While n fluid steps are performed corresponding to one structural step,

Structure FEM−Mesh   Interface Mesh   Fluid LBM−Grid

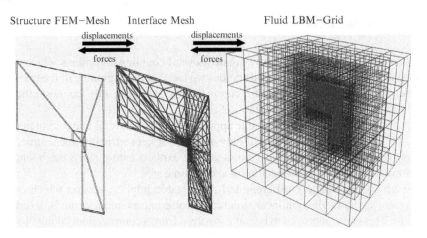

Fig. 4 Coupling using interface mesh.

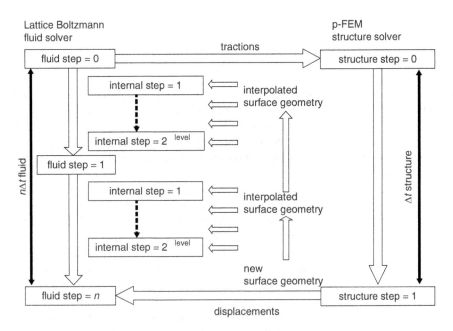

Fig. 5 Explicit coupling algorithm with multilevel nested time-stepping.

further sub-steps (called fluid-internal steps) need to be performed in each of these fluid steps, in order to propagate the fluid unknowns $f_I(t, \mathbf{x})$ through the hierarchical, non-uniform grid defined by the LBM solver (see Sect. 2). After the setup process (Sect. 5.2), the algorithm follows these steps:

1. The fluid solver computes the traction vector on the interface mesh points.
2. The tractions are handed through the interface mesh to the structural solver.
3. The structure integrates these tractions in its load vector either the standard way or with a load conservative scheme [45]. The structural solver calculates the displacements which are then exchanged through the interface mesh.
4. The fluid solver performs an interpolation of the positions of the interface mesh in time and calculates the solution.
5. Step 4 is repeated for all fluid-internal steps.
6. Step 5 is repeated for the number n of fluid-subcycling steps. The fluid stresses of the subcycling steps plus the fluid internal steps are averaged.

Thus, the algorithm is a staggered coupling algorithm with a subiteration for the fluid with the specialty that the fluid-solver needs to perform fluid-internal steps on its subgrids. As such, the presented algorithm is only conditionally stable and first order accurate in time. It was first reported by [19] and has been used by a countless number of authors since then in different contexts. Review articles such as e.g. [20, 61–63] have summarized and analyzed a large number of possible algorithms for the partitioned solution of coupled mechanical systems and classified the scheme utilized in this article as a *conventional serial staggered* (*CSS*) procedure. Their analysis extends far beyond the version of the *CSS* procedure presented here. Some of these algorithms were implemented and evaluated in terms of their energy conservation properties in the context of coupling to a lattice Boltzmann method in [45].

However, and although subject to improvement, the simple version utilized in this paper suffices to compute all FSI examples presented below.

6 Boundary Conditions

6.1 Kinematic Boundary conditions

The kinematic boundary conditions for the LBM-solver are imposed by the displacements and velocities at the surface of the structure. They can be obtained in a straight forward manner by the usual postprocessing-step for Finite Element Methods in structural dynamics.

In the LBM-flow solver, however, macroscopic flow quantities are gained via forming the corresponding moments of the microscopic flow quantities, i.e. the particle distribution functions. In turn, macroscopic boundary conditions can only be set implicitly via the particle distribution functions. A well known and simple way to introduce no-slip walls is the so-called bounce back scheme. It allows spatial second order accuracy if the boundary is aligned with one of the lattice vectors $\hat{\mathbf{e}}_i = \mathbf{e}_i/c$ and is first order accurate otherwise. To maintain second order accuracy for arbitrarily shaped, moving boundaries we use the modified bounce back scheme developed in [3, 51] for the velocity boundary conditions.

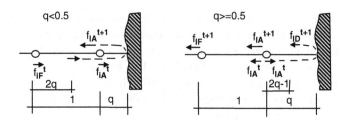

Fig. 6 Interpolations for second order bounce back scheme.

In Fig. 6 two cases along a link i can be identified:

(a) wall–node distance $q_i < 0.5 |\mathbf{e}_i \Delta t|$ and
(b) wall–node distance $q_i \geq 0.5 |\mathbf{e}_i \Delta t|$.

The modified bounce back scheme is

$$f_{IA}^{t+1} = (1 - 2q) \cdot f_{iF}^t + 2q \cdot f_{iA}^t - 2\rho w_i \frac{\mathbf{e}_i \mathbf{u}_w}{c_s^2}, \quad 0.0 < q < 0.5, \tag{14}$$

$$f_{IA}^{t+1} = \frac{2q - 1}{2q} \cdot f_{IA}^t + \frac{1}{2q} \cdot f_{iA}^t - \rho w_i \frac{\mathbf{e}_i \mathbf{u}_w}{q c_s^2}, \quad 0.5 \leq q \leq 1.0. \tag{15}$$

The recovery of second order accuracy even for curved boundaries is clearly demonstrated in [26]. A detailed discussion of LBE boundary conditions can be found in [30]. In contrast to the simple bounce-back scheme, the use of these interpolation based no-slip boundary conditions results in a notable mass loss across the no-slip lines. Yet, the results obtained with bounce-back were inferior which highlights the importance of a proper geometric resolution of the flow domain. Pressure boundary conditions are implemented by setting the incoming distributions to [79]

$$f_I(t + \Delta t, \mathbf{x}) = -f_i(t, \mathbf{x}) + f_I^{eq}(p_0, \mathbf{u}(t_B, \mathbf{x}_B)) + f_i^{eq}(p_0, \mathbf{u}(t_B, \mathbf{x}_B)), \tag{16}$$

where p_0 is the given pressure, $t_B = t + \frac{1}{2}\Delta t$, $\mathbf{x}_B = \mathbf{x} + \frac{1}{2}\mathbf{e}_i$, (f_I, f_i) are an anti-parallel pair of distributions with velocities $\mathbf{e}_i = -\mathbf{e}_I$, and f_I is the incoming and f_i the outgoing distribution function value. $\mathbf{u}$ is obtained by extrapolation.

Activation/Deactivation of new fluid nodes
If a new fluid node is created due to the moving structure, linear inter- or extrapolation (depending on the geometrical configuration) are used to compute the local velocity at the concerned node. A local Poisson type iteration described in [57] is then applied locally to compute the corresponding pressure and the higher order moments.

6.2 Force Boundary Conditions

Force boundary conditions are imposed onto the structural boundary by the fluid.

There are two possibilities to evaluate forces on boundaries using the LB Method. The first one is the momentum-exchange based method and the second one is the pressure/stress integration based method. A detailed description and comparison of both methods can be found in [58] and the details for the implementation in 2D can be found in [27, 45]. The boundary of the structure is given by the interface mesh consisting of flat triangles in 3D as discussed in Sect. 5.

The momentum exchange method works well if one is only interested in the value of the integrated forces acting on structural elements, but it leads to noisy results for the computation of the forces if the boundary of the elements of the structure or the interface mesh have a size comparable to the grid distance of the fluid mesh. Incorrect local forces are created if too few links contribute to the force computation. In this case the stress integration method is favorable.

An advantage of the LBM compared to conventional CFD solvers is the local availability of the stress tensor. The stress tensor $S_{\alpha\beta}$ with scalar pressure is:

$$S_{\alpha\beta} = -c_s^2 \rho \delta_{\alpha\beta} + \sigma_{\alpha\beta}. \tag{17}$$

$\sigma_{\alpha\beta}$ is computed from the non equilibrium part of the distributions and δ is the Kronecker-Delta.

$$\sigma_{\alpha\beta} = \left(1 - \frac{\Delta t}{2\tau}\right) \sum_{i=1}^{b-1} f_i^{neq} \left(e_{i\alpha}e_{i\beta} - \frac{1}{D}\mathbf{e}_i\mathbf{e}_i\delta_{\alpha\beta}\right). \; brium \; with \; the \; resisting \; force \tag{18}$$

In three dimensions, 256 different cases emerge of how a point may lie inside the grid. These cases conform with the Marching-Cubes (3-D) algorithm [55] which was developed for computer graphics to compute isosurfaces from Cartesian grids (see Fig. 7).

Two methods for the force evaluation were implemented. The first method we term stress integration method. It computes and selects the corresponding marching cubes cases. Then the stresses are inter- and/ or extrapolated onto the boundary via recovering the stresses on the nodes inside the structure as described in [28,45]. The second method sets the stresses on the boundary point (i.e. on the interface mesh) equal to the ones at the nearest neighboring node in normal direction. This is also termed next-neighbor interpolation [33].

Once the tractions are evaluated on the structural boundary they need to be integrated in the structural load vector. Again, there are two possibilities, both of which were already described in [45] and repeated here briefly for completeness. The first possibility is to directly integrate the transferred tractions in the deformed configuration. This possibility assumes a p-continuous load function across element boundaries. The second possibility is to apply a composed integration scheme. Kinks in the load function can then be taken into account. Details for the convergence rate of such a scheme were also given in [71].

Fig. 7 Representation of the 15 base configurations of the Marching Cube algorithm.

7 Experimental benchmark

In [45] the numerical benchmark (see Figs. 8 and 9) of the DFG research unit 493 [81] was studied with the two-dimensional version of VIRTUALFLUIDS. Additionally, we verified the methodology against the experimental benchmark described in [32], where the results of our approach are not only compared to the experimental values, but also put into context to the results obtained by other research groups.

In contrast to the numerical benchmark, the flag is very thin and an additional mass is attached at its end to add inertia to the system and thereby maintain stable, periodic motions. The cylinder may rotate freely around its center. Two different cases ($Re = 140$ and $Re = 190$) are considered each of which excites a different structural mode.

The cylinder is made of aluminium with a density of $\rho = 2828[\frac{kg}{m^3}]$ while the membrane and the attached mass are made of steel with a density of $\rho = 7855[\frac{kg}{m^3}]$. Young's modulus of the steel was measured to be $E = 200[\frac{kN}{m\,m^2}]$.

The fluid used in the experiment is Polyethyleneglycolsyrup with a density of $1050[\frac{kg}{m^3}]$ and a kinematic viscosity of $0.000164[\frac{m^2}{s}]$.

The center of the cylinder is situated at a distance of $55[m\,m]$ from the inlet (9). The origin of the coordinate system is set at the center of the cylinder. The channel is $338[m\,m]$ long and $240[m\,m]$ high. The gravitional force acts along the x-axis.

The gravity is equal to $9.81[\frac{m}{s^2}]$ and is scaled to the LB system using Eq. (13). In comparison to the numerical benchmark, the stiffness (EI) of the flag is reduced by a factor of $1/1000$ and the structural mass is reduced by a factor of $1/700$, while the fluid mass is approximately the same. The system is, therefore, badly conditioned and a partitioned solution approach is very prone to instabilities.

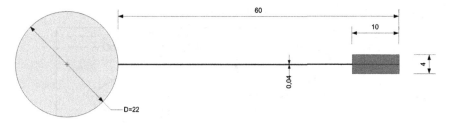

Fig. 8 Geometrical definition of the structure in *mm* [31].

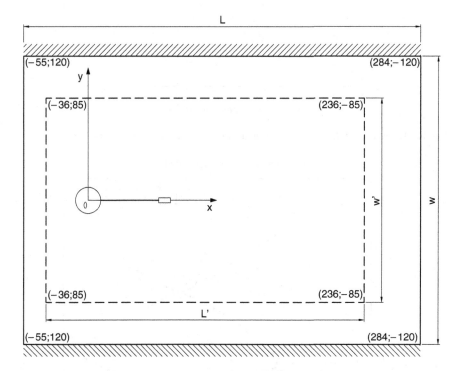

Fig. 9 Physical domain and measurement domain (dashed) in *mm* [31].

On coarse LB grids, a stable simulation can only be obtained if the mass of the membrane is increased. For the grid with 340×240 nodes the mass of the membrane needed to be increased by a factor of four. For a higher resolution of the fluid grid with 1360×960 nodes, it was not necessary to increase the mass to obtain a stable simulation. For 100 hours of computational time, 1.6 seconds realtime could be simulated for the grid with 240 nodes in the height. This might seem little, but is considered to be quite fast compared to other methods. For the grid with 480 nodes per height, the time was 245 hours for 1.6 seconds realtime on a 2.0 GHz AMD Opteron 64-Bit Processor. The Mach number in the simulation was equal to 0.05.

The close agreement of the computed trajectories at the end point of the flag with the ones measured in the experiments is depicted in Fig. 10 and Fig. 11. For a direct

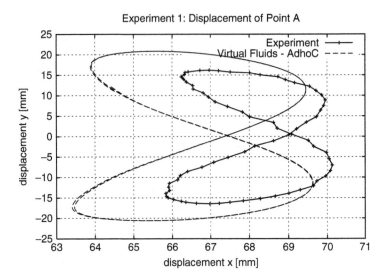

Fig. 10 Trajectories of the endpoint compared with experimental data for $Re = 140$.

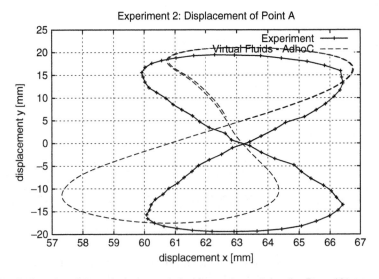

Fig. 11 Trajectories of the endpoint compared with experimental data for $Re = 190$.

comparison to the results obtained by other groups see [32] which is presented in a different chapter of this book.

The corresponding frequencies of the trajectories are given in Table 1 where the experimental results were taken as the reference for the computation of the relative deviation.

The low Reynolds number benchmark can be reproduced very well. Only the higher Reynolds number benchmark exhibits a notable deviation in the frequency of

Table 1 Experiments: computed versus measured eigenfrequencies.

case	experiment 1: $Re = 140$	experiment 2: $Re = 190$
measured eigenfrequency [Hz]	6.38	13.58
computed eigenfrequency [Hz]	6.71	16.70
rel. deviation [%]	5.17	22.98

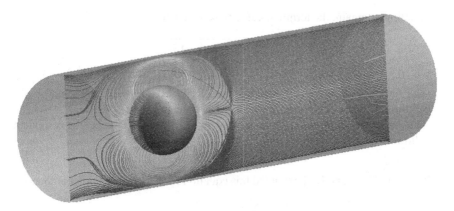

Fig. 12 Screenshot of a sphere in channel.

about 22[%]. Additionally, a small unsymmetry is present in the trajectories as the zero crossing is slightly offset. It is pointed out that other methods, such as an ALE approach did have severe difficulties with the mesh deformation in this benchmark due to the fact that the end mass exhibits a rotation of more than 45 degrees at each cycle. This is no issue in our approach.

8 Validation of 3D-FSI: The falling sphere

We now set out to extend the methodology to three dimensions and evaluate its performance against a three-dimensional numerical testcase consisting of a falling sphere in a pipe (see Fig. 12). As only rigid body motions are involved, the fluid solver VIRTUALFLUIDS was coupled with the dynamics engine PHYSICSENGINE (PE) [40] to simulate the fluid-structure interactions.

The benchmark is defined as follows: a sphere of radius r and density ρ_s falls in a fluid of density ρ_f and viscosity ν under the action of gravity. When the sphere reaches its terminal velocity, its weight F_G and the statical buoyancy F_A are in equilibrium with the resisting force F_D.

$$F_G = \frac{4}{3}\pi r^3 \rho_s g, \tag{19}$$

$$F_A = \frac{4}{3}\pi r^3 \rho_f g, \tag{20}$$

$$F_D = \frac{1}{2}C_{DW}\rho_f \pi r^2 U^2. \tag{21}$$

In an infinite domain, the terminal velocity is then defined as:

$$U = \sqrt{\frac{4}{3}\frac{(\rho_s - \rho_f)}{\rho_f}\frac{2r}{C_{DW}}g}. \tag{22}$$

The terminal velocity is evaluated iteratively, as the drag coefficient C_{DW} depends on the Reynolds number $Re = Ud/\nu$ defined for the sphere.

The drag coefficient C_D for laminar flow in the Stokes regime can be computed as follows:

$$C_D = 24/Re. \tag{23}$$

Schiller and Naumann [67] extended this equation for higher Reynolds numbers.

$$C_D = \frac{24}{Re}(1 + 0.15Re^{0.687}). \tag{24}$$

In case the sphere is falling in a finite domain represented by a circular pipe, additional corrections have to be considered in order to take into account the wall effects. These were given by Fayon und Happel [18] as a function of a perturbation expansion of $\lambda = r_{sphere}/r_{cylinder}$:

$$C_{DW} = C_D + \frac{24}{Re}(K - 1), \tag{25}$$

with

$$K = \frac{1 - 0.75857\lambda^5}{1 - 2.105\lambda + 2.0865\lambda^3 - 1.7068\lambda^5 + 0.72603\lambda^6}. \tag{26}$$

Eq. (25) is valid for $\lambda < 0.6$ and $Re < 50$ and can predict the drag coefficient with $\pm 5[\%]$ accuracy.

8.1 Fixed sphere

A sphere of diameter D is submerged into a channel of cylindrical shape and diameter $2D$. The sphere is fixed at the centerline of the cylinder and situated at a distance of $2.4D$ measured from the inlet of the channel. To simulate a moving sphere, no-slip boundary conditions are specified on the sphere while velocity boundary conditions are defined at the inlet of the channel as well as at its walls. The outlet of the channel is supplied with a fixed pressure boundary condition with $p = 0$.

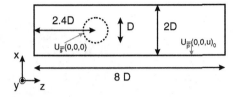

Fig. 13 Numerical simulation setup.

Table 2 Deviation from the theoretical value of the drag coefficient acting on the sphere for different mesh resolutions considering $Re = 1$.

domain size	$u_0\ [m\ s^{-1}]$	$v\ [m^2\ s^{-1}]$	$D/\Delta x$	C_{DW} [-]	rel. deviation [%]
$30^2 \times 120$	0.0111	0.166	15	141.19	2.3
$60^2 \times 240$	0.0055	0.166	30	142.00	1.7
$90^2 \times 360$	0.0037	0.166	45	142.30	1.5

Table 3 Deviation from the theoretical value of the drag coefficient acting on a regular moving sphere for different mesh resolutions considering $Re = 1$.

domain size	$u_0\ [m\ s^{-1}]$	$v\ [m^2\ s^{-1}]$	$D/\Delta x$	C_{DW} [-]	rel. deviation [%]
$30^2 \times 120$	0.0111	0.166	15	139.53	3.4
$60^2 \times 240$	0.0055	0.166	30	141.64	1.9
$90^2 \times 360$	0.0037	0.166	45	142.38	1.4

Figure 13 depicts the domain schematically. The drag forces are computed via the momentum exchange method. The simulation was computed with three different resolutions for Reynolds number $Re = 1$. The reference value for the drag coefficient is $c_{d,W} = 144.48$. Results are given in Table 2. The computed drag coefficient converges to the reference values as the mesh resolution increases.

8.2 Moving sphere

Having shown that a fixed sphere can be computed with a deviation from the theoretical value of the drag coefficient of approximately two percent, the same setup is now utilized for the simulation of a moving sphere. Therefore, no-slip boundary conditions are applied on the walls of the channel as well as at its inlet and outlet. As before, the resulting drag coefficient is measured.

The sphere is submerged into the fluid domain and positioned in the center of the cylinder at a distance of one sphere diameter from the top of the cylinder. After 500 iterations in the coarse setup and 4000 iterations in the fine setup the drag coefficient converges. The results are depicted in Table 3.

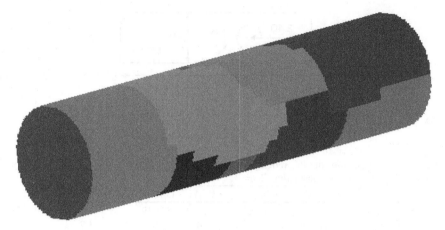

Fig. 14 Partitioning of the fluid domain for the parallel computation mit 14 subdomains.

8.3 Falling Sphere

The falling sphere in a fluid was computed with different mesh resolutions and different sphere densities. The simulation parameter for this test case are defined as follows: $r_{Cylinder} = 0.1[m]$, $r_{Sphere} = 0.05[m]$, $v = 0.01[\frac{m^2}{s}]$ and $\rho_f = 1000[\frac{kg}{m^3}]$. The length of the pipe is $L = 0.8[m]$. It is thus sufficiently long such that the terminal velocity is reached before the sphere hits the bottom. No-slip boundary conditions are specified on the cylinder's wall. Gravity is taken into account by applying a volume force on the fluid (see Eq. (13)). The boundary condition on the sphere's wall is defined via the modified bounce-back rule of second order accuracy described in Eq. (15) to take into account the sphere's velocity. As a the computation is very time consuming and huge amount of memory is needed for the high resolution configuration, this testcase was simulated with the parallel version (see Fig. 14) of VIRTUALFLUIDS. This extension is described in the PhD thesis of Freudiger [24]. In the highest resolution the computation had 37 million nodes[2].

For the grid and Mach number parameter study the density of the sphere was set to $\rho_s = 2000[\frac{kg}{m^3}]$ and the density of the fluid was set to $\rho_s = 1000[\frac{kg}{m^3}]$. The viscosity in the LB grid was set to 0.1666. The terminal velocity of the sphere was determined iteratively from the analytical solution to be: $u_{ref} = 0.0906$. In this case the Reynolds number is $Re = 0.91$. The resulting forces at the sphere fluctuated around 0.1[%] from their theoretical value. The sphere's velocity was measured for different mesh resolutions. The values are listed in Table 4.

The deviation of the sphere's velocity from the theoretical value for different Reynolds numbers (i.e. different densities of the sphere) is given in Table 5. The

---

[2] This is equivalent to about seven hundred million degrees of freedom in the D3Q19 model which was applied here.

Table 4 Deviation of the sphere's velocity from its theoretical value for different mesh resolutions.

domain size	$D/\Delta x$	$u\,[m\,s^{-1}]$	rel. deviation [%]
$30^2 \times 120$	15	0.09080	0.2
$60^2 \times 240$	30	0.09215	1.7
$90^2 \times 360$	45	0.09186	1.4
$120^2 \times 480$	60	0.09190	1.4
$150^2 \times 600$	75	0.09180	1.3
$180^2 \times 720$	90	0.09177	1.3
$210^2 \times 840$	105	0.09174	1.2

Table 5 Deviation of the sphere's terminal velocity from the theoretical value for different Reynolds numbers.

ρ_s	Re	$u_{analytical}\,[m\,s^{-1}]$	$u_{computed}\,[m\,s^{-1}]$	rel. deviation [%]
1200	0.18	0.0184	0.0187	1.6
1500	0.46	0.0457	0.0468	2.4
2000	0.91	0.0906	0.0896	1.1
4000	2.65	0.2653	0.2740	3.3
6000	4.34	0.4340	0.4500	3.7
8000	5.98	0.5977	0.6250	4.5
10000	7.58	0.7578	0.7800	2.9

grid resolution was $30^2 \times 120$ nodes. The deviation w.r.t. the theoretical value is in the uncertainty range of Eq. (25).

9 Plate in a cross flow

Bathe and Ledezma suggest a plate in a cross flow for an FSI benchmark with geo-metrically non-linear structural deformations [1, 2]. This benchmark measures the displacements of the top of the plate at different Reynolds numbers ranging from $Re = 500$ to $Re = 5000$. In [1, 2], the fluid was simulated with an Arbitrary-Lagrangian-Eulerian (ALE) formulation coupled to a Lagrangian description of the structure. The computation results in a quasi-stationary deflection of the plate.

For the coupling of the lattice Boltzmann solver VIRTUALFLUIDS with the p-FEM solver ADHOC the case $Re = 500$ and $Re = 2500$ is considered. Additionally, a laminar test case at $Re = 10$ is investigated.

The configuration parameters are as follows: The plate has a Young's modulus of $E = 7 \times 10^{10}[\frac{N}{m^2}]$ with a Poisson's ratio of $v_s = 0.3$ while the thickness of the plate is $h = 0.00125[m]$. The fluid domain is block shaped. The bottom as well as the rectangular plate have no-slip boundary conditions. The velocity at the inlet is defined via Eq. (27) and the outlet has a constant pressure of $p = 0[Pa]$.

$$v(z) = \frac{3V}{2H^2}(2Hz - z^2). \tag{27}$$

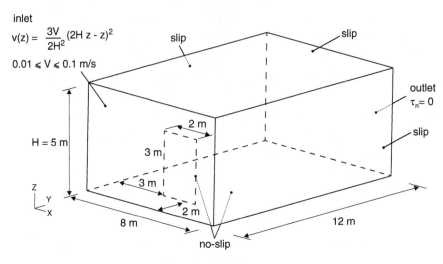

Fig. 15 Domain with integrated rectangular plate [1].

Slip boundary conditions are imposed as depicted in Fig. 15. The fluid has a density of $\rho = 1000[\frac{kg}{m^3}]$ and a dynamic viscosity of $\mu = 0.1[\frac{kg}{ms}]$. The kinematic viscosity thus amounts to $\nu = 0.0001[\frac{m^2}{s}]$.

The characteristic length in the Reynolds number is the height of the channel $H = 5[m]$. The plate was discretized by $10 \times 1 \times 18$ p-elements with the Ansatz space $S_{ps}^{p_\xi, p_\eta, p_\zeta}(\Omega_{st}^h)$ and $p = 3$. The maximum element aspect ratio was $1 : 160$. The fluid domain was computed with a uniform grid and a local refinement around the plate as given in Fig. 16. The Mach number was $Ma = 0.05$.

In order to obtain a stationary fluid solution, a low Reynolds number of $Re = 10$ was chosen. Generally, the force acting on the structure can be estimated as given in [5]:

$$F_d = \frac{1}{2}C_d A\rho v^2. \tag{28}$$

This is an approximation based on Stokes law. C_d is the drag coefficient, A is the area perpendicular to the flow, ρ is the density of the fluid and v^2 is the free field velocity of the fluid. The crucial point here is that C_d is not only dependent on the shape of the object but also on the Reynolds number. It is tabulated for standard cases e.g. in [60]. For a sphere at $Re = 1$ the drag coefficient is $C_d \approx 140$ while at $Re \geq 1000$ it decreases to around 0.47. For a plate in cross flow at $Re \geq 1000$, $C_d \approx 1$. To the author's knowledge, the case of a clamped plate subject to a cross flow at $Re = 10$ is not tabulated. It was therefore determined by a fully three-dimensional computation via CFX where the discretization was adapted to the

Fig. 16 Grid refinement around the plate domain.

pressure field in three steps. This resulted in a discretization of 2 016 708 tetrahedral elements with 399 399 nodes. A view of the pressure field and its discretization as computed by CFX is depicted in Fig. 17(a). The pressure field as computed by LBM is depicted in Fig. 17(b).

From the steady state computation with CFX, the drag coefficient was determined to be $C_d = 9.46$. A rough estimate of the expected deflection at the top of the plate can be computed via application of the force given by Eq. (28) to a clamped beam. Under the assumption of small displacements and a constant load, the analytical solution simply is:

$$u_y(A) = \frac{qL^4}{8EI};\tag{29}$$

where the load is $q = \dfrac{F_d}{A}$, the height of the beam is $L = 3[m]$. The stiffness EI is the plate stiffness and can be computed by:

$$EI = \frac{Et^3}{12(1 - \nu^2)}.\tag{30}$$

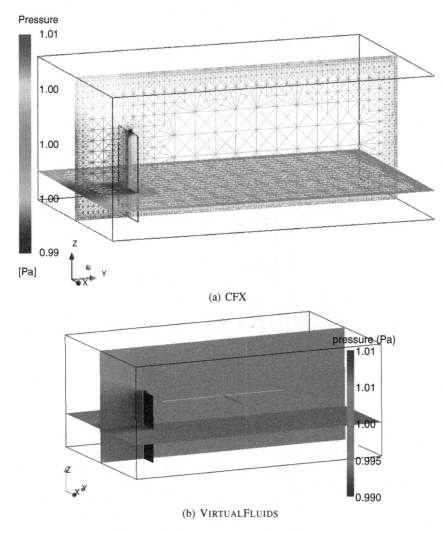

(a) CFX

(b) VIRTUALFLUIDS

Fig. 17 Pressure field of plate in cross flow computed by two different methods: (a) CFX: Visualization with CFX-postprocessor. (b) VIRTUALFLUIDS.

The first study is to evaluate the simplified stress computation and the correct detection of the neighbouring nodes next to the structure according to its normal. In order to achieve a range of different deflections at $Re = 10$, the benchmark was computed with three different inflow velocities $v_1 = 2.0 \times 10^{-4}[\frac{m}{s}]$, $v_2 = 2.0 \times 10^{-3}[\frac{m}{s}]$ and $v_3 = 2.0 \times 10^{-2}[\frac{m}{s}]$ where the fluid viscosity was adjusted accordingly. The deflection is expected to be in the range of the prediction via Eq. (28) and Eq. (29) and scale quadratically with the inflow velocity as long as the deflection remains within the limits of geometrical linearity.

To reach a steady state quickly, the mass of the structure was chosen to be $\rho_s = 1.0 \times 10^6 [\frac{kg}{m^3}]$ and the first eigenfrequency of the structure was damped out with a stiffness proportional damping. The Mach number was chosen to be $Ma = 0.05$. The results lie within the expected range of accuracy and are given in Table 6. The transient behaviour over time is depicted in Fig. 18.

In a second step, the Reynolds number was raised to $Re = 500$ by means of increasing the inflow velocity to $v = 1.0 \times 10^{-2} [\frac{m}{s}]$, while all other parameters were kept constant. Unlike at the Reynolds number of $Re = 10$, the Reynolds number of $Re = 500$ leads to transient behaviour in the fluid. Vortices detach at the tip of the plate and disperse into the domain. A snapshot of the streamlines is depicted in Fig. 20 clearly demonstrating these vortices. However, they only have a minor effect on the structure which settles at a quasi stationary deflection of $u_y(A) = 8.97 \times 10^{-2} [m] \pm 5.64 \times 10^{-4} [m]$ on a uniform grid. With a refinement of one level

Table 6 Displacement of the rectangular plate for $Re = 10$.

domain size	$u\,[m/s]$	$v\,[\frac{m^2}{s}]$	displacement cantilever $[m]$	displacement $[m]$
$96 \times 144 \times 60$	0.0002	0.0001	$1.52e-4$	$1.656e-4$
$96 \times 144 \times 60$	0.002	0.001	$1.52e-2$	$1.645e-2$
$96 \times 144 \times 60$	0.02	0.01	1.52	1.235
$192 \times 288 \times 120$	0.0002	0.0001	$1.52e-4$	$1.690e-4$
$192 \times 288 \times 120$	0.002	0.001	$1.52e-2$	$1.683e-2$
$192 \times 288 \times 120$	0.02	0.01	1.52	1.249

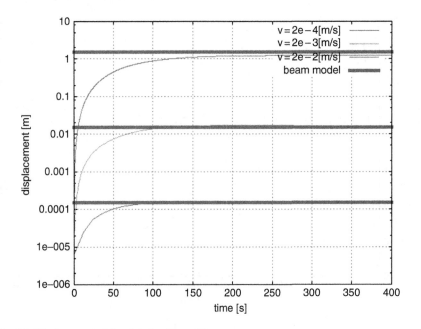

Fig. 18 Displacement of the plate for $Re = 10$.

and two levels the mean deflections are $8.91m \times 10^{-2}[m]$ and $8.93 \times 10^{-2}[m]$. The grid refinement is depicted in Fig. 16

By contrast, Bathe and Ledezma [1] compute the deflection at Point A to be $u_y(A) = 6.6 \times 10^{-2}[m]$. In order to clarify the deviation, the benchmark was recomputed with Ansys-CFX Multiphysics Version 12. The explicit coupling algorithms in Ansys-CFX Multiphysics lead to instabilities for this benchmark which is why a two-way Fluid-Structure interaction with an implicit coupling needed to be run. The implicit coupling required an underrelaxation in the displacements with a factor of $\omega = 0.15$. The time step was chosen to be $\Delta t = 2.5[s]$ for fluid and structure alike. The structure was discretized with 1536 hexahedral elements. It was integrated with a Newmark [59] method with a numerical damping where $\beta = 0.49$ and $\gamma = 0.9$. The fluid was discretized with 569 402 tetrahedral Finite Volume elements with 102 384 nodes. The complete time history of the deflection of point A as computed by VIRTUALFLUIDS-ADHOC (LBM-p-FEM) is directly compared to the time history of Point A obtained by the computation with Ansys-CFX Multiphysics in Fig. 19.

A direct comparison as shown in Fig. 19 is very strict. Different starting values, for example, will almost always lead to different time-histories before the system has locked in. This difference diminishes as time procedes. The deflection at point A is thus compared in its last oscillation only and amounts to $u_y(A) = 9.43 \times 10^{-2}[m] \pm 1.0 \times 10^{-3}[m]$ for the Ansys-CFX Multiphysics computation. It is thus

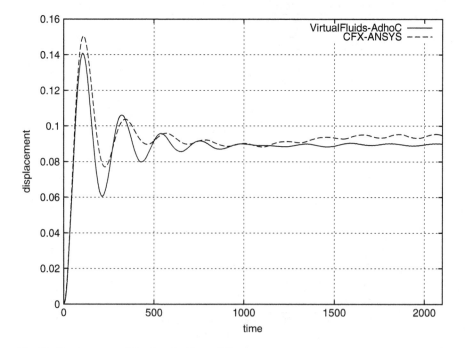

Fig. 19 Displacement of the plate for $Re = 500$.

Fig. 20 Screen shots of the rectangular plate for $Re = 500$ (streamlines, vorticity).

4.9% larger in the mean and a little less than twice as large with respect to the oscillation amplitude as compared to the solution computed by VIRTUALFLUIDS-ADHOC (LBM-*p*-FEM).

These differences might be due to the fact that LBM discretization is still too compressible considering the intermediate Mach number of $Ma = 0.05$ as compared to the incompressible nature of the fluid discretization utilized in CFX. These lower deflections might also partly be attributed to the simpler and less exact next neighbor force transfer method chosen in the LBM-solver. However, the results agree quite well considering the complicated physics involved and the *completely different approaches* used for this comparison. The reason why both computations, (VIRTUALFLUIDS-ADHOC) as well as Ansys-CFX Multiphysics deviate from the results published in [1, 2] could not be detected and the authors are in contact with Prof. Bathe in an effort to gain more insight into this issue.

In a third test the Reynolds number was raised to $Re = 2500$ by increasing the fluid velocity at the inflow to $v = 5.0 \times 10^{-2} [\frac{m}{s}]$ while all other fluid and structural discretization parameters were kept constant w.r.t. the test case $Re = 500$. The resulting deflection at point A is depicted in Fig. 21(a) and the streamlines are depicted in Fig. 21(b). The mean deflection for this case amounts to

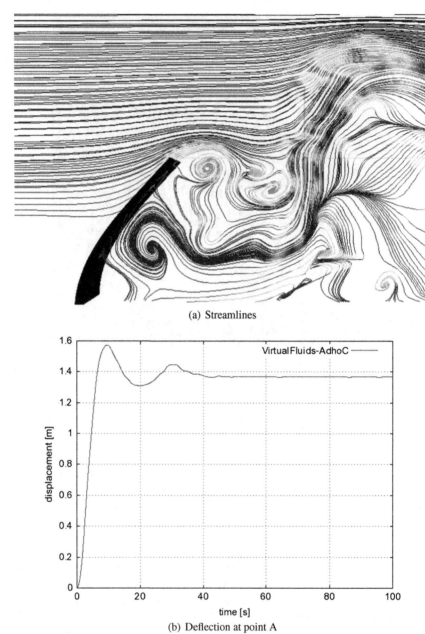

(a) Streamlines

(b) Deflection at point A

Fig. 21 Plate in cross flow at $Re = 2500$.

$u_y(A) = 1.36[m]$. As in the previous case where $Re = 500$, Bathe and Ledezma [1] again arrive at a smaller deflection and give $u_y(A) \approx 1.0[m]$[3].

It is pointed out that the explicit coupling setup VIRTUALFLUIDS-ADHOC proved to be very efficient also in three dimensions. A typical computation for this plate in a cross flow benchmark (e.g. the computation of the time history as shown in Fig. 19) with ANSYS MULTIPHYSICS took approximately four days when computed in parallel with four processes on a modern 2GHz Intel Xenon E5405 four processor CPU's. The same task took only one day with the serial versions of VIRTUALFLUIDS-ADHOC. The proposed setup is therefore by more than one magnitude faster.

10 Beyond benchmarking

10.1 Very large structural deflection

As demonstrated by benchmark of a falling sphere in Sect. 8, it comes as no surprise that the ability to compute very large displacements without the necessity to remesh can also be taken advantage of when coupling to flexible structures. As shown in Fig. 22 it is possible, for example, to let the structural component of the experimental benchmark described in Sect. 7 fall freely through the fluid domain. This can be realized via releasing the Dirichlet boundary conditions at the center of the cylinder. It is stressed that, once the boundary conditions are correctly imposed and the coupling setup is validated and verified, these large deflections impose virtually no additional difficulty to this type of coupling.

10.2 A free surface example

This example is to point out a typical benefit of a partitioned as opposed to a monolithic solution approach. The computation of other types of surface coupled problems are possible with only minor changes. Figure 23 shows the coupling between a tower modeled by high-order FEM hit by a free surface wave. Again, the tower motion was computed by ADHOC and the fluid by VIRTUALFLUIDS.

---

[3] Part of the problem here is that such comparative studies are painfully time consuming, even on the most modern machines. The last run with Ansys-CFX Multiphysics for the case $Re = 500$, for example, took 4 days even though it was computed in parallel using one Intel Xeon E5405 processors with four cores at 2.0[GHz]. Many runs were required to tune the simulation parameters such as a good underrelaxation factor, the proper time step, tight but not too tight convergence values, a fine but computable spatial resolution of the domain, etc..

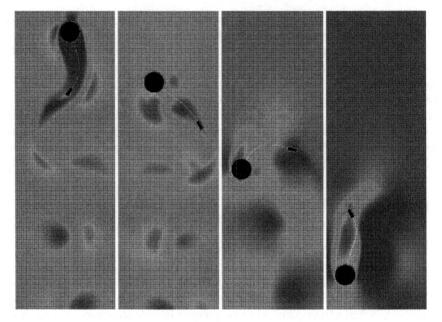

Fig. 22 Very large displacement FSI: free falling experimental benchmark[5] ($Re = 140$).

10.3 Towards realistic engineering problems

Large parts of this work dealt with the verification and validation of examples which are interesting for research purposes in order to gain confidence in the involved codes and the newly developed methods. However, having achieved this it is of interest as well to attempt to compute examples of industrial scale. The Millau bridge fits into this class of problems. This example is to serve as a motivation for future development and to discuss the principle applicability of the setup to problems of this size.

The Millau bridge is the world largest cable-stayed road-bridge spanning the valley of the river Tan in southern France [84]. A picture of the bridge in construction is depicted in Fig. 24. It was constructed using a timed shifting method. It is prone to wind induced vibrations especially in the construction phase when the cantilever arm of the bridge is not yet supported by the subsequent bridge pier.

The computational model of this most critical situation is depicted in Fig. 25. The structure was discretized by 239 Hexahedral elements of order 4 in all directions while 2.85×10^6 nodes were used for the LBM computation in VIRTUALFLUIDS. Figure 25 depicts this computational setup, the geometry of the bridge and the valley. The slices show the magnitude of the velocity vector. The Reynolds number was set to $Re = 1.0 \times 10^6$ and a Smagorinsky turbulence model [75] was applied. For examples of such size, three main problems arise: (a) The spatial resolution

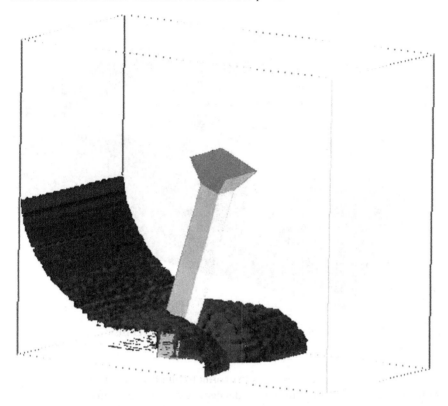

Fig. 23   Tower hit by a wave.

Fig. 24   Millau Bridge in construction [84].

Fig. 25 Millau Bridge modeled by ADHOC coupled to VIRTUALFLUIDS.

necessary to resolve the eddies leads to enormous numbers of degrees of freedoms. Although 2.85×10^6 nodes for the model depicted in Fig. 25 seems a lot but this is not enough. With the computational domain having a hight of $400[m]$ the resolution along this height is only about 200 nodes resulting in approximately one node for every two meters or about 2.5 over the height of the cross section of the bridge. One way to reach an adequate resolution around the bridge is to utilize local refinement. Another possibility to model eddies which stem from obstacles with the same size as the grid resolution is to apply a cascaded LBM model; see for example [25] in which a vortex shedding behind a cylinder with the radius equal to the grid spacing is accurately modeled. (b) The structural model might yet be too simple. For a realistic modeling, structural cable elements would be required. Generally, depending on the level of detail required, the structural solver, too, reaches its limits in terms of computational complexity. In an effort to reduce the structural complexity of the model, the bridge deck and pillars are currently modeled as one solid body. This can lead to an overestimated rotational stiffness due to an overestimated shear modulus. The computational models are currently refined step by step in order to be able to tackle this class of engineering problems.

11 Conclusions

This contribution presents an explicit coupling of a Lattice Boltzmann Method to a high-order structural discretization for bi-directional Fluid-Structure interaction with very large displacements. Based on the work of Krafczyk [46,47], it was jointly developed in the course of the work for the DFG-Research Unit 493 'Fluid-Structure Interaction: Modeling, Simulation, Optimization' Previous contributions in to this project were presented in [27, 29, 44, 45, 71].

In all examples, the presented setup turned out to be well suited for this type of Fluid-Structure interaction. Within this setup it was possible to:

- Use an efficient, explicit overall coupling approach in a straight forward manner,
- verify *and* validate the results against purely computational and experimental benchmarks, respectively,
- compute examples where the structure undergoes *very* large deflections, only limited by the size of the computational domain,
- extend the approach to three dimensions and apply the methodology to examples of large scale.

Moreover, this work has applied the proposed methods to realistic engineering problems. In this context, benchmarks with turbulent flow are subject of current investigations.

Acknowledgements Financial support of the first authors of each research group (S.Geller and S. Kollmannsberger) by the German Research Foundation in the framework of the Research Unit 493 *Fluid-Struktur-Wechselwirkung* is greatfully acknowledged. The authors would also like to thank K. Iglberger for providing the PE-PHYSICSENGINE.

Appendix

Transformation matrix **M**

$$
\begin{bmatrix}
1\cdot & (1 & 1 & 1 & 1 & 1 & 1 & 1 & 1 & 1 & 1 & 1 & 1 & 1 & 1 & 1 & 1 & 1 & 1) \\
c^2\cdot & (-1 & 0 & 0 & 0 & 0 & 0 & 0 & 1 & 1 & 1 & 1 & 1 & 1 & 1 & 1 & 1 & 1 & 1) \\
c^4\cdot & (1 & -2 & -2 & -2 & -2 & -2 & -2 & 1 & 1 & 1 & 1 & 1 & 1 & 1 & 1 & 1 & 1 & 1) \\
c\cdot & (0 & 1 & -1 & 0 & 0 & 0 & 0 & 1 & -1 & 1 & -1 & 1 & -1 & 1 & -1 & 0 & 0 & 0 & 0) \\
c^3\cdot & (0 & -2 & 2 & 0 & 0 & 0 & 0 & 1 & -1 & 1 & -1 & 1 & -1 & 1 & -1 & 0 & 0 & 0 & 0) \\
c\cdot & (0 & 0 & 0 & 1 & -1 & 0 & 0 & 1 & -1 & -1 & 1 & 0 & 0 & 0 & 0 & 1 & -1 & 1 & -1) \\
c^3\cdot & (0 & 0 & 0 & -2 & 2 & 0 & 0 & 1 & -1 & -1 & 1 & 0 & 0 & 0 & 0 & 1 & -1 & 1 & -1) \\
c\cdot & (0 & 0 & 0 & 0 & 0 & 1 & -1 & 0 & 0 & 0 & 0 & 1 & -1 & -1 & 1 & 1 & -1 & -1 & 1) \\
c^3\cdot & (0 & 0 & 0 & 0 & 0 & -2 & 2 & 0 & 0 & 0 & 0 & 1 & -1 & -1 & 1 & 1 & -1 & -1 & 1) \\
c^2\cdot & (0 & 2 & 2 & -1 & -1 & -1 & -1 & 1 & 1 & 1 & 1 & 1 & 1 & 1 & 1 & -2 & -2 & -2 & -2) \\
c^4\cdot & (0 & -2 & -2 & 1 & 1 & 1 & 1 & 1 & 1 & 1 & 1 & 1 & 1 & 1 & 1 & -2 & -2 & -2 & -2) \\
c^2\cdot & (0 & 0 & 0 & 1 & 1 & -1 & -1 & 1 & 1 & 1 & 1 & -1 & -1 & -1 & -1 & 0 & 0 & 0 & 0) \\
c^4\cdot & (0 & 0 & 0 & -1 & -1 & 1 & 1 & 1 & 1 & 1 & 1 & -1 & -1 & -1 & -1 & 0 & 0 & 0 & 0) \\
c^2\cdot & (0 & 0 & 0 & 0 & 0 & 0 & 0 & 1 & 1 & -1 & -1 & 0 & 0 & 0 & 0 & 0 & 0 & 0 & 0) \\
c^2\cdot & (0 & 0 & 0 & 0 & 0 & 0 & 0 & 0 & 0 & 0 & 0 & 0 & 0 & 0 & 0 & 1 & 1 & -1 & -1) \\
c^2\cdot & (0 & 0 & 0 & 0 & 0 & 0 & 0 & 0 & 0 & 0 & 0 & 1 & 1 & -1 & -1 & 0 & 0 & 0 & 0) \\
c^3\cdot & (0 & 0 & 0 & 0 & 0 & 0 & 0 & 1 & -1 & 1 & -1 & -1 & 1 & -1 & 1 & 0 & 0 & 0 & 0) \\
c^3\cdot & (0 & 0 & 0 & 0 & 0 & 0 & 0 & -1 & 1 & 1 & -1 & 0 & 0 & 0 & 0 & 1 & -1 & 1 & -1) \\
c^3\cdot & (0 & 0 & 0 & 0 & 0 & 0 & 0 & 0 & 0 & 0 & 0 & 1 & -1 & -1 & 1 & -1 & 1 & 1 & -1)
\end{bmatrix}
$$

Orthogonal basis vectors $\{\Phi_i, i = 0, \ldots, b-1\}$

$$\Phi_{0,\alpha} = 1, \quad \Phi_{1,\alpha} = \mathbf{e}_\alpha^2 - c^2, \quad \Phi_{2,\alpha} = 3(\mathbf{e}_\alpha^2)^2 - 6\mathbf{e}_\alpha^2 c^2 + c^4, \tag{31}$$

$$\Phi_{3,\alpha} = e_{\alpha x}, \quad \Phi_{5,\alpha} = e_{\alpha y}, \quad \Phi_{7,\alpha} = e_{\alpha z}, \tag{32}$$

$$\Phi_{4,\alpha} = (3\mathbf{e}_\alpha^2 - 5c^2)e_{\alpha x}, \quad \Phi_{6,\alpha} = (3\mathbf{e}_\alpha^2 - 5c^2)e_{\alpha y},$$

$$\Phi_{8,\alpha} = (3\mathbf{e}_\alpha^2 - 5\,c^2)e_{\alpha z}, \tag{33}$$

$$\Phi_{9,\alpha} = 3e_{\alpha x}^2 - \mathbf{e}_\alpha^2, \quad \Phi_{11,\alpha} = e_{\alpha y}^2 - e_{\alpha z}^2, \tag{34}$$

$$\Phi_{13,\alpha} = e_{\alpha x}e_{\alpha y}, \quad \Phi_{14,\alpha} = e_{\alpha y}e_{\alpha z}, \quad \Phi_{15,\alpha} = e_{\alpha x}e_{\alpha z}, \tag{35}$$

$$\Phi_{10,\alpha} = (2\mathbf{e}_\alpha^2 - 3\,c^2)(3e_{\alpha x}^2 - \mathbf{e}_\alpha^2), \quad \Phi_{12,\alpha} = (2\mathbf{e}_\alpha^2 - 3\,c^2)(e_{\alpha y}^2 - e_{\alpha z}^2), \tag{36}$$

$$\Phi_{16,\alpha} = (e_{\alpha y}^2 - e_{\alpha z}^2)e_{\alpha x}, \quad \Phi_{17,\alpha} = (e_{\alpha z}^2 - e_{\alpha x}^2)e_{\alpha y},$$

$$\Phi_{18,\alpha} = (e_{\alpha x}^2 - e_{\alpha y}^2)e_{\alpha z}. \tag{37}$$

References

1. K.J. Bathe and G.A. Ledezma. Benchmark problems for incompressible fluid flows with structural interactions. *Computers and Structures*, 85:628–644, 2007.
2. K.J. Bathe and H. Zhang. A mesh adaptivity procedure for CFD & fluid-structure interactions. *Computers and Structures*, 87:604–617, 2009.
3. M. Bouzidi, M. Firdaouss, and P. Lallemand. Momentum transfer of a Boltzmann-Lattice fluid with boundaries. *Physics of Fluids*, 13(11):3452–3459, 2001.
4. M. Brenk, H.-J. Bungartz, M. Mehl, and T. Neckel. Fluid-Structure Interaction on Cartesian Grids: Flow Simulation and Coupling Environment. In *Fluid-Structure Interaction, Modelling, Simulation and Optimisation*, volume 53 of *Lecture Notes in Computational Science and Engineering*, pages 294–335. Springer, 2006.
5. Encyclopedia Britannica. fluid mechanics. *Encyclopedia Britannica, http://www.britannica.com/EBchecked/topic/211272/fluid-mechanics*, 2009.
6. S. Chapman and T.G. Cowling. *The Mathematical Theory of Non-Uniform Gases*. Cambridge University Press, New York, 1990.
7. J. Chung and G.M. Hulbert. A time integration algorithm for structural dynamics with improved numerical dissipation: the generalized-α method. *Journal of Applied Mechanics, Transactions ASME*, 60:371–375, 1993.
8. B. Crouse. *Lattice-Boltzmann Strömungssimulation auf Baumdatenstrukturen*. PhD thesis, Lehrstuhl für Bauinformatik, Fakultät für Bauingenieur- und Vermessungswesen, Technische Universität München, 2003.
9. B. Crouse, E. Rank, M. Krafczyk, and J. Tölke. A LB-based approach for adaptive flow simulations. *International Journal of Modern Physics B*, 17:109–112, 2003.
10. D. d'Humières. Generalized lattice-Boltzmann equations. In B. D. Shizgal and D. P. Weave, editors, *Rarefied Gas Dynamics: Theory and Simulations*, volume 159 of *Prog. Astronaut. Aeronaut.*, pages 450–458, Washington DC, 1992. AIAA.
11. D. d'Humières, I. Ginzburg, M. Krafczyk, P. Lallemand, and L. Luo. Multiple-relaxation-time lattice Boltzmann models in three dimensions. *Philosophical Transactions: Mathematical, Physical and Engineering Sciences*, 360(1792):437–451, 2002.
12. A. Düster, H. Bröker, H. Heidkamp, U. Heißerer, S. Kollmannsberger, Z. Wassouf, R. Krause, A. Muthler, A. Niggl, V. Nübel, M. Rücker, and D. Scholz. *AdhoC^4 – User's Guide*. Lehrstuhl für Bauinformatik, Technische Universität München, 2004.
13. A. Düster, H. Bröker, and E. Rank. The p-version of the finite element method for three-dimensional curved thin walled structures. *International Journal for Numerical Methods in Engineering*, 52:673–703, 2001.
14. A. Düster, L. Demkowicz, and E. Rank. High order finite elements applied to the discrete Boltzmann equation. *International Journal for Numerical Methods in Engineering*, 67:1094–1121, 2006.
15. A. Düster, S. Hartmann, and E. Rank. p-fem applied to finite isotropic hyperelastic bodies. *Computer Methods in Applied Mechanics and Engineering*, 192:5147–5166, 2003.

16. A. Düster, A. Niggl, and E. Rank. Applying the *hp-d* version of the FEM to locally enhance dimensionally reduced models. *Computer Methods in Applied Mechanics and Engineering*, 196:3524–3533, 2007.
17. A. Düster, D. Scholz, and E. Rank. *pq*-Adaptive solid finite elements for three-dimensional plates and shells. *Computer Methods in Applied Mechanics and Engineering*, 197:243–254, 2007.
18. A. Fayon and J. Happel. Effect of a cylindrical boundary on fixed rigid sphere in a moving fluid. *AIChE journal*, 6:55–58, 1960.
19. C. Felippa, K. Park, and J. DeRuntz. Stabilization of staggered solution procedures for fluid-structure interaction analysis. In T. Blytschko and T. Geers, editors, *Computational Methods for Fluid-Structure Interaction Problems*, volume 26, pages 95–124. American Society of Mechanical Engineers, New York, 1977.
20. C. Felippa, K. Park, and C. Farhat. Partitioned analysis of coupled mechanical systems. *Computer Methods in Applied Mechanics and Engineering*, 190:3247–3270, 2001.
21. Z.G. Feng and E.E. Michaelides. The immersed boundary-lattice Boltzmann method for solving fluid-particles interaction problems. *Journal of Computational Physics*, 195:602–628, 2004.
22. O. Filippova and D. Hänel. Boundary-fitting and local grid refinement for LBGK models. *International Journal of Modern Physics C*, 8:1271–1279, 1998.
23. Message Passing Interface Forum. MPI: A Message-Passing Interface Standard. *International Journal of Supercomputer Applications*, 8(3/4), 1994.
24. S. Freudiger. *Entwicklung eines parallelen, adaptiven, komponentenbasierten Strmungskerns fr hierarchische Gitter auf Basis des Lattice Boltzmann Verfahrens*. PhD thesis, Technische Universität Braunschweig, 2009.
25. M. Geier. De-aliasing and stabilization formalism of the cascaded lattice Boltzmann automaton for under-resolved high reynolds number flow. *International Journal for Numerical Methods in Engineering*, 56:1249–1254, 2008.
26. S. Geller, M. Krafczyk, J. Tölke, S. Turek, and J. Hron. Benchmark computations based on Lattice-Boltzmann, Finite Element and Finite Volume Methods for laminar Flows. *Computers & Fluids*, 35:888–897, 2006.
27. S. Geller, J. Tölke, and M. Krafczyk. Lattice Boltzmann Methods on Quadree-Type Grids for Fluid-Structure Interaction. In H.J. Bungartz and M. Schäfer, editors, *Fluid-Structure Interaction, Modelling, Simulation and Optimisation*, volume 53 of *Lecture Notes in Computational Science and Engineering*, pages 270–293. Springer, 2006.
28. S. Geller, J. Tölke, M. Krafczyk, S. Kollmannsberger, A. Düster, and E. Rank. A coupling algorithm for high order solids and lattice Boltzmann fluid solvers. In P. Wesseling, E. Onate, and J. Periaux, editors, *Proceedings of the European Conference on Computational Fluid Dynamics, ECCOMAS CFD 2006*, TU Delft, The Netherlands, 2006.
29. S. Geller, J. Tölke, M. Krafczyk, D. Scholz, A. Düster, and E. Rank. Simulation of bidirectional fluid-structure interaction based on explicit coupling approaches of Lattice Boltzmann and p-FEM solvers. In *Proceedings of the Int. Conf. on Computational Methods for Coupled Problems in Science and Engineering*, Santorini, Greece, 2005.
30. I. Ginzburg and D. d'Humières. Multi-reflection boundary conditions for lattice Boltzmann models. *Physical Review E*, 68:066614, 2003.
31. J.P. Gomes and H. Lienhart. Experimental study in a fluid-structure interaction reference test case. In H.J. Bungartz and M. Schäfer, editors, *Fluid-Structure Interaction, Modelling, Simulation and Optimisation*, volume 53 of *Lecture Notes in Computational Science and Engineering*, pages 294–335. Springer, 2006.
32. J.P. Gomes and H. Lienhart. Experimental study in a fluid-structure interaction reference test case. In H.J. Bungartz and M. Schäfer, editors, *Fluid-Structure Interaction*, Lecture Notes in Computational Science and Engineering. Springer, 2010. to appear.
33. A. Halfmann. *Ein geometrisches Modell zur numerischen Simulation der Fluid-Struktur-Interaktion windbelasteter, leichter Flächentragwerke*. PhD thesis, Lehrstuhl für Bauinformatik, Fakultät für Bauingenieur- und Vermessungswesen, Technische Universität München, 2002.

34. K. Han, Y.T. Feng, and D.R.J. Owen. Coupled lattice Boltzmann and discrete element modelling of fluid-particle interaction problems. *Computers & Structures*, 85:1080–1088, 2007.

35. K. Han, Y.T. Feng, and D.R.J. Owen. Coupled lattice Boltzmann method and discrete element modelling of particle transport in turbulent fluid flows. *International Journal for Numerical Methods in Engineering*, 72:1111–1134, 2007.

36. U. Heißerer, S. Hartmann, A. Düster, W. Bier, Z. Yosibash, and E. Rank. *p*-fem for finite deformation powder compaction. *Computer Methods in Applied Mechanics and Engineering*, 197:727–740, 2008.

37. U. Heißerer, S. Hartmann, A. Düster, and Z. Yosibash. On volumetric locking-free behavior of p-version finite elements under finite deformations. *Communications in Numerical Methods in Engineering*, 24:1019–1032, 2008.

38. J. Hron and S. Turek. A monolithic FEM solver for ALE formulation of fluid structure interaction with configurations for numerical benchmarking. In *Proceedings of the International Conference on Computational Methods for Coupled Problems in Science and Engineering*, Santorini, 2005.

39. B. Hübner, E. Walhorn, and D. Dinkler. A monolithic approach to fluid-structure interaction using space-time finite elements. *Computer Methods in Applied Mechanics and Engineering*, 193:2087–2104, 2004.

40. K. Iglberger. pe – physics engine. http://www10.informatik.uni-erlangen.de/~klaus/, 2009.

41. K. Iglberger, N. Thürey, and U. Rüde. Simulation of moving particles in 3d with the lattice Boltzmann method. *Computers & Mathematics with Applications*, 55:1461–1468, 2008.

42. M. Junk, A. Klar, and L. S. Luo. Asymptotic analysis of the lattice Boltzmann equation. *Journal of Computational Physics*, 210:676, 2005.

43. S. Kollmannsberger, A. Düster, and E. Rank. FSI based on bidirectional coupling of high order solids to a Lattice Boltzmann Method. In *Proceedings of PVP 2006*, 11th International Symposium on Emerging Technologies in Fluids, Structures, and Fluid/Structure Interactions, within the ASME Pressure Vessel and Piping Conference, Vancouver, B.C, Canada, 2006.

44. S. Kollmannsberger, S. Geller, A. Düster, J. Tölke, M. Krafczyk, and E. Rank. Fluid-Structure Interaction based on Lattice Boltzmann and *p*-FEM: Verification and Validation. In *Proceedings of the International Conference on Computational Methods for Coupled Problems in Science and Engineering*, Ischia Island, Italy, 2009.

45. S. Kollmannsberger, S. Geller, A. Düster, J. Tölke, C. Sorger, M. Krafczyk, and E. Rank. Fixed-grid Fluid-Structure interaction in two dimensions based on a partitioned Lattice Boltzmann and *p*-FEM approach. *International Journal for Numerical Methods in Engineering*, 79:817–845, 2009.

46. M. Krafczyk. *Gitter-Boltzmann-Methoden: Von der Theorie zur Anwendung*. Postdoctoral thesis, Lehrstuhl für Bauinformatik, Fakultät für Bauingenieur- und Vermessungswesen, Technische Universität München, 2001.

47. M. Krafczyk, J. Toelke, E. Rank, and M. Schulz. Two-dimensional simulation of fluid-structure interaction using lattice-boltzmann methods. *Computers and Structures*, 79:2031–2037, 2001.

48. U. Küttler and W. Wall. Vector extrapolation for strong coupling fluid-structure interaction solvers. *Journal of Applied Mechanics*, 2(76), 2009.

49. A. Ladd. Numerical simulations of particulate suspensions via a discretized Boltzmann equation. Part 1: theoretical foundations. *Journal of Fluid Mechanics*, 271:285, 1994.

50. A. Ladd. Numerical simulations of particulate suspensions via a discretized Boltzmann equation. Part 2: numerical results. *Journal of Fluid Mechanics*, 271:311, 1994.

51. P. Lallemand and Luo L.-S. Lattice boltzmann method for moving boundaries. *Journal of Computational Physics*, 184:406–421, 2003.

52. P. Lallemand and L. Luo. Theory of the lattice Boltzmann method: Dispersion, dissipation, isotropy, Galilean invariance, and stability. *Physical Review E*, 61(6):6546–6562, 2000.

53. P. LeTallec and J. Mouro. Fluid structure interaction with large structural displacements. *Computer Methods in Applied Mechanics and Engineering*, 190:3039–3067, 2001.

54. R. Löhner, J.R. Cebral, F.F. Camelli, J.D. Baum, E.L. Mestreau, and O.A. Soto. Adaptive embedded/immersed unstructured grid techniques. *Archives Of Computational Methods In Engineering*, 14:279–301, 2007.
55. W.E. Lorensen and H.E. Cline. A Discontinuous Galerkin Method for the Navier-Stokes Equations. *ACM SIGGRAPH Computer Graphics archive*, 21:163–169, 1987.
56. X.J. Luo. *An automatic adaptive directional variable p-version method in 3D curved domains*. PhD thesis, Rensselaer Polytechnic Institute, Troy, New York, 2005.
57. R. Mei, L.S. Luo, P. Lallemand, and D. d'Humières. Consistent Initial Conditions for LBE Simulations. *Computers & Fluids*, 35:855–862, 2006.
58. R. Mei, D. Yu, W. Shyy, and L Lou. Force evaluation in the lattice Boltzmann method involving curved gemoetry. *Physical Review E*, 65:041203, 2002.
59. N.M. Newmark. A numerical method for structural dynamics. *Journal of Engineering Mechanics (ASCE)*, 85:67–94, 1959.
60. H. Oertel, Böhle M., and U. Dohrmann. *Strömnungsmechanik*. Vierweg+Teubner, 5. edition, 2009.
61. K. Park and C. Felippa. Partitioned analysis of coupled systems. In T. Blytschko and T. Hughes, editors, *Computational Methods for Transient Analysis*, chapter 3, pages 157–219. North-Holland, Amsterdam-New York, 1984.
62. S. Piperno and C. Farhat. Partitioned procedures for the transient solution of coupled aroelastic problems – part II: Energy transfer analysis and three dimensional applications. *Computer Methods in Applied Mechanics and Engineering*, 190:3147–3170, 2001.
63. S. Piperno, C. Farhat, and B. Larrouturou. Partitioned procedures for the transient solution of coupled aroelastic problems – part I: Model problem, theory and two-dimensional application. *Computer Methods in Applied Mechanics and Engineering*, 124:79–112, 1995.
64. Y.H. Qian, D. d'Humières, and P. Lallemand. Lattice BGK models for Navier-Stokes equations. *Europhysics Letters*, 17(6):479–484, 1992.
65. E. Rank, A. Düster, V. Nübel, K. Preusch, and O.T. Bruhns. High order finite elements for shells. *Computer Methods in Applied Mechanics and Engineering*, 194:2494–2512, 2005.
66. M. Rheinländer. A consistent grid coupling method for lattice-boltzmann schemes. *Journal of Statistical Physics*, 121:49–74, 2005.
67. L. Schiller and A.Z. Naumann. Über die grundlegenden berechnungen bei der schwerkraftaufbereitung. *Journal of Verein Deutscher Ingenieure*, 77:318–320, 1933.
68. D. Scholz. *An anisotropic p-adaptive method for linear elastostatic and elastodynamic analysis of thin-walled and massive structures*. Dissertation, Lehrstuhl für Bauinformatik, Fakultät für Bauingenieur- und Vermessungswesen, Technische Universität München, 2006.
69. D. Scholz, A. Düster, and E. Rank. Model-adaptive fluid-structure interaction using high order structural elements. In *Proceedings of the Int. Conf. on Computational Methods for Coupled Problems in Science and Engineering*, Santorini, Greece, 2005.
70. D. Scholz, A. Düster, and E. Rank. Model-adaptive structural FEM computations for fluid-structure interaction. In *Proceedings of the Third M.I.T. Conference on Computational Fluid and Solid Mechanics*, Cambridge, USA, 2005.
71. D. Scholz, S. Kollmannsberger, A. Düster, and E. Rank. Thin Solids for Fluid-Structure Interaction. In H.J. Bungartz and M. Schäfer, editors, *Fluid-Structure Interaction, Modelling, Simulation and Optimisation*, volume 53 of *Lecture Notes in Computational Science and Engineering*, pages 294–335. Springer, 2006.
72. X. Shi and S.P. Lim. A LBM-DLM/FD method for 3D fluid-structure interactions. *Journal of Computational Physics*, 226:2028–2043, 2007.
73. X. Shi and N. Phan-Thien. Distributed Lagrange multiplier/fictitious domain method in the framework of lattice Boltzmann method for fluid-structure interactions. *Journal of Computational Physics*, 206:81–94, 2005.
74. Xing Shi and Siak Piang Lim. A LBM-DLM/FD method for 3D fluid-structure interactions. *Journal of Computational Physics*, 226:2028–2043, 2007.
75. J. Smagorinsky. General circulation experiments with the primitive equations – i the basic experiment. *Monthly Weather Review*, 91:99–164, 1963.

76. Y. Sui, Y.-T. Chew, P. Roy, and H.-T. Low. A hybrid immersed-boundary and multi-block lattice Boltzmann method for simulating fluid and moving-boundaries interactions. *International Journal for Numerical Methods in Engineering*, 53:1727–1754, 2007.

77. B.A. Szabó and I. Babuška. *Finite element analysis*. John Wiley & Sons, 1991.

78. B.A. Szabó, A. Düster, and E. Rank. The p-version of the Finite Element Method. In E. Stein, R. de Borst, and T. J. R. Hughes, editors, *Encyclopedia of Computational Mechanics*, volume 1, chapter 5, pages 119–139. John Wiley & Sons, 2004.

79. N. Thürey. *A single-phase free-surface Lattice-Boltzmann Method*. PhD thesis, Lehrstuhl f"ur Systemsimulation (Informatik 10), Universität Erlangen-Nürnberg, 2003.

80. S. Turek and J. Hron. Proposal for Numerical Benchmarks for Fluid-Structure Interaction between an Elastic Object and Laminar Incompressible Flow. In H.J. Bungartz and M. Schäfer, editors, *Fluid-Structure Interaction, Modelling, Simulation and Optimisation*, volume 53 of *Lecture Notes in Computational Science and Engineering*, pages 371–385. Springer, 2006.

81. S. Turek and J. Hron. Numerical Benchmarks for Fluid-Structure Interaction between an Elastic Object and Laminar Incompressible Flow. In H.J. Bungartz and M. Schäfer, editors, *Fluid-Structure Interaction*, Lecture Notes in Computational Science and Engineering. Springer, 2010. to appear.

82. E. Walhorn, A. Kölke, B. Hübner, and Dinkler D. Fluidstructure coupling within a monolithic model involving free surface flows. *Computers & Structures*, 83:2100–2111, 2005.

83. W. Wall, P. Gammnitzer, and A. Gerstenberger. A strong coupling partitioned approach for fluid-structure interaction with free surfaces. *Computers & Fluids*, 36:169–183, 2007.

84. Wikipedia. Viaduc de millau. http://de.wikipedia.org/wiki/Viaduc_de_Millau.

85. D. Yu, R. Mei, and W. Shyy. A multi-block lattice Boltzmann method for viscous fluid flows. *International Journal for Numerical Methods in Fluids*, 39:99–120, 2002.

An XFEM Based Fixed-Grid Approach for 3D Fluid-Structure Interaction

W.A. Wall, A. Gerstenberger, U. Küttler, and U.M. Mayer

Abstract This paper gives an overview on our recent research activities on a fixed grid fluid-structure interaction scheme that can be applied to the interaction of most general structures with incompressible flow. The developed approach is based on an eXtended Finite Element Method (XFEM) based strategy to allow moving interfaces on fixed Eulerian fluid grids. The enriched Eulerian fluid field and the Lagrangian structural field are partitioned and iteratively coupled using Lagrange multiplier techniques for non-matching grids. The approach allows the simulation of the interaction of thin and bulky structures exhibiting large deformations. Extensions towards automatic adaptivity and fluid boundary layer meshes are sketched and the principle applicability to contact simulations of submerged structures are presented.

1 Introduction

Fluid-structure interaction is of great relevance in many fields of engineering as well is in the applied sciences. Hence, the development and application of respective simulation approaches has gained great attention over the past decades. Some current endeavors in this field are: the advancement from special purpose or special problem to quite general approaches; the desire to capture very general and complex systems; and the exigent need of robust high quality approaches even for such complex cases, *i.e.* approaches that have the potential to turn over from being a challenging and fascinating research topic to a development tool with real predictive capabilities [63]. Often, when interaction effects are essential this comes along with large structural deformations.

A sketch of the general problem of the interaction of a flow field and a flexible structure is shown in Fig. 1. The interface Γ^i separates the structural domain Ω^s

W.A. Wall, A. Gerstenberger, U. Küttler, and U.M. Mayer
Institute for Computational Mechanics, Technische Universität München, Boltzmannstr. 15, 85747 Garching, Germany
e-mail: wall@lnm.mw.tum.de

H.-J. Bungartz et al. (eds.), *Fluid Structure Interaction II*, Lecture Notes in Computational Science and Engineering 73, DOI 10.1007/978-3-642-14206-2_12, © Springer-Verlag Berlin Heidelberg 2010

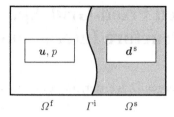

Fig. 1 FSI setup: Interface Γ^{i} separates fluid field Ω^{f} and structural field Ω^{s}.

from the fluid domain Ω^{f}. Most research and commercial codes that are available for simulations of the interaction of flows and flexible, often thin-walled, structures are based on the Arbitrary Lagrangian Eulerian (ALE) method. These approaches go back to early works like [7, 8, 16, 30, 32, 52]. The essential feature of ALE based methods is that the fluid field is formulated and solved on a deforming grid. This fluid grid deforms with the structure at the interface and then the fluid grid is smoothed within the fluid domain. Within the last fifteen years, our research group contributed a number of different ALE-based schemes and has applied them to a variety of problems, see e. g. [19–21, 35–37, 48, 64, 65].

But even the most advanced and best cultured ALE based scheme comes to its limits where only re-meshing helps to prevent severe fluid mesh distortion. In such situations, one might be tempted to turn over to approaches that work with a fixed grid. Here, the interface is described either explicitly, using some kind of Lagrangian interface markers or a Lagrangian structural discretization, or implicitly, using e. g. level-set functions on a fixed fluid grid. Changing properties and discontinuities in the fluid solution have to be taken care of with modifications on the fluid equations and/or fluid discretization.

Prominent fixed grid methods for incompressible flow include the Immersed Boundary (IB) method [54, 55] and its many derivations [38, 40, 46]. It is capable to simulate thin and deformable boundaries and fully fledged, deformable 3d structures submersed in incompressible flow [66, 68]. An approach with many similarities to the IB method is the so called Distributed Lagrange Multiplier / Fictitious Domain (DLM/FD) method [26, 27], which has since been extended to simulate thin, deformable structural surfaces [3, 14, 42] as well as to flexible and fully fledged structures [67]. Both methods have in common a Lagrangian mesh for the structure moving through the fluid mesh and forcing the fluid material inside the structure to deform as the structure. Other fixed-grid or Cartesian-grid methods are, among others, [11], [17] and [43].

In [23], we listed a number of requirements that fixed grid methods should fulfill to be applicable to general structures interacting with incompressible flow:

- the physical and fictitious flow region are completely decoupled such that no artificial energy gain or loss occurs across the interface,
- the fluid-structure coupling takes place only along the interface,
- there is no mesh size dependency between fluid and structure mesh, and
- it should be possible to turn off most parts of the fictitious domain in order to reduce memory and performance consumption.

Until a few years ago, many fixed grid methods in literature did not fulfill one or several of above listed requirements. For instance, in most DLM/FD and IB approaches to date, if used with volume occupying structures, the coupling takes place in the entire structure domain between the fictitious flow field and the structure domain. Since the structure movement is tied to the fictitious flow, it is forced to deform in an incompressible way and artificial viscosity may modify the real physics of the structural domain. Of course the error made by such coupling might not be prohibitive high in specific situations, for example, if incompressible biophysical materials are modelled, however, it still generates unwanted dependencies between the fluid and the structure mesh and adds additional sources of errors for an already complex physical problem.

In an attempt to meet all of the above requirements, we proposed a partitioned iterative coupling scheme between a standard Lagrangian structural description and an Eulerian formulation for the fluid in [23] that uses features of the eXtended Finite Element Method (XFEM) [6,50] and Lagrange multiplier/Mortar techniques [9,10,53]. With such an XFEM based approach for moving boundaries on fixed fluid grids, in principal all of the listed requirements can be addressed, which is demonstrated within this review. Because of the attractive features of XFEM, XFEM-based techniques have become popular for fixed-grid FSI methods and are meanwhile an very active research area, see e. g. [39,61,69].

The paper is structured as follows: The general FSI problem is stated in Sect. 2 and the intermediate interface field as an external reference is defined. The introduction of the intermediate interface field as an external reference allows us to derive the XFEM fluid problem and its coupling to the interface in Sect. 3 separately from the coupling between structure and interface. The interface-structure coupling and the overall solution approach is then reviewed in Sect. 4. Extensions to improve boundary layer reslution by means of boundary layer meshes combined with adaptivity and an outlook towards contact of submerged structures is shown in Sect. 5.

2   Statement of Coupled Fluid-Structure Problem

A general fluid-structure interaction problem statement consists of the description of fluid and solid fields, appropriate fluid-structure interface conditions at the common interface and conditions for the remaining boundaries, respectively.

2.1   Single Fields

For the fluid domain Ω^f with the position vector x, the Navier-Stokes equations for incompressible flow in Eulerian form is stated as

$$\rho^f \frac{\partial u}{\partial t} = -\rho^f u \cdot \nabla u + \nabla \cdot \sigma^f + b^f \qquad \text{in } \Omega^f, \qquad (1)$$

$$\nabla \cdot u = 0 \qquad \text{in } \Omega^f. \qquad (2)$$

Here, the partial time derivative of the fluid velocity u times the fluid density ρ^f is balanced by the convection term, the divergence of the Cauchy stress tensor σ^f and external, velocity independent volumetric forces b^f. For brevity, the volumetric forces are omitted in the subsequent derivation, however they are included in the actual implementation and their derivation requires no additional difficulties. For incompressible flow, the Cauchy stress tensor σ^f is split into pressure p and a deviatoric viscous stress tensor τ^f

$$\sigma^f = -pI + \tau^f. \tag{3}$$

For simplicity of this presentation, we use the Newtonian material law, which defines the viscous stress as a product of the dynamic viscosity denoted as μ and the strain rate tensor ε

$$\tau^f = 2\mu\varepsilon = \mu(\nabla u^f + (\nabla u^f)^T). \tag{4}$$

Dirichlet and Neumann boundary conditions at $\Gamma^{f,D}$ and $\Gamma^{f,N}$ are given as

$$u^f = \hat{u}^f \quad \text{in } \Gamma^{f,D} \quad \text{and} \quad \sigma^f \cdot n^f = \hat{h}^f \quad \text{in } \Gamma^{f,N}, \tag{5}$$

and are applied as usual. Variables with superscript ^ denote prescribed conditions for the fluid computation. Without consideration of a moving interface, the weak form of the Navier-Stokes equation in Eulerian form, after integration by parts, becomes

$$\left(v, \rho^f \frac{\partial u}{\partial t}\right)_{\Omega^f} + \left(v, \rho^f u \cdot \nabla u\right)_{\Omega^f} + \left(\nabla \cdot v, \sigma^f\right)_{\Omega^f}$$
$$+ \left(q, \nabla \cdot u\right)_{\Omega^f} - \left(v, \hat{h}^f\right)_{\Gamma^{f,N}} = 0. \tag{6}$$

For our implementation, we use linear or quadratic equal-order shape functions for velocity and pressure. Such a formulation is known to show instabilities for two reasons – dominating convection and (inf-sup) unstable pairs of velocity and pressure shape functions. The stabilization parameters used throughout this work are discussed and defined in [23].

In most applications, the structure is described using a Lagrangian description, where the material derivative becomes a simple partial derivative with respect to time, such that the momentum equation becomes

$$\rho^s \frac{\partial^2 d^s}{\partial t^2} = \nabla \cdot \sigma^s + \rho^s b^s. \tag{7}$$

The displacement d^s is defined as the difference between the current position x and the initial position X and b^s is an external acceleration field acting on the structural domain. Structural velocity u^s and acceleration a^s are defined as

$$a^s = \frac{\partial u^s}{\partial t} = \frac{\partial^2 d^s}{\partial t^2} \quad \text{or} \quad a^s = \dot{u}^s = \ddot{d}^s.$$

In the large deformation case it is common to describe the constitutive equation using a stress-strain relation based on the Green-Lagrange strain tensor E and the 2. Piola-Kirchhoff stress tensor $S(E)$ as a function of E. The 2. Piola-Kirchhoff stress can be obtained from the Cauchy stress σ as

$$S = J F^{-1} \cdot \sigma \cdot F^{-T} \quad \text{with} \quad E = \frac{1}{2}(F^T \cdot F - I), \quad F = \frac{\partial x}{\partial X},$$

where J denotes the determinant of the deformation gradient tensor F. Dirichlet and Neumann conditions are defined as

$$d^s = \hat{d}^s \quad \text{in } \Gamma^{s,D} \quad \text{and} \quad \sigma^s \cdot n^s = \hat{h}^s \quad \text{in } \Gamma^{s,N}. \tag{8}$$

For transient simulations, initial conditions are required for structural displacements. The weak from after integration by parts without consideration of the fluid-structure coupling is

$$(\delta d^s, \rho^s \ddot{d}^s)_{\Omega^s} + (\nabla \delta d^s, \sigma^s)_{\Omega^s} - (\delta d^s, \hat{h}^s)_{\Gamma^{s,N}} = 0. \tag{9}$$

As demonstrated in Sect. 4, the structural equation is not affected by the fixed-grid fluid treatment and complex material and geometric behavior can be treated as usual. In particular, the equation is discretized using different (hybrid-mixed) element techniques in space and direct time integration schemes. For the numerical results presented in this article we use the St.-Venant-Kirchhoff material law, *i.e.* a linear relation between S and E, for simplicity. All non-linearities are handled via a Newton-Raphson scheme. As it will become clear in the next section, the structural formulation is by no means restricted to non-linear elasticity. Any material model described in the Lagrangean formulation can be used without modification of the fixed-grid FSI scheme presented in this article.

2.2 Fluid-Structure Interface Conditions

Fluid-structure interaction is a *surface*-coupled problem with dynamic and kinematic coupling conditions only along the interface. For our applications, we assume no mass flow normal to the interface. Consequently, the normal velocities at the interface have to match as

$$u^f \cdot n^f = -\dot{d}^s \cdot n^s \qquad \text{in } \Gamma^i. \tag{10}$$

Note the opposite sign due to the different normal vector n^f and n^s for fluid and structural domain, respectively. If viscous fluids are considered, there is usually also a matching condition for the tangential velocities, which can be combined with the equation above to obtain no-slip coupling conditions as

$$u^f = \dot{d}^s \qquad \text{in } \Gamma^i. \tag{11}$$

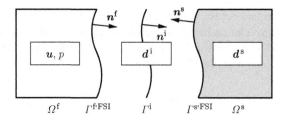

Fig. 2 3-field setup: fluid field Ω^{f}, interface Γ^{i} and structural field Ω^{s} along with respective domain normals and variables.

The force equilibrium requires the surface traction to be equal as

$$\boldsymbol{\sigma}^{\mathrm{f}} \cdot \boldsymbol{n}^{\mathrm{f}} = -\boldsymbol{\sigma}^{\mathrm{s}} \cdot \boldsymbol{n}^{\mathrm{s}} \qquad \text{in } \Gamma^{\mathrm{i}}. \tag{12}$$

A three-field setup as introduced in [23], where fluid and structural fields and an independent interface field are treated separately, has proven to be very valuable in separating formulation and implementation of the fixed-grid FSI approach. This setup – along with the respective variables living on these fields or interfaces – is shown in Fig. 2. Similar ideas for coupling 2 fields using an intermediate reference surface ("frame"), named as localized Lagrange multiplier method, can be found e. g. in [53] and references therein. For the 3-field problem, the interface condition Eq. (11) is modified as

$$\boldsymbol{u}^{\mathrm{f}} = \dot{\boldsymbol{d}}^{\mathrm{i}}, \tag{13}$$

$$\boldsymbol{d}^{\mathrm{i}} = \boldsymbol{d}^{\mathrm{s}}. \tag{14}$$

In other words, if we constrain the velocity of both fields independently to be the same as the interface velocity, then the original matching condition (11) is fulfilled. Let $\boldsymbol{\lambda}$ and $\boldsymbol{\mu}$ be two traction fields on the interface Γ^{i} such that

$$\boldsymbol{\sigma}^{\mathrm{f}} \cdot \boldsymbol{n}^{\mathrm{f}} = \boldsymbol{\lambda}, \tag{15}$$

$$\boldsymbol{\sigma}^{\mathrm{s}} \cdot \boldsymbol{n}^{\mathrm{s}} = \boldsymbol{\mu}. \tag{16}$$

Then the interface traction balance (12) can be stated as

$$\boldsymbol{\lambda} = -\boldsymbol{\mu}. \tag{17}$$

Since we generally assume that neither fluid nor structural discretization match the interface mesh, we need to couple three non-fitting meshes. The single fields can be described separately using the coupling of each field to the interface. During this derivation, the interface serves as a reference field with given position and velocity in time. In the final FSI system, depending on the Lagrange multiplier technique and the monolithic or partitioned solution approach, the interface unknowns remain in the system of equations or can be condensed out.

3 Moving fluid boundaries and interfaces on fixed Eulerian grids

3.1 From Explicit to Implicit Surfaces

The key element of a fixed-grid FSI approach is to properly define moving interfaces on a fixed Eulerian fluid grid. For that purpose, we define a domain Ω that contains the fluid domain Ω^f completely and extends into the structural domain Ω^s. The interface between fluid and structure now becomes an internal interface that separates Ω into two subdomains Ω^+ and Ω^-, where Ω^+ corresponds to the *physical* fluid domain Ω^f and Ω^- is the remaining domain filling Ω. Hence, the flow field in Ω^- is entirely *fictitious* with no physical meaning to the FSI problem and should be removed from the system of equations.

As requested in the introduction, the main tasks are to decouple the physical flow in Ω^+ from the fictitious flow in Ω^- and, using the 3-field setup, to employ the interface conditions between the interface and the physical flow field in Ω^+. The re-formulation of the explicit fluid surface as an embedded interface is depicted in Fig. 3.

The jump in the velocities $[\![u]\!]$ between the physical values u^+ and the void (u^-) can be expressed as

$$[\![u]\!] = u^+ - \underbrace{u^-}_{=0} = \hat{u}^i \quad \text{in } \Gamma^i, \tag{18}$$

where the jump height equals the interface velocity $\hat{u}^i$ from the kinematic fluid-interface coupling condition Eq. (13). The interface velocity $\hat{u}^i$ is assumed to be

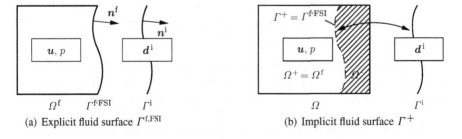

(a) Explicit fluid surface $\Gamma^{f,FSI}$ (b) Implicit fluid surface Γ^+

Fig. 3 Fluid part of the FSI problem: Γ^i divides the computational fluid domain Ω into a physical fluid domain Ω^+ and the fictitious domain Ω^-.

given for the moment. Likewise, one can identify a jump in the stress field, which results in a surface traction vector field λ as

$$[\![\sigma \cdot n]\!] = \sigma^+ \cdot n - \underbrace{\sigma^- \cdot n}_{=0} = \lambda \quad \text{in } \Gamma^i. \tag{19}$$

Initial conditions and Dirichlet and Neumann boundary conditions are only required for Ω^+ as no flow is modelled in Ω^-. Likewise, fluid body forces b, if present, are only applied to Ω^+.

For simplicity of presentation, the one-step-θ time discretization is employed in the following. The time-discrete weak form without interface condition is developed by testing with velocity and pressure test functions v and q, respectively

$$\left(v, \rho^f u\right)_\Omega + \Delta t \theta \left\{ \left(v, \rho^f u \cdot \nabla u - \nabla \cdot \sigma^f\right)_\Omega + \left(q, \nabla \cdot u\right)_\Omega \right\}$$
$$= \left(v, \rho^f u^n + \Delta t (1-\theta) \rho^f a^n\right)_\Omega. \tag{20}$$

The acceleration of the old time step is denoted as a^n. The stabilization terms are omitted to better clarify the basic principles of this approach.

3.2 Moving Interfaces by the XFEM

As should be clear by now, in the proposed fixed grid scheme, the fluid-structure interface is generally not aligned with fluid element surfaces. Consequently, we need to find a way to represent a jump in the primary variables, namely the velocity and pressure, within the elements. Furthermore, also the derivatives have to be discontinuous, since the stress field is discontinuous, too.

The eXtended Finite Element Method has been proposed in [6, 50] to represent discontinuities in the Finite Element approximation. For that purpose, the finite element shape functions are extended or enriched by using additional degrees of freedom combined with known solution or special enrichment functions. Applied to the velocity field, an enriched velocity approximation can be defined as

$$u^h(x,t) = \sum_I N_I(x)\tilde{u}_I + \sum_J N_J(x)\psi(x,t)\check{u}_J + \ldots . \tag{21}$$

Here, $\tilde{u}_I$ represent the standard nodal degrees of freedom at node I, while additional degrees of freedom $\check{u}_I$ multiplied by a properly chosen enrichment function $\psi(x,t)$ are used to enhance the solution. The superscript h indicates the discretized field function.

The choice of the enrichment function depends on the character of the interface conditions. In the fixed-grid fluid setup, there is a jump from the physical field in Ω^+ to zero velocity and pressure in Ω^-. Likewise, derived variables like stress tensors

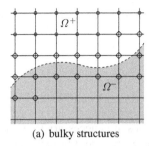
(a) bulky structures

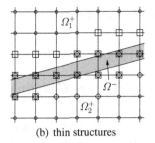
(b) thin structures

Fig. 4 Enrichment for bulky and thin structures: For thin structures, two overlapping enrichments ($\diamondsuit$ and $\square$) represent the discontinuity, while away from the surface, standard degrees of freedom ($\circ$) are used in Ω^+. Uncut elements within the fictitious domain Ω^- get no degrees of freedom.

resulting from velocity derivatives and pressure are only present in the physical flow field, whereas in Ω^- they are zero as no flow occurs. Hence, for our purpose the enrichment function is defined as a step function $\check{\psi}(x,t)$ [13].

$$\check{\psi}(x,t) = \begin{cases} +1 & \text{in } \Omega^+ \\ 0 & \text{in } \Omega^- \end{cases}. \tag{22}$$

Along Γ^i, both velocity and pressure are discontinuous and enriched with $\check{\psi}$.

Several interfaces within one fluid element can be handled without problems, which allows to treat thin structures within an FSI approach on "large" fluid elements. If several interfaces intersect the local support of unknowns, multiple enrichments can be used, using one enrichment for each interface as depicted in Fig. 4. Higher order shape functions are constructed with no additional effort making this approach capable for consistent low and high order finite element approximations. Elements divided by discontinuities are subdivided into subdomains and numerical integration of the weak form is performed over sub-domains [50]. The subdivision is performed using a constrained Delaunay algorithm. The search to find candidates of intersecting background fluid elements and interface elements is accelerated using parallel search tree techniques. For a detailed description of this process for higher order elements with curved interfaces, see [45].

If the interface moves over time, the enriched approximation functions $P(x,t) = N_I(x)\psi(x,t)$ at n and $n + 1$ are generally different near the interface. Mesh topology and fluid node positions do not change. Nodes that happen to be in the structural domain at time step n and are in the fluid domain at the new time level $n + 1$ have to be treated properly as no old solution is available. Our current approach is inspired by the Ghost fluid methods [1, 17, 31], which are popular for Finite Difference and Finite Volume methods. Here, values at the old time level are constructed by extrapolation of each sides interface velocity, which in our case is given by the movement of the Lagrangian interface mesh. After extrapolation, a projection step generates a discrete incompressible velocity field at the old time level. This has proven to be a robust method, however, the temporal accuracy for moving interfaces has to be investigated in more detail. Other ways to deal with this situation are presented in [12] and [69].

3.3 Fluid-Interface Coupling

Due to the non-fitting fluid and interface meshes, it is necessary to weakly enforce the fluid-interface conditions Eq. (13). In our work, two different ways of weakly enforcing the interface constraints on the background grid have been developed and implemented. In the first approach that has been used in the initial 2D implementation [23], the fluid-interface condition is enforced weakly at t^{n+1} at the t^{n+1} location of the interface Γ^i by testing the condition with a test function $\delta\lambda(x)$ along the interface as

$$
\begin{aligned}
\left(v, \rho^f u\right)_\Omega &+ \Delta t \theta \Big\{ \left(v, \rho^f u \cdot \nabla u\right)_\Omega + \left(\nabla \cdot v, -pI + \tau^f\right)_\Omega + \left(q, \nabla \cdot u\right)_\Omega \\
&- \left(v, \lambda\right)_{\Gamma+} - \left(\delta\lambda, u - u^{i,n+1}\right)_{\Gamma+} \Big\} \\
&= \left(v\rho^f, u^n + \Delta t(1-\theta)a^n\right)_\Omega .
\end{aligned}
\tag{23}
$$

The Lagrange multiplier field and the corresponding test functions are discretized along the interface Γ^i, which establishes an interface mesh along Γ^i. Similarly, we can define discrete interface velocities $u^{i,h}$ and displacements $d^{i,h}$. The choice of appropriate approximations for λ can not be made without consideration of the underlying fluid discretization. Inappropriate choices can lead to leakage, oscillations (zero-energy modes) or locking due to violation of the inf-sup condition, see e. g. [34, 39, 49]. For the 2D implementation in [23], we used bi-quadratic ansatz functions N_I for fluid velocity and pressure and linear ansatz function N_I^i for interface velocities and the Lagrange multiplier field. The Lagrange multiplier nodes where placed at the intersection of fluid element edges and structural surface element edges.

This approach has been used successfully for 2D problems and its accuracy has been shown in [22, 23]. However, it is quite hard to construct stable interface Lagrange multiplier spaces for 3D models. Furthermore, when thinking about large-scale computations as typically encountered for 3D fluid problems, the parallel interface mesh generation can become costly and solving the resulting saddle-point problem in Eq. (23) poses additional problems for the parallel iterative solution of the linearized fluid system. Hence, we proposed an alternative approach in [24] that is simpler to implement in parallel and does not result in a saddle point problem like the Lagrange multiplier approach. It also does not require user-provided stabilization parameters for the interface constraint like Nitsches method [4,15,18,28,29,51], an alternative emerging approach without saddle point structure.

The key point of the new method is the replacement of the vector field λ along the interface with an additional stress field $\bar{\sigma}$ as independent unknown within intersected fluid elements. Hence, we have three primary unknown fields, an elementwise continuous velocity u and pressure p and an elementwise discontinuous Cauchy stress $\bar{\sigma}$ and their corresponding test functions v, q and $\bar{\gamma}$, respectively. An extra strain rate balance equation defines the 6 unknowns of the symmetric stress tensor field uniquely. The resulting task is therefore: find u, p and $\bar{\sigma}$ such that

$$
\begin{aligned}
& \left(v, \rho^{\mathrm{f}} u\right)_{\Omega} + \Delta t \theta \big\{ (v, \rho^{\mathrm{f}} u \cdot \nabla u)_{\Omega} + \left(\nabla v, -p I + \tau\right)_{\Omega} + (q, \nabla \cdot u)_{\Omega} \\
& - (\bar{\gamma}, \bar{\varepsilon} - \varepsilon)_{\Omega} - \left(\bar{\gamma} \cdot n^{\mathrm{f}}, u - u^{\mathrm{i}}\right)_{\Gamma^{\mathrm{i}}} - \left(v, \bar{\sigma} \cdot n^{\mathrm{f}}\right)_{\Gamma^{\mathrm{i}}} \big\} \\
& = \left(v \rho^{\mathrm{f}}, u^n + \Delta t(1 - \theta) \dot{u}^n\right)_{\Omega} + \Delta t \theta \left(v, \hat{h}\right)_{\Gamma_{\mathrm{N}}}.
\end{aligned}
\tag{24}
$$

Dirichlet and Neumann conditions away from the interface are employed as usual. Note that we intentionally do not replace τ in the fluid momentum equation with the primary stress unknowns. This way, the (stabilized) fluid formulation remains unchanged. Such a replacement, however, might become an option, if complex fluid materials are modelled as, e. g. in [5]. The two strain rates $\bar{\varepsilon}$ and ε and the velocity-dependent stress tensor τ are computed from the primary unknowns

$$
\bar{\varepsilon} = \frac{1}{2\mu}(\bar{\sigma} + p I),
\tag{25}
$$

$$
\varepsilon = \frac{1}{2}(\nabla u + (\nabla u)^{\mathrm{T}}),
\tag{26}
$$

$$
\tau = 2\mu\varepsilon.
\tag{27}
$$

The stress is approximated with element-wise discontinuous shape functions and can be condensed out on the element level! For uncut elements, the stress equation decouples from the momentum and mass continuity equations and can be omitted, altogether. The resulting global system contains only velocity and pressure unknowns and has no saddle-point characteristic. For an in-depth discussion of this approach including benchmark computations and numerical convergence analysis, see [24]. Exemplarily, two 3D benchmark computations from [24] are shown in Fig. 5. Computations showed optimal performance and convergence rates for all

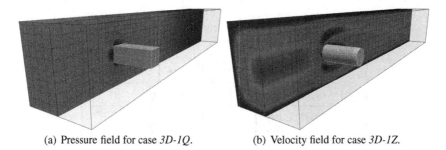

(a) Pressure field for case *3D-1Q*.   (b) Velocity field for case *3D-1Z*.

Fig. 5 3D benchmark cases *3D-1Q* and *3D-1Z*.

common element shapes, namely equal order, linear and quadratic hexahedral and tetrahedral elements.

An important property of the new embedded Dirichlet approach with beneficial consequences for the FSI coupling is that the interface discretization can be chosen *arbitrarily*. This allows to choose the interface mesh to be identical to the structure surface mesh. Hence, no weak coupling between interface and structure mesh is required and the weak interface-structure coupling as presented in [23] can be replaced by matching nodes algorithms.

Both fluid-interface coupling approaches in combination with the XFEM formulation allow to decouple the physical domain from the fictitious domain Ω^- and a weak coupling between fluid flow and interface movement. They are now used in an FSI approach as presented in the next section.

4   Fluid-Structure Coupling

In this section, we combine both fields in one coupled system. Before the discrete system can be established, the interface velocity u^i has to be discretized in time, since interface displacement d^i and velocity u^i are connected by the simple differential equation $\dot{d}^i = u^i$. For simplicity of presentation, we chose the one-step-θ method for the interface movement where the time-discrete form of $\dot{d}^i = u^i$ becomes

$$\frac{d^{i,n+1} - d^{i,n}}{\Delta t} = \theta^i u^{i,n+1} + (1 - \theta^i) u^{i,n}. \tag{28}$$

For second order accuracy, θ^i has to be chosen as $\theta^i = 0.5$ resulting in the second order trapezoidal rule. However, it is noted in [19] that the trapezoidal rule might lead to oscillations, since no damping is included.

The fully coupled system using three distinct meshes in Ω, Ω^s and Γ^i, respectively, is shown in Fig. 6. The three meshes are required, if the original fluid-interface coupling with Lagrange multiplier λ as described in the previous section is employed. The structure domain is coupled via a Mortar formulation to the interface

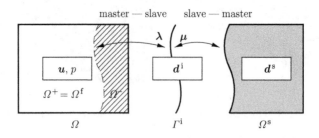

Fig. 6 Coupled problem with 3 fields: fluid-interface coupling with Lagrange multiplier approach.

using $(\delta\mu, d^s - d^i)_{\Gamma^i}$. The interface hosts both Lagrange multiplier fields, which makes it the slave side for both couplings — the fluid-interface coupling and the structure-interface coupling.

$$\delta P^f = \left(v\rho^f, \frac{\partial u}{\partial t} + u \cdot \nabla u\right)_\Omega + (\nabla v, -pI + \tau)_\Omega + (q, \nabla \cdot u)_\Omega$$
$$- (v, \lambda)_{\Gamma^i} - (\delta\lambda, u - \dot{d}^i)_{\Gamma^i}, \tag{29a}$$

$$\delta W^i = (\delta d^i, \lambda + \mu)_{\Gamma^i} \tag{29b}$$

$$\delta W^s = \left(\delta d^s, \rho^s \frac{\partial^2 d^s}{\partial t^2}\right)_{\Omega^s} + (\nabla\delta d^s, \sigma^s)_{\Omega^s} - (\delta d^s, \mu)_{\Gamma^i} - (\delta\mu, d^s - d^i)_{\Gamma^i}. \tag{29c}$$

Correct scaling of coupling matrices using Eq. 28 allows to solve for velocities in the fluid field and displacements on the interface mesh and the structural domain in one monolithic system. After linearization, the final monolithic system at Newton-Raphson iteration step k is of the form

$$\begin{bmatrix} \mathbf{F}_{uu} & \mathbf{F}_{up} & \mathbf{M}_\lambda^T & 0 & 0 & 0 \\ \mathbf{F}_{pu} & \mathbf{F}_{pp} & 0 & 0 & 0 & 0 \\ \mathbf{M}_\lambda & 0 & 0 & \mathbf{D}_\lambda & 0 & 0 \\ 0 & 0 & \mathbf{D}_\lambda^T & 0 & \mathbf{D}_\mu^T & 0 \\ 0 & 0 & 0 & \mathbf{D}_\mu & 0 & \mathbf{M}_\mu \\ 0 & 0 & 0 & 0 & \mathbf{M}_\mu^T & \mathbf{S}_{d^s d^s} \end{bmatrix}_k \begin{bmatrix} \Delta\mathbf{u} \\ \Delta\mathbf{p} \\ \Delta\mathbf{l} \\ \Delta\mathbf{d}^i \\ \Delta\mathbf{m} \\ \Delta\mathbf{d}^s \end{bmatrix} = - \begin{bmatrix} \mathbf{r}_u \\ \mathbf{r}_p \\ \mathbf{r}_\lambda \\ \mathbf{r}_{d^i} \\ \mathbf{r}_\mu \\ \mathbf{r}_{d^s} \end{bmatrix}_k . \tag{30}$$

Here, $\mathbf{S}_{d^s d^s}$ represents the linearized structural system matrix and $\mathbf{M}_\mu$ and $\mathbf{D}_\mu$ the standard interface-structure surface Mortar matrices. $\mathbf{M}_\lambda$ and $\mathbf{D}_\lambda$ are the previously defined Mortar matrices between intersected fluid elements and the interface mesh. The Lagrange multiplier increments are denoted as $\Delta\mathbf{l}$ and $\Delta\mathbf{m}$. Note that we did not separate structure and fluid unknowns into degrees of freedom in the interior of each domain and surface degrees of freedom. Of course, the coupling matrices influence only surface degrees of freedom of the structure and degrees of freedom of intersected fluid elements. Note further that the necessary scaling factors to couple displacements and velocities at the interface are assumed to be contained within the coupling matrices and the interface residual terms. However, the scaling terms are exactly the same as in ALE coupling approaches and can essentially be copied from the appropriate publications, see e. g. [37].

In the monolithic case, that system is solved as given in Eq. (30). For a partitioned approach, the lower right corner describing the interface-structure coupling can be replaced by a partitioned Dirichlet-Neumann scheme, where the structural displacement at the surface forms a Dirichlet condition for the interface-fluid system, while the resulting fluid surface traction from $\lambda = -\mu$ is applied as Neumann condition on the structure. The details of this approach can be found in [23]. For the dynamic FSI problems involving incompressible fluid flow and lightweight structures considered here, we employ an iterative staggered scheme based on [47, 48, 64], where each field is solved implicitly and an iterative procedure over the fields ensures

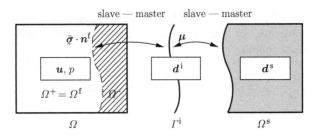

Fig. 7 Coupled problem with 3 fields: fluid-interface coupling with hybrid stress approach.

convergence for the interface conditions at the new time step level $n + 1$. In [20] it has been shown that such strong coupling schemes are necessary for these cases. As proposed in [48,64], we use Aitken's acceleration scheme for vector sequences [33] to obtain the relaxation parameter ω_i.

The monolithic system has two times a saddle point structure due to the two Lagrange multiplier fields λ and μ. However, even for the partitioned FSI coupling, the fluid-interface system still has a saddle point structure, which hamperes the intended solution of large fluid systems on parallel machines.

With the new embedded Dirichlet approach, the Lagrange multiplier field λ is replaced by the hybrid stress formulation. Since the stress field is defined on the background grid, the background fluid could be called the slave side of the fluid-interface coupling. As mentioned in the previous section, the master field for the fluid-interface coupling can be chosen arbitrarily and we choose the interface mesh to be the same as the structure surface, which results in a node-matching FSI approach between interface mesh and structure surface mesh. This improved setup is shown in Fig. 7. The corresponding weak form can be written as

$$\delta P^{\mathrm{f}} = \left(v\rho^{\mathrm{f}}, \frac{\partial u}{\partial t} + u \cdot \nabla u\right)_{\Omega} + \left(\nabla v, -pI + \tau\right)_{\Omega} + \left(q, \nabla \cdot u\right)_{\Omega}$$
$$- \left(\bar{\gamma}, \bar{\varepsilon} - \varepsilon\right)_{\Omega} - \left(v, \bar{\sigma} \cdot n^{\mathrm{f}}\right)_{\Gamma^{\mathrm{i}}} - \left(\bar{\gamma} \cdot n^{\mathrm{f}}, u - \dot{d}^{\mathrm{i}}\right)_{\Gamma^{\mathrm{i}}}, \tag{31a}$$

$$\delta W^{\mathrm{i}} = \left(\delta d^{\mathrm{i}}, \bar{\sigma} \cdot n^{\mathrm{f}} + \mu\right)_{\Gamma^{\mathrm{i}}}, \tag{31b}$$

$$\delta W^{\mathrm{s}} = \left(\delta d^{\mathrm{s}}, \rho^{\mathrm{s}} \frac{\partial^2 d^{\mathrm{s}}}{\partial t^2}\right)_{\Omega^{\mathrm{s}}} + \left(\nabla \delta d^{\mathrm{s}}, \sigma^{\mathrm{s}}\right)_{\Omega^{\mathrm{s}}} - \left(\delta d^{\mathrm{s}}, \mu\right)_{\Gamma^{\mathrm{i}}} - \left(\delta \mu, d^{\mathrm{s}} - d^{\mathrm{i}}\right)_{\Gamma^{\mathrm{i}}}. \tag{31c}$$

The new hybrid fluid-interface coupling does not suffer from the saddle point structure. Here, the monolithic system with condensed element stresses for intersected elements can be stated as

$$\begin{bmatrix} \mathbf{F}_{uu} + \mathbf{C}_{uu} & \mathbf{F}_{up} + \mathbf{C}_{up} & \mathbf{C}_{ud^{\mathrm{i}}} & 0 & 0 \\ \mathbf{F}_{pu} & \mathbf{F}_{pp} & 0 & 0 & 0 \\ \mathbf{C}_{d^{\mathrm{i}}u} & \mathbf{C}_{d^{\mathrm{i}}p} & \mathbf{C}_{d^{\mathrm{i}}d^{\mathrm{i}}} & \mathbf{D}_{\mu}^{\mathrm{T}} & 0 \\ 0 & 0 & \mathbf{D}_{\mu} & 0 & -\mathbf{M}_{\mu} \\ 0 & 0 & 0 & -\mathbf{M}_{\mu}^{\mathrm{T}} & \mathbf{S}_{d^{\mathrm{s}}d^{\mathrm{s}}} \end{bmatrix}_{k} \begin{bmatrix} \Delta \mathbf{u} \\ \Delta \mathbf{p} \\ \Delta \mathbf{d}^{\mathrm{i}} \\ \Delta \mathbf{m} \\ \Delta \mathbf{d}^{\mathrm{s}} \end{bmatrix} = - \begin{bmatrix} \mathbf{r}_u + \mathbf{r}_u^{\mathrm{c}} \\ \mathbf{r}_p \\ \mathbf{r}_{d^{\mathrm{i}}} + \mathbf{r}_{d^{\mathrm{i}}}^{\mathrm{c}} \\ \mathbf{r}_{\mu} \\ \mathbf{r}_{d^{\mathrm{s}}} \end{bmatrix}_{k}. \tag{32}$$

The matrices $\mathbf{C}_{\bullet\bullet}$ and force terms $\mathbf{r}_{\bullet}^c$ stem from the stress condensation process, which is described in detail in [24]. Again the time scaling terms are included in coupling matrices and residuals.

The advantage of being able to chose the interface discretization identical to the structure surface discretization is two-fold: For a partitioned scheme, where the linear system in Eq. (32) is split around the Lagrange multiplier μ as usual, the structure-interface coupling is a node-matching scheme, because for matching interface-structure surfaces with matching interface-structure shape functions we have

$$\Delta \mathbf{d}^i = \mathbf{D}_\mu^{-1} \mathbf{M}_\mu \Delta \mathbf{d}^s = \mathbf{I} \Delta \mathbf{d}^s = \Delta \mathbf{d}^s, \tag{33}$$

with $\mathbf{I}$ being the identity matrix. This essentially replaces the weak interface-structure coupling with a matching-nodes coupling approach. For the monolithic scheme, the identical discretizations remove the need for an extra interface mesh. It can be removed from the global system, which leaves the following system to be solved

$$\begin{bmatrix} \mathbf{F}_{uu} + \mathbf{C}_{uu} & \mathbf{F}_{up} + \mathbf{C}_{up} & \mathbf{C}_{ud^i} \\ \mathbf{F}_{pu} & \mathbf{F}_{pp} & \mathbf{0} \\ \mathbf{C}_{d^i u} & \mathbf{C}_{d^i p} & \mathbf{S}_{d^s d^s} + \mathbf{C}_{d^i d^i} \end{bmatrix}_i \begin{bmatrix} \Delta \mathbf{u} \\ \Delta \mathbf{p} \\ \Delta \mathbf{d}^s \end{bmatrix} = - \begin{bmatrix} \mathbf{r}_u + \mathbf{r}_u^c \\ \mathbf{r}_p \\ \mathbf{r}_{d^s} + \mathbf{r}_{d^i}^c \end{bmatrix}_i . \tag{34}$$

As said before, the coupling terms influence only surface degrees of freedom of the structure and degrees of freedom of intersected fluid elements.

Two-dimensional examples for the fluid-interface coupling via Lagrange multiplier λ are given in [22,23,62]. The new hybrid approach allows now fully coupled three-dimensional stationary and fully transient simulations. In this presentation, we constrain ourself to an illustrative example result of a steady-state 3D FSI simulation. In this example, a flexible structure (Poison ratio $\nu = 0.48$, Young's modulus $E = 90\,\mathrm{N/m^2}$) is deformed due to a flow through a channel with dimensions $0.5 \times 1.0 \times 3.0\mathrm{m}$. The fully coupled FSI simulation is performed until steady-state is reached. Top and bottom channel walls have zero (no-slip) velocity prescribed, the inflow from the right is prescribed by a parabolic velocity condition. The Reynolds number based on the mean inflow velocity and the channel height is 16. The stationary equilibrium solution is shown in Fig. 8. A transient simulation is given together with a contact simulation in Sect. 5.2. A detailed description of the 3D FSI approach using the new embedded Dirichlet formulation will be the topic of an upcoming paper [25].

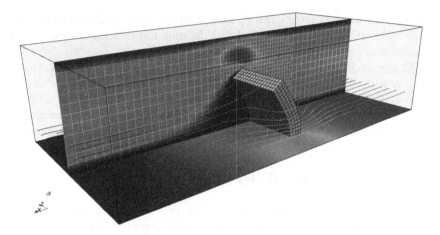

Fig. 8 Stationary flow field through a channel with a flexible structure. The parabolic inflow (right) and the wall boundaries are standard Dirichlet boundary conditions. The outflow boundary (left) is of Neumann (zero traction) type. The structure is fixed at the lower wall. The fluid-structure interface is modeled using the proposed embedded Dirichlet approach.

5 Enhancements towards complex FSI problems

5.1 Adaptivity vs. Boundary Layer Meshes

'Pure' fixed-grid methods in principle allow for unlimited deformation of the structure. In addition, no extra computational costs for mesh movement and mesh smoothing is required. However, unlike in ALE methods, an adequate (boundary layer) mesh can rarely be constructed *a priori*, since the movement of the fluid-structure interface may vary much more than ALE methods would allow. Hence, for pure fixed-grid methods special care is necessary to create an appropriate – meaning an sufficiently fine – mesh that allows for reliable simulation of complex problems.

In [22], we propose and discuss two different techniques to improve this situation. The first approach is a rather straightforward usage of adaptivity. It is based on local, adaptive mesh refinement and coarsening scheme combined with an error-estimator and/or mesh refinement heuristics like the distance to structure surfaces. Our second proposal to improve the accuracy/efficiency of fixed-grid methods is to use a hybrid approach combining fixed-grid and ALE techniques. It essentially adds a surface layer of deformable fluid elements with an ALE formulation to the structural surface. Such a fluid patch would capture the near-surface flow efficiently with an appropriate boundary layer mesh, which is then coupled to the fixed Eulerian background mesh. A sketch of this setup using the original Lagrange multiplier for the fluid-interface coupling is shown in Fig. 9. The coupled fixed fluid - moving fluid system can then efficiently be solved monolithically as demonstrated in [22].

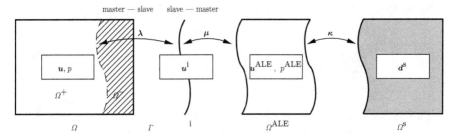

master — slave   slave — master

Fig. 9 Boundary layer mesh approach: a fixed fluid grid and a deforming ALE fluid grid are coupled using the XFEM/Lagrange Multiplier approach. The moving ALE grid can in turn be coupled to the structure field. Shown are also variables living on each domain and the Lagrange multiplier fields.

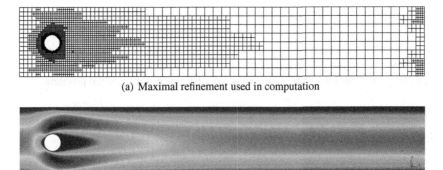

(a) Maximal refinement used in computation

(b) x-component of the velocity field

Fig. 10 CFD benchmark: Finest, adapted mesh used to calculate lift and drag values on the cylinder. It consists of 284 linear Lagrange multiplier elements and 2740 quadratic 9-node fluid elements. Fig. 10(b) shows the x-component of the velocity field. The structure is not displayed.

Our new hybrid fluid-interface coupling would simplify the fixed-fluid moving-fluid coupling, as the interface mesh is not required anymore.

The following example shall further clarify the principle difference between automatic adaptivity and the boundary layer mesh approach. In Fig. 10, the flow around a cylinder has been resolved by using Cartesian subdivision of elements with hanging nodes as described in [22]. Due to the circular shape, it is not possible to generate a mesh that optimally resolves only the strong gradient normal to the cylinder surface. Figure 10 shows the finest mesh that we used to study this problem. With this, we could already achieve very good agreement with the averaged results given in [59].

In contrast, with a boundary layer mesh, instead of refining the mesh near the cylinder surface as before, we used the proposed hybrid approach. The final solution can be seen in Fig. 11. The velocity component u_x of the fluid solution is shown over the whole fluid domain in Fig. 11(a) and near the interface in Fig. 11(b). The surface

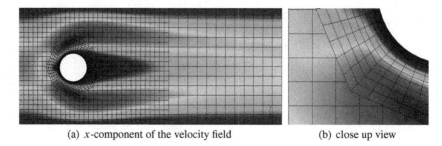

(a) x-component of the velocity field   (b) close up view

Fig. 11 CFD benchmark using the boundary layer mesh approach: The initial mesh was refined in the inflow region and in addition, a fluid mesh patch surrounding the structure is applied.

fitted mesh resembles the boundary layer near the cylinder surface much closer and the boundary layer can be resolved much more efficiently. Of course, the boundary layer mesh approach can be combined with the aforementioned h-adaptivity. For high Reynolds number flows, the boundary layer mesh approach should outperform adaptivity by element subdivision, especially for three-dimensional problems.

5.2 Contact of Submerged Structures

Contact of submerged structures can be observed in many FSI problems. For instance, in bio-physical applications, closing valves, collapsing veins, interaction of blood cells are a few examples, where numerical tools could provide valuable insight. With ALE-methods, the deforming mesh essentially does not allow two objects to come into full contact. With special techniques, however, even with ALE fluid methods simplified FSI contact is possible, see *e.g.* [58, 60]. Fixed-grid methods, on the other hand, have no deforming mesh between such objects and potentially open up an entire new field that can be treated by numerical tools. Recent examples for fixed-grid FSI simulations including contact of submerged structures can be found in [2, 41, 57].

In [44], the presented XFEM-FSI approach has been combined with a new contact formulation based on dual-Lagrange multipliers and a so-called primal-dual active set strategy for contact constraint enforcement as described in [56]. From an algorithmic point of view, the key element of the contact formulation is the complete condensation of the contact Lagrange multipliers within the structural formulation. Consequently, the structure block can be extended with a contact formulation without interfering with the FSI formulation. To our knowledge, [44] presents the first fluid-structure-interaction approach including advanced approaches for FSI *and* contact.

A preliminary example of such FSI contact is shown in Fig. 12. A body force pulls a little, soft block towards a large stiff block, that has a fixed bottom surface. At the same time, a growing inflow at the top of the fluid domain pushes the little

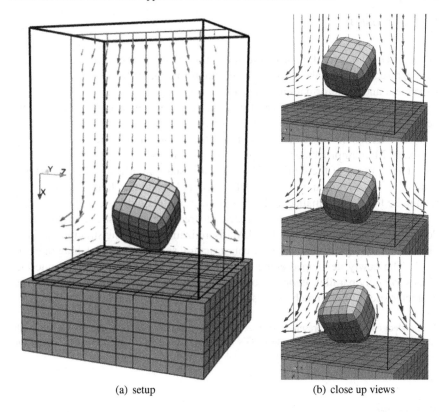

(a) setup   (b) close up views

Fig. 12 Setup of contact example and 3 snapshots of the first contact between a soft body and a relatively stiff block.

block additionally in x-direction towards the wall. Two exits at the lower end of the channel allow the flow to exit the domain. The three little pictures in Fig. 12(b) show (from top to bottom): how the fluid is pushed out of the closing gap between the little and the large block, the moment of largest deformation of the little block at first contact, and the bouncing back of the small block. The same simulation without fluid flow (not shown here) allows the block to bounce back many times to approximately the same height from where it started. Including the FSI simulation, the movement is damped by the fluid viscosity, such that it comes to rest on the lower block quickly. Since no frictional contact between both objects is employed, eventually, the little block is pushed sideways out of the fluid domain. Note that in this simulation we have not employed any special wet or dry contact physics and that this simulation is not properly resolved to be quantitatively accurate. The example shall just demonstrate the increased flexibility that comes with fixed grid methods for FSI.

6 Conclusion

Fixed grid methods for fluid-structure interaction are subject of a growing number of current research undertaken. The ultimate goal is to remove the burden of fluid mesh movement and, if deformation of the structure becomes excessive, remeshing. For that purpose, we worked out a list of minimal requirements that have to be fulfilled before fixed grid methods can match or surpass ALE based methods.

In an attempt to meet these requirements, we proposed a new XFEM based approach. Main improvements compared to most IB and DLM/FD methods are the complete removal of the fictitious domain and a sharp interface description. Physical and fictitious domains are completely decoupled and the fluid unknowns from within the fictitious domain could be removed from the fluid system of equation.

Due to the use of a three-field approach, the treatment of the structure is independent of any internals of the fluid XFEM discretization. From the structural point of view, the coupling exists solely between structural surface and a matching interface mesh. The newly developed embedded Dirichlet technique for the fluid-interface allows to apply parallel, iterative solution techniques by avoiding the saddle point characteristics of non-matching mesh coupling.

Acknowledgements The present study is supported by grant WA1521/1 of the "Deutsche Forschungsgemeinschaft" (DFG) within the DFG's *Forschergruppe 493* "FSI: Modelling, Simulation, and Optimization". This support is gratefully acknowledged.

References

1. Arienti, M., Hung, P., Morano, E., Shepherd, J.E.: A level set approach to Eulerian-Lagrangian coupling. Journal of Computational Physics **185**(1), 213–251 (2003)
2. Astorino, M., Gerbeau, J.F., Pantz, O., Traoré, K.F.: Fluid-structure interaction and multi-body contact: Application to aortic valves. Computer Methods in Applied Mechanics and Engineering **198**(45-46), 3603–3612 (2009)
3. Baaijens, F.P.T.: A fictitious domain/mortar element method for fluid-structure interaction. International Journal for Numerical Methods in Fluids **35**(7), 743–761 (2001)
4. Becker, R., Burman, E., Hansbo, P.: A nitsche extended finite element method for incompressible elasticity with discontinuous modulus of elasticity. Computer Methods in Applied Mechanics and Engineering **198**(41-44), 3352–3360 (2009)
5. Behr, M.A., Franca, L.P., Tezduyar, T.E.: Stabilized finite element methods for the velocity-pressure-stress formulation of incompressible flows. Computer Methods in Applied Mechanics and Engineering **104**(1), 31–48 (1993)
6. Belytschko, T., Black, T.: Elastic crack growth in finite elements with minimal remeshing. International Journal for Numerical Methods in Engineering **45**(5), 601–620 (1999)
7. Belytschko, T., Kennedy, J.M.: Computer models for subassembly simulation. Journal of Nuclear Engineering and Design **49**, 17–38 (1978)
8. Belytschko, T., Kennedy, J.M., Schoeberle, D.: Quasi-Eulerian finite element formulation for fluid structure interaction. Journal of Pressure Vessel Technology **102**, 62–69 (1980)
9. Bernardi, C., Maday, Y., Patera, A.T.: Asymptotic and numerical methods for partial differential equations with critical parameters, vol. 384, chap. Domain decomposition by the mortar element method, pp. 269–286. Reidel, Dordrecht (1993)

10. Bernardi, C., Maday, Y., Patera, A.T.: A new nonconforming approach to domain decomposition: the mortar element method. Nonlinear partial differential equations and their applications **299**, 13–51 (1994)
11. Cirak, F., Radovitzky, R.: A Lagrangian-Eulerian shell-fluid coupling algorithm based on level sets. Computers & Structures **83**(6-7), 491–498 (2005)
12. Codina, R., Houzeaux, G., Coppola-Owen, H., Baiges, J.: The fixed-mesh ale approach for the numerical approximation of flows in moving domains. Journal of Computational Physics **228**(5), 1591–1611 (2009)
13. Daux, C., Moës, N., Dolbow, J., Sukumar, N., Belytschko, T.: Arbitrary cracks and holes with the extended finite element method. International Journal for Numerical Methods in Engineering **48**(12), 1741–1760 (2000)
14. De Hart, J., Peters, G.W., Schreurs, P.J., Baaijens, F.P.: A two-dimensional fluid-structure interaction model of the aortic value. Journal of Biomechanics **33**(9), 1079–1088 (2000)
15. Dolbow, J., Harari, I.: An efficient finite element method for embedded interface problems. International Journal for Numerical Methods in Engineering **78**(2), 229–252 (2009)
16. Donéa, J., Fasoli-Stella, P., Giuliani, S.: Lagrangian and Eulerian finite element techniques for transient fluid-structure interaction problems. In: Trans. 4th Int. Conf. on Structural Mechanics in Reactor Technology (1977)
17. Fedkiw, R.P., Aslam, T., Merriman, B., Osher, S.: A non-oscillatory Eulerian approach to interfaces in multimaterial flows (the ghost fluid method). Journal of Computational Physics **152**(2), 457–492 (1999)
18. Fernández-Méndez, S., Huerta, A.: Imposing essential boundary conditions in mesh-free methods. Computer Methods in Applied Mechanics and Engineering **193**(12-14), 1257–1275 (2004)
19. Förster, C., Wall, W.A., Ramm, E.: On the geometric conservation law in transient flow calculations on deforming domains. International Journal for Numerical Methods in Fluids **50**(12), 1369–1379 (2006)
20. Förster, C., Wall, W.A., Ramm, E.: Artificial added mass instabilities in sequential staggered coupling of nonlinear structures and incompressible viscous flows. Computer Methods in Applied Mechanics and Engineering **196**(7), 1278–1293 (2007)
21. Gee, M.W., Küttler, U., Wall, W.A.: Truly monolithic algebraic multigrid for fluid–structure interaction. International Journal for Numerical Methods in Engineering **submitted** (2009)
22. Gerstenberger, A., Wall, W.A.: Enhancement of fixed-grid methods towards complex fluid-structure interaction applications. International Journal for Numerical Methods in Fluids **57**(9), 1227–1248 (2008)
23. Gerstenberger, A., Wall, W.A.: An extended finite element method / Lagrange multiplier based approach for fluid-structure interaction. Computer Methods in Applied Mechanics and Engineering **197**(19-20), 1699–1714 (2008)
24. Gerstenberger, A., Wall, W.A.: An embedded Dirichlet formulation for 3d continua. International Journal for Numerical Methods in Engineering DOI: 10.1002/nme.2755 (2009)
25. Gerstenberger, A., Wall, W.A.: An fixed-grid approach for 3d fluid-structure interaction **in preparation** (2010)
26. Glowinski, R., Pan, T.W., Hesla, T.I., Joseph, D.D.: A distributed Lagrange multiplier / fictitious domain method for particulate flows. International Journal of Multiphase Flow **25**(5), 755–794 (1999)
27. Glowinski, R., Pan, T.W., Periaux, J.: A fictitious domain method for external incompressible viscous flow modeled by Navier-Stokes equations. Computer Methods in Applied Mechanics and Engineering **112**(1-4), 133–148 (1994)
28. Hansbo, A., Hansbo, P.: An unfitted finite element method, based on nitsche's method, for elliptic interface problems. Computer Methods in Applied Mechanics and Engineering **191**(47-48), 5537–5552 (2002)
29. Hansbo, A., Hansbo, P.: A finite element method for the simulation of strong and weak discontinuities in solid mechanics. Computer Methods in Applied Mechanics and Engineering **193**(33-35), 3523–3540 (2004)

30. Hirth, C., Amsden, A.A., Cook, J.: An Arbitrary Lagrangian-Eulerian computing method for all flow speeds. Journal of Computational Physics **14**, 227–253 (1974)
31. Hong, J.M., Shinar, T., Kang, M., Fedkiw, R.: On boundary condition capturing for multiphase interfaces. Journal of Scientific Computing **31**(1), 99–125 (2007)
32. Hughes, T.J., Liu, W.K., Zimmermann, T.: Lagrangian-Eulerian finite element formulation for incompressible viscous flows. Computer Methods in Applied Mechanics and Engineering **29**, 329–349 (1981)
33. Irons, B.M., Tuck, R.C.: A version of the Aitken accelerator for computer iteration. International Journal for Numerical Methods in Engineering **1**(3), 275–277 (1969)
34. Ji, H., Dolbow, J.E.: On strategies for enforcing interfacial constraints and evaluating jump conditions with the extended finite element method. International Journal for Numerical Methods in Engineering **61**(14), 2508–2535 (2004)
35. Küttler, U., Förster, C., Wall, W.A.: A solution for the incompressibility dilemma in partitioned fluidstructure interaction with pure Dirichlet fluid domains. Computational Mechanics **38**(4–5), 417–429 (2006)
36. Küttler, U., Wall, W.A.: Fixed-point fluid-structure interaction solvers with dynamic relaxation. Computational Mechanics **43**(1), 61–72 (2008)
37. Küttler, U., Wall, W.A.: Vector extrapolation for strong coupling fluid-structure interaction solvers. Journal of Applied Mechanics **76**(2) (2009)
38. Lee, L., LeVeque, R.J.: An immersed interface method for incompressible Navier-Stokes equations. SIAM Journal on Scientific Computing **25**(3), 832–856 (2003)
39. Legay, A., Chessa, J., Belytschko, T.: An Eulerian-Lagrangian method for fluid-structure interaction based on level sets. Computer Methods in Applied Mechanics and Engineering **195**(17-18), 2070–2087 (2006)
40. LeVeque, R.J., Calhoun, D.: Cartesian grid methods for fluid flow in complex geometries. In: L.J. Fauci, S. Gueron (eds.) Computational Modeling in Biological Fluid Dynamics, vol. 124, pp. 117–143. IMA Volumes in Mathematics and its Applications, Springer-Verlag (2001)
41. van Loon, R., Anderson, P., van de Vosse, F.: A fluid-structure interaction method with solid-rigid contact for heart valve dynamics. Journal of Computational Physics **217**(2), 806–823 (2006)
42. van Loon, R., Anderson, P.D., Baaijens, F.P., van de Vosse, F.N.: A three-dimensional fluid-structure interaction method for heart valve modelling. Comptes Rendus Mecanique **333**(12), 856–866 (2005)
43. Löhner, R., Baum, J.D., Mestreau, E., Sharov, D., Charman, C., Pelessone, D.: Adaptive embedded unstructured grid methods. International Journal for Numerical Methods in Engineering **60**(3), 641–660 (2004)
44. Mayer, U., Popp, A., Gerstenberger, A., Wall, W.A.: 3d fluid-structure interaction based on a combined XFEM/FSI and dual mortar contact approach. Computational Mechanics **accepted** (2009)
45. Mayer, U.M., Gerstenberger, A., Wall, W.A.: Interface handling for three-dimensional higher-order XFEM computations in fluid-structure interaction. International Journal for Numerical Methods in Engineering **47**(7), 846–869 (2009)
46. Mittal, R., Iaccarino, G.: Immersed boundary methods. Annual Review of Fluid Mechanics **37**(1), 239–261 (2005)
47. Mok, D.P.: Partitionierte Lösungsansätze in der Strukturdynamik und der Fluid-Struktur-Interaktion. Tech. Rep. PhD Thesis, Report No. 36, Institute of Structural Mechanics, University of Stuttgart (2001)
48. Mok, D.P., Wall, W.A.: Partitioned analysis schemes for the transient interaction of incompressible flows and nonlinear flexible structures. In: W. Wall, K.U. Bletzinger, K. Schweizerhof (eds.) Trends in Computational Structural Mechanics, pp. 689–698. CIMNE: Barcelona (2001)
49. Moës, N., Béchet, E., Tourbier, M.: Imposing Dirichlet boundary conditions in the extended finite element method. International Journal for Numerical Methods in Engineering **67**(12), 1641–1669 (2006)
50. Moës, N., Dolbow, J., Belytschko, T.: A finite element method for crack growth without remeshing. International Journal for Numerical Methods in Engineering **46**(1), 131–150 (1999)

51. Nitsche, J.: Über ein Variationsprinzip zur Lösung von Dirichlet-Problemen bei Verwendung von Teilräumen, die keinen Randbedingungen unterworfen sind. Abh. Math. Sem. Univ. Hamburg **36**, 915 (1971)
52. Noh, W.: CEL: A time-dependent two-space-dimensional coupled Eulerian-Lagrangian code. In: B. Alder, S. Fernbach, M. Rotenberg (eds.) Methods in Computational Physics, vol. 3, pp. 117–179. Academic Press: New York (1964)
53. Park, K.C., Felippa, C.A., Ohayon, R.: Partitioned formulation of internal fluid-structure interaction problems by localized Lagrange multipliers. Computer Methods in Applied Mechanics and Engineering **190**(24-25), 2989–3007 (2001)
54. Peskin, C.S.: Numerical analysis of blood flow in the heart. Journal of Computational Physics **25**(3), 220–252 (1977)
55. Peskin, C.S.: The immersed boundary method. Acta Numerica **11**(1), 479–517 (2002)
56. Popp, A., Gee, M.W., Wall, W.A.: A finite deformation mortar contact formulation using a primal-dual active set strategy. International Journal for Numerical Methods in Engineering **79**(11), 1354–1391 (2009)
57. Diniz dos Santos, N., Gerbeau, J.F., Bourgat, J.F.: A partitioned fluid-structure algorithm for elastic thin valves with contact. Computer Methods in Applied Mechanics and Engineering **197**(19-20), 1750–1761 (2008)
58. Sathe, S., Tezduyar, T.: Modeling of fluidstructure interactions with the spacetime finite elements: contact problems. Computational Mechanics **43**(1), 51–60 (2008)
59. Schäfer, M., Turek, S.: Flow Simulation with High-Performance Computers II, Notes on Numerical Fluid Mechanics, vol. 52, article Benchmark Computations of Laminar Flow Around a Cylinder, pp. 547–566. Vieweg (1996)
60. Tezduyar, T.E., Sathe, S.: Modelling of fluid-structure interactions with the space-time finite elements: Solution techniques. International Journal for Numerical Methods in Fluids **54**(6-8), 855–900 (2007)
61. Wagner, G.J., Ghosal, S., Liu, W.K.: Particulate flow simulations using lubrication theory solution enrichment. International Journal for Numerical Methods in Engineering **56**(9), 1261–1289 (2003)
62. Wall, W.A., Gamnitzer, P., Gerstenberger, A.: Fluidstructure interaction approaches on fixed grids based on two different domain decomposition ideas. International Journal of Computational Fluid Dynamics **22**(6), 411–427 (2008)
63. Wall, W.A., Gerstenberger, A., Gamnitzer, P., Förster, C., Ramm, E.: Large deformation fluid-structure interaction – advances in ALE methods and new fixed grid approaches. In: H.J. Bungartz, M. Schäfer (eds.) Fluid-Structure Interaction: Modelling, Simulation, Optimisation, LNCSE, vol. 53, pp. 195–232. Springer Verlag (2006)
64. Wall, W.A., Mok, D.P., Ramm, E.: Partitioned analysis approach of the transient coupled response of viscous fluids and flexible structures. In: W. Wunderlich (ed.) Solids, Structures and Coupled Problems in Engineering, Proc. ECCM '99. Munich (1999)
65. Wall, W.A., Ramm, E.: Fluid-structure interaction based upon a stabilized (ALE) finite element method. In: S. Idelsohn, E. Oñate, E. Dvorkin (eds.) Computational Mechanics New Trends and Applications, Proc. 4th World Congress on Computational Mechanics - Buenos Aires. CIMNE, Barcelona (1998)
66. Wang, X., Liu, W.K.: Extended immersed boundary method using FEM and RKPM. Computer Methods in Applied Mechanics and Engineering **193**(12-14), 1305–1321 (2004)
67. Yu, Z.: A DLM/FD method for fluid/flexible-body interactions. Journal of Computational Physics **207**(1), 1–27 (2005)
68. Zhang, L., Gerstenberger, A., Wang, X., Liu, W.K.: Immersed finite element method. Computer Methods in Applied Mechanics and Engineering **193**(21-22), 2051–2067 (2004)
69. Zilian, A., Legay, A.: The enriched space-time finite element method (EST) for simultaneous solution of fluid-structure interaction. International Journal for Numerical Methods in Engineering **75**(3), 305–334 (2008)

Fluid-Structure Interaction in the Context of Shape Optimization and Computational Wind Engineering

M. Hojjat, E. Stavropoulou, T. Gallinger, U. Israel, R. Wüchner, and K.-U. Bletzinger

Abstract Within this contribution, an integrated concept for the shape optimal design of light-weight and thin-walled structures like shells and membranes subject to fluid flow is presented. The Nested Analysis and Design approach is followed and a partitioned FSI simulation for the state analysis is embedded. The gained modularity allows for the adaption of the single ingredients to various technical applications by choosing appropriate coupling algorithms for the solution of the coupled problem and the sensitivity analysis as well as different strategies to describe the shapes to be optimized. A non-matching grid capability at the coupling interface supports this flexibility. The focus here is on problems of aeroelasticity in the field of Computational Wind Engineering. To ensure reliable results, investigations on the correct modeling as well as goal-oriented benchmarking are carried out. Moreover, special emphasis is given to the appropriate combination of different approaches for shape description in establishing the closed design cycle. Finally, the success of the overall solution and optimization strategy is demonstrated with an example of a hybrid, light-weight structure, subject to turbulent wind flow.

1 Introduction

In modern structural design, there is a strong tendency to build light structures with very efficient load-carrying behavior. This is motivated by the optimal use of material and their intended use (retractable roofs, airplanes, etc.). The design of light-weight structures is a challenging task in structural engineering and demands for additional investigations and in-depth analysis of effects which are of minor importance in the design of standard structures. This is due to the fact that in case of structures subject to loads caused by surrounding fluid flow, lightness and slenderness make structures more sensitive to flow-induced deformations and vibrations.

M. Hojjat, E. Stavropoulou, T. Gallinger, U. Israel, R. Wüchner, and K.-U. Bletzinger
Lehrstuhl für Statik, Technische Universität München, Germany
e-mail: {hojjat,stavropoulou,gallinger,israel,wuechner,bletzinger}@bv.tum.de

H.-J. Bungartz et al. (eds.), *Fluid Structure Interaction II*, Lecture Notes
in Computational Science and Engineering 73, DOI 10.1007/978-3-642-14206-2_13,
© Springer-Verlag Berlin Heidelberg 2010

These deformations may change the flow field considerably which results in an interaction between the structure and the flow field. Typical examples in civil engineering are wind-induced effects on thin shells like chimneys, extremely light membrane structures like wide-span roofs and severe oscillations of slender bridges like the famous Tacoma Narrows bridge disaster. In these cases, the isolated consideration of the structure, i.e. reducing the complex physical problem with aeroelastic effects by finding appropriate assumptions about the fluid load, involves the risk of ignoring effects which are relevant for the final design. Therefore, the reliable and verified numerical simulation of the surface-coupled problem including the fluid-structure interactions (FSI) is necessary and the modeling of the coupling, as well as the single fields is a crucial part of the design procedure. Hence, the goal-oriented and detailed benchmarking of all parts of the numerical simulations is indispensable.

Due to the interaction of the single fields, determination of a good and "optimal" structure is far from trivial and hardly possible only on the basis of experience. Therefore there is a strong need for an overall methodology to improve and optimize the chosen designs. Since the constituents of the numerical wind tunnel, i.e. light thin-walled structures undergoing large displacements and highly turbulent wind flows, are very complex problems, a partitioned solution scheme is advantageous because it allows to use the best-suited simulation technology for each problem. This modular and flexible software concept for the solution of FSI problems needs to be integrated in a shape optimization approach which leads to a Nested Analysis and Design (NAND) concept. In particular, the strong coupling between the physical fields necessitates a coupled sensitivity analysis which is solved by a staggered scheme. Finally, this integrated design process enables the simulation and optimization of technically relevant free-form structures.

2 Design and optimization

The final goal of every structural design process is to create a structure which can reliably withstand the external loadings, being at the same time efficient in terms of material use. To determine the final structural layout, methods of structural optimization can be applied. Typically, optimized structures are free-form shells or membranes since they are very light with regard to their load bearing capacity. In cases of structures interacting with surrounding fluid flows, the complete coupled system must be considered in the optimization procedure which leads to a multiphysics optimization approach [2, 12, 16, 25, 26]. In general, shape optimization in the respective single fields fluid and structure [1,5,7,29,36] constitute valuable parts and form a basis for the optimization in fluid-structure interaction problems.

2.1 Design workflow and optimization loop

From a methodological point of view, it is important to distinguish between partitioned (or staggered) and monolithic solution strategies for the simulation of fluid-structure interactions and the optimization problem. In contrast to a monolithic approach, a partitioned solution scheme in the analysis of the FSI problem separates the two coupled physical fields and the whole system is solved by subsequent single field solutions and appropriately tailored coupling strategies to represent the complete system [13]. In the development of optimization strategies a similar situation exists: In the Simultaneous Analysis and Design (SAND) approach all physical equations are solved together with the optimization problem at once (i.e. it is solved for the state and optimization variables at the same time), whereas in Nested Analysis and Design (NAND), the complete Multidisciplinary Design Optimization problem is decomposed into the ingredients of optimizer, state analysis and sensitivity analysis.

In this contribution, a completely modular software framework for gradient based shape optimization considering FSI, which guarantees the flexibility for various technically relevant applications, is presented. The resulting workflow is depicted in Fig. 1. Besides the links between the different modules, the corresponding variables are introduced: The general chain of shape parameterizations is visualized and the dependency of the finite element nodal positions $\mathbf{x}$ from the form finding parameters $\mathbf{y}$ and the design variables which are typically the CAD parameters $\mathbf{s}$ can be seen. Therefore, during the sensitivity analysis, the gradient of the objective function f with respect to the shape design variables $\mathbf{s}$ is computed by expanding it to the sensitivities of the individual modules. Different possibilities of shape definition are introduced in Sect. 2.2.

2.2 Shape representation

In optimization, a meaningful definition of design variables is of utmost importance. In the case of shape optimization, there are several alternatives to represent the shape changes which differ significantly in the numerical effort and the design freedom.

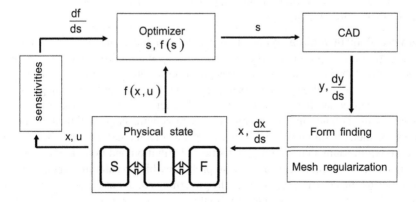

Fig. 1 Design optimization: workflow and respective shape variables.

Fig. 2 Representation of shape, CAGD based (left), parameter-free based (middle) and form-finding (right).

According to the desired task, the best suited method must be applied: in preliminary design stages, it is typically needed to have a big design space, whereas in a nearly finished structural design process a more detailed shape optimization with only a very limited change in the concept is desired. This has a huge impact in the shape description to be used and also on the technical realization. In Fig. 2 the different possibilities are displayed.

CAGD based The current standard in structural shape optimization is the separation of the geometric model and the analysis model, which is the case in all CAGD (Computer Aided Geometric Design) based shape optimizations. Here, the design variables are selected geometry parameters of the CAGD-model which leads usually to a low number of optimization variables and renders this approach rather inexpensive with regard to the optimization. The linking of the locations of the FE nodes to the design elements is established by means of mathematical relations (e.g. B-Splines, Bézier curves or Coons patches).

Finite element based The typical restriction in shape variations caused by the use of CAGD is canceled by directly taking the nodal positions of the finite element mesh as design variables. This so-called parameter-free or FE-based geometric modeling approach has the maximum design freedom but is linked to a huge numerical effort and therefore demands for suitable adjoint formulations in the sensitivity analysis. Moreover, some further numerical problems occur which demand for filtering and regularization strategies [7].

Numerical form finding Within thin-walled structures, the ones in membrane action are the most efficient structures which can be built. To reach this desirable load-carrying behavior mechanical design criteria can be exploited via numerical form finding methods. As an example, the Updated Reference Strategy (URS) [4, 5, 42] can be applied as an effective and efficient way to compute free form shapes of membranes and (in combination with general shape optimization procedures) of shells within the design process. Physically, the task of form finding is to find an equilibrium shape for a given stress distribution and boundary conditions. Mathematically, it represents an inverse problem which results in singular expressions during the numerical solution of the governing equations for equilibrium. This basic relationship is the principle of virtual work which is modified in the URS by a homotopy mapping to overcome the singularity in the original and singular virtual work expression ($\delta W_\sigma = 0$). The stabilizing term δW_S is formulated in terms of PK2 stresses S rather than in Cauchy stresses σ and fades out as the solution is approached and therefore the original, unmodified solution is received. The weak form of this stabilized form finding method which has to be discretized by finite elements states as:

$$\delta W_\lambda = \lambda \, \delta W_\sigma + (1 - \lambda) \, \delta W_S,$$

$$= \lambda \left[h \int_A det\mathbf{F} \left(\sigma \cdot \mathbf{F}^{-T} \right) : \delta \mathbf{F} \, dA \right] + (1 - \lambda) \left[h \int_A (\mathbf{F} \cdot \mathbf{S}) : \delta \mathbf{F} \, dA \right] = 0.$$

(1)

As can be observed in this equation, the design variables under consideration are the anisotropy and distribution of pre-stress together with the edge cable forces and the positions of the supports. These (usually few) parameters define naturally real free-form surfaces by enforcing the equilibrium condition which is also related to the determination of minimal surfaces. Hence, a good reduction of design parameters compared to CAGD-free methods can be realized by at the same time using a non-parametric shape definition which in turn allows to obtain mechanically motivated, real free-form shapes. Remarkably, the form finding process includes the solution of nonlinear equations which has impact on the evaluation of sensitivities.

2.3 Mesh regularization

The quality of the solution in finite element and finite volume method strongly depends on the domain discretization and therefore having a good mesh is strongly desired. However, many of the mesh generation techniques are not able to provide a discretization with good element shapes in the whole domain, especially in the case of complex geometries. Moreover, even if the quality of the generated mesh is acceptable, in several finite element applications which deal with varying geometries, during computation, the shape of elements might get distorted. For instance, in shape optimization problems and large deformation fluid-structure interaction simulations the retaining of the initial discretization in not guaranteed.

In this work, a global method which regularizes the finite element mesh to a desired condition is presented. In this method, an artificial stress is applied on the surface or on the volume mesh and a global linear system of equilibrium equations is solved. The applied stress adapts each element towards a desired predefined template geometry and at the end a globally smooth mesh is achieved. In this way, both shape and size of each element is controlled.

This method is closely related to the form finding method in terms of solving Eq. (1). In the case of regularization, there are some additional aspects that should be considered. Firstly, the reference geometry has to be changed such that the applied

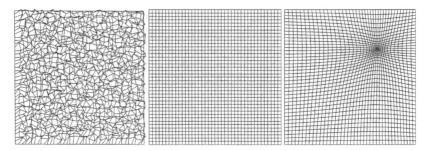

Fig. 3 Local refinement in case of singularity.

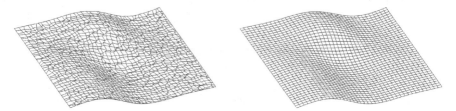

Fig. 4 Regularization of a curved distorted grid.

prestress brings the elements to their desired shapes. Secondly, the applied prestress has to control the size of each element and at the end the resulting discretization has to preserve the domain geometry.

In order to control the shape of elements their reference geometry is assumed to be the desired one, hence the applied prestress modifies the elements towards the given shape. When the control of the size of the elements is of interest, as it is in the case of singularities, the applied prestress should be adapted accordingly. In Fig. 3 a square template is used for all elements whereas the prestress is adjusted according to the distance to the singular point.

Applying regularization on a mesh, the shape of the surface or volume has to be preserved. For parameter-free geometries, this can be done in an approximation manner, since the exact geometry is not known. However, in shape optimization this approximation is of no harm, since the optimization is the driving process which determines the final shape. In this work, the shape is preserved by allowing modifications of surface nodes only in the in-plane direction. Hence, a reduced system of equations is solved. In Fig. 4 the distorted mesh of a curved surface on the right is regularized while preserving the geometry.

2.4 Physical state

During optimization computation of the objective function involves the evaluation of the actual physical state. In the present work, due to the coupling between the structure and the fluid fields, a fluid-structure interaction analysis is required and a three field analysis including structure, fluid and fluid mesh is performed.

In this chapter, the mesh equation is included in the fluid equation and for the solution of the two remaining equations -fluid and structural-, a partitioned solution strategy is applied. A detailed description of the two fields is given in the following sections and the treatment of the interface in Sect. 2.4.3.

Within this work, the interaction between the fluid and the mesh fields is represented using an ALE approach in which the deformation of the structures wetter surface is used to update the boundaries of the computational domain of the fluid and the nodes of the corresponding mesh are fixed to the material points.

2.4.1 Structure - Shells and membranes

In the context of large deformations, where the geometrical nonlinearity is not negligible, the following boundary value problem has to be considered:

$$\rho\frac{\partial^2 \mathbf{d}}{\partial t^2} = \nabla(\mathbf{F}\cdot\mathbf{S}) + \rho\mathbf{f} \quad \text{in} \quad \Omega_S, \tag{2}$$

with initial and boundary conditions, $\mathbf{d}(x,t) = \mathbf{d}_\Gamma(t)$, $\boldsymbol{\sigma}(x,t) = \boldsymbol{\sigma}_\Gamma(t)$ on Γ, where $\mathbf{F}$, $\mathbf{S}$ and $\mathbf{f}$ are the material deformation gradient, the second Piola-Kirchhoff stress tensor and the volume forces on Ω_S, respectively.

The second Piola-Kirchhoff stress tensor $\mathbf{S}$ is given via an appropriate constitutive relation, e.g. for St. Venant-Kirchhoff material with $\mathbf{C}$ being the constitutive tensor and $\mathbf{E}$ representing the Green-Lagrange strain:

$$\mathbf{S} = \mathbf{C} : \mathbf{E}, \qquad \mathbf{E} = \frac{1}{2}(\mathbf{F}^T \cdot \mathbf{F} - \mathbf{I}). \tag{3}$$

The computational domain is discretized and the above equations are solved with the finite element method, in a Lagrangian description, using the isoparametric element concept. For the time integration the generalized-α method is used [10] and in the case of stationary problems the inertia term vanishes and Eq. (2) reduces to a static equilibrium equation.

In the modeling of thin-walled structures, membrane or shell elements are used for discretization of the finite element model. Within this work, for the shell applications a 7-parameter shell element is used. Furthermore, special care is given to prevent locking phenomena by applying the well-known ANS and EAS methods [3]. For membranes, the thickness is assumed relatively small and three translational degrees of freedom are assumed per node. Shape of these structures is obtained with form-finding, described in Sect. 2.2 [4, 42].

Both, shell and membrane elements produce reduced structural models with two dimensional representation which can describe the three-dimensional physical properties by introducing mechanical assumptions for the thickness direction. With this way, more efficient computations are performed and the accuracy of the model is significantly improved. Due to this reduced order modeling of the structure special care has to be given to the interface and an additional operation has to be included in the structural part to transfer information between the 2D structure and the 3D fluid model [6].

More precisely, in the case of shells, the interface surface is found by moving the 2 dimensional surface of the structure half of the thickness normal to the surface in both sides. On these two moved surfaces, the exchange of data is performed consistently with respect to shell theory: The displacements and rotation calculated in the structure are transformed to displacements for the fluid and the normal and shear forces from the fluid are transformed to normal forces and moments applied on the shell. In membranes, the thickness is considered to be negligible. So, in both fields the structure is represented with two surfaces which coincide.

2.4.2 Fluid - Atmospheric wind

In the case of structural design, since almost all civil engineering structures are exposed to the atmospheric wind flow, a prediction of the wind loads applied on them is needed. Specially, in the case of light-weight structure design, consideration of the wind effects on the structure has a significant importance. In order to achieve a good prediction of the wind loads, one should take into account main characteristics of the atmospheric boundary layer (ABL). ABL has a complex form in terms of velocity and turbulence intensity profiles, and their variations in time, which are dependent on many parameters such as geographic location and type of surrounding urban. A robust wind model should characterize these physical issues in a set of few easy-to-handle model parameters. Furthermore, large structure size, low air viscosity and high wind velocity lead to very big Reynolds numbers (several millions) and subsequently large computation times. Therefore, computational wind engineering (CWE) simulation has to be efficient with respect to the computation time. In this section, governing equations and some of ABL modeling aspects with focus on fluid-structure interaction application is presented.

State equations and turbulence models Considering physical properties of air and the typical Mach number in wind engineering problems, simulation of wind flow can be considered as an incompressible Newtonian problem. Therefore, one can find state variable fields (pressure and velocity) by solving the momentum and continuity equations (Navier-Stokes) on a computational domain with appropriate boundary conditions. Using finite volume method, momentum equations can be discretized and linearized. In the case of transient simulation, this can be done by using an iterative algorithm such as PISO (Pressure Implicit Split-Operation) [19]. For steady-state models, SIMPLE algorithm follows a similar approach to PISO, neglecting the time-derivative term in the momentum equation.

Despite the fact that LES is known to be more accurate than RANS, taking into account the required computational effort, RANS is rather a more realistic technique for solving this specific type of FSI-CWE problems. However, reviewing the literature [24, 32] shows that a relatively high order of accuracy can be achieved using RANS models in wind flow simulations. Particularly, since within this work, the optimization is realized with steady-state assumption, time-averaged pressures calculated by RANS are totally sufficient and quantities such as the peak pressure, which can be captured only by more flexible models like LES are not of interest.

Reynolds averaged models are based on splitting field variables to the sum of the mean value and the fluctuating part. This leads to an additional momentum term in Navier-Stokes equations:

$$u = \bar{u} + u, \tag{4}$$

$$\frac{\partial \bar{u}_i}{\partial t} + \bar{u}_j \frac{\partial \bar{u}_i}{\partial x_j} = -\frac{1}{\rho} \frac{\partial \bar{p}}{\partial x_i} + \nu \frac{\partial^2 \bar{u}_i}{\partial x_k^2} - \frac{\partial \overline{u_i' u_j'}}{\partial x_j}, \qquad \frac{\partial \bar{u}_i}{\partial x_i} = 0. \tag{5}$$

The aim of Reynolds averaged methods is to model (in contrast to resolve) this momentum term. Turbulent viscosity models propose dividing it into an isotropic part which will be added to the pressure field and an anisotropic part which is considered as an additional viscosity, added to the molecular viscosity.

$$\frac{\partial \bar{u}_i}{\partial t} + \bar{u}_j \frac{\partial \bar{u}_i}{\partial x_j} = -\frac{1}{\rho} \frac{\partial (\bar{p} + \frac{2}{3} k_e)}{\partial x_i} + \frac{\partial}{\partial x_j} \left(2(\nu + \nu_T) \bar{S}_{ij} \right). \tag{6}$$

RANS models are classified by the number of extra transport equations that have to be solved in order to obtain the turbulence viscosity value. Two widely-used two-equation RANS models are $k - \varepsilon$ and $k - \omega$, in which ν_T is modeled solving transport equations of k (turbulence kinetic energy) and ε (dissipation rate) for $k - \varepsilon$, and k and $\omega \equiv \frac{\varepsilon}{k}$ for $k - \omega$.

$k - \omega$ shows a better performance comparing to $k - \varepsilon$ in large pressure gradients and near-wall flows [41], whereas $k - \varepsilon$ gives a more realistic estimation of the turbulence viscosity for free-stream flows and small pressure gradients. $k - \omega$ SST model [27] combines strengths of both models by using a linear combination of ε and ω transport equations. Coefficients of this mixed transport equation are calculated as follows:

$$X_{k-\omega SST} = X_{k-\omega} F_1 + X_{k-\epsilon} (1 - F_1). \tag{7}$$

Blending function (F_1) gives the emphasis to $k - \omega$ results near the wall, and directs the solution more to the $k - \varepsilon$ results far from the wall.

The nature of wind is time varying and a transient modeling technique is necessary to simulate wind behavior. Although RANS is based on time averaging, it is possible to use it for transient flows. The so-called URANS (Unsteady RANS) model divides the time domain into time-steps and within each time-step, steady RANS assumptions are applied. Time-dependent variables are estimated by common time-integration techniques. Validity of URANS and the proper choice of time scales in it for different applications are still open questions in turbulence modeling.

Wind simulation In CWE, the geometry of the structure is built as an obstacle in a discretized computational domain. Solving the turbulent flow around it, pressure values on the surface of the structure are calculated. In this type of problems, there are some modeling aspects to be considered, few of which are as follows:

- Boundary conditions: Assuming a straight wind flow, the following boundary conditions need to be specified. Top and side surfaces are treated as free-slip or symmetric or periodic boundaries [17]. Outlet pressure can be set to a constant value. The challenging part is to prescribe the fields (pressure, velocity and turbulence intensity) at the inlet, as similar as possible to the real wind conditions. Furthermore, at the bottom of the computational domain boundary, a rough wall boundary shall be used, rather than a fully smooth no-slip wall [8]. This is due to the natural roughness objects existing on the surface of the earth such as grass, trees, other buildings, etc. In the following, the proper choice of inlet fields is described considering a rough wall boundary layer.
- The domain size: Size of the computational domain should be much larger than the structure. This prevents the disturbance of the fluid boundaries on the obstacle. For different fields of application, in literature, there exist suggestions for the proper CWE domain size. Franke et al. [14] proposes a distance of 2-8, 5, 6 and 15 (normalized by the structure size) for the front, side, top and back between the obstacle and the computational domain, based on blockage percentage.

Atmospheric boundary layer Basically, a turbulent wall boundary layer consists of a linear near-wall (viscous) sublayer and a logarithmic sublayer. Due to the large Reynolds number, in ABL the viscous sublayer is very thin and therefore almost all structures are located in the logarithmic region. The mean velocity profile in logarithmic region of a smooth wall boundary is described as $\bar{u} = \frac{u_\tau}{\kappa} \ln y^+ + B$, where u_τ is the friction velocity, $\kappa = 0.41$ is *von Kàrmàn* constant, y^+ is the distance to the wall in inner coordinates and $B = 5.2$ is a universal constant. Existence of roughness at the wall surface does not change the general form of the velocity profile, but just the multiplier of y^+ in the mentioned formula [37]. Therefore, Inserting B and the effect of roughness inside the "ln" expression, the velocity profile in a rough boundary layer would be $\bar{u} = \frac{u_\tau}{\kappa} \ln \left(\frac{y}{y_0} \right)$, where y_0 is the *roughness length* and is a function of viscosity, u_τ and k_s (sand roughness height). Assuming this velocity profile, Richards and Hoxey [30] suggest the following inlet conditions for atmospheric wind velocity, kinetic energy and dissipation rate profiles:

$$\bar{u}(z) = \frac{u^*}{\kappa} \ln \left(\frac{z + z_0}{z_0} \right), \qquad k = \frac{u^{*2}}{\sqrt{C_\mu}}, \qquad \varepsilon = \frac{u^{*3}}{k(z + z_0)}, \qquad (8)$$

where $u^* = \left(\kappa u_{ref} \right) \mathbf{div} \ln \left(\frac{z_{ref}}{z_0} \right)$ and $C_\mu = 0.09$. $\bar{u}_{ref}$ is a known wind velocity at the height z_{ref}, and z_0 is the surface roughness length.

Moving boundaries In case of moving boundaries, the mesh field has to be deformed and the mesh fluxes have to be corrected, respectively. There are different algorithms of calculating this mesh motion. For instance, H. Jassak and Z. Tukovic [20] determine the mesh field motion by solving the Laplace equation with variable diffusion on elements. One of the strengths of this method is that less distortion in the mesh close to the moving boundary occurs. In the present work, this mesh motion algorithm is applied to calculate the movement of the boundaries due to the structural displacement in fluid structure interaction analysis as well as in update of the design in the optimization loop.

2.4.3 Coupling

In order to establish equilibrium between the fluid and the structure field at the common interface Γ^{FS}, certain coupling conditions have to be fulfilled. These are the continuity of displacements and surface tractions, respectively:

$$d_S^{\Gamma^{FS}} = d_F^{\Gamma^{FS}}, \quad n_S^{\Gamma^{FS}} \cdot \tau_S^{\Gamma^{FS}} = n_F^{\Gamma^{FS}} \cdot \tau_F^{\Gamma^{FS}}. \tag{9}$$

The equilibrium state of a coupled system is reached, if for a certain interface displacement $d^{\Gamma^{FS}}$ the field and interface residua tend to zero:

$$R_S(t, d^{\Gamma^{FS}}) = 0, \quad R_F(t, d^{\Gamma^{FS}}) = 0, \quad R^{\Gamma}(t, d^{\Gamma^{FS}}) = 0. \tag{10a}$$

Coupled problem in operator form The solution of the coupled problem is based on a partitioned Dirichlet-Neumann approach. This means that the fluid and structure fields are solved as separate partitions and the interaction is captured by influencing the specific boundary conditions. The fluid field is solved due to a given interface displacement and is therefore the Dirichlet partition. The structure field is solved due to given boundary forces and is therefore the Neumann partition. To develop algorithms for the implicit solution of the coupled problem, the problem is denoted in its operator form. Because the used algorithms rely on black-box field solutions, the corresponding field operators are introduced. The fluid field is denoted by the field operator $\mathscr{F}$ and the structure field by the field operator $\mathscr{S}$. Due to the nonlinearity of the field and the coupled problem, the solution process involves subiterations within as well as between the fields. Typically, each iteration consists of the following steps:

1. Apply a displacement $\mathbf{d}^{\Gamma}$ onto the fluid interface Γ. The fluid domain Ω^F is deformed and the mesh fluxes are calculated.
2. Solve the fluid problem on the deformed state using an ALE-approach, leading to velocity $\mathbf{v}$ and pressure $\mathbf{p}$. Calculate the surface forces f^{ext} at the interface: $f^{ext} = \mathscr{F}(\mathbf{d}^{\Gamma})$.
3. Apply the surface forces f^{ext} onto the structure interface. Solve the structure problem for the corresponding displacements $\tilde{\mathbf{d}}$ and evaluate them at the interface: $\tilde{\mathbf{d}}^{\Gamma} = \mathscr{S}(f^{ext}) = \mathscr{S}(\mathscr{F}(\mathbf{d}^{\Gamma})) = \mathscr{S} \circ \mathscr{F}(\mathbf{d}^{\Gamma})$.

The nonlinear coupled problem is therefore given by $\tilde{\mathbf{d}}^{\Gamma} = \mathscr{S} \circ \mathscr{F}(\mathbf{d}^{\Gamma})$, and its solution is found if $\tilde{\mathbf{d}}^{\Gamma} = \mathbf{d}^{\Gamma}$. The residuum at the interface $\mathbf{R}^{\Gamma}(\mathbf{d}^{\Gamma})$ is defined as:

$$\mathbf{R}^{\Gamma}(\mathbf{d}^{\Gamma}) = \mathscr{S} \circ \mathscr{F}(\mathbf{d}^{\Gamma}) - \mathbf{d}^{\Gamma} = \tilde{\mathbf{d}}^{\Gamma} - \mathbf{d}^{\Gamma}. \tag{11}$$

Solution algorithms for coupled problems There exists a huge variety of algorithms for the partitioned solution of the coupled problem, e.g. fixed-point [22, 43], vector extrapolation [23, 33], quasi-Newton [11, 39], inexact Newton-Krylov [28] or Multigrid/Multilevel methods. One of the greatest advantages of a partitioned approach is its modularity. Depending on the objective of a coupled simulation, the computational framework can be easily adapted for the specific needs of the single-field solvers, what leads to a great flexibility and a broad application range. But if a coupled solution algorithm is chosen, which necessitates complicated changes and enhancements to the single-field solution method, this approach is not suitable. Therefore, within this project the emphasis is placed on methods, which rely on black-box field solvers. There exist mainly two approaches, which fulfill this requirement, namely fixed-point iteration in combination with Aitken's Δ^2-method [22, 43] and a quasi-Newton method with reduced order modeling of the field behavior [11].

Fixed-point iteration: The basic form of a fixed-point method is given by:

$$x^{k+1} = \Phi(x^k), \quad k = 1, ..., n. \tag{12}$$

Operator Φ evaluates the solution of step number $k+1$, being given the last iteration solution, x^k. Considering a coupled problem at time $n+1$, the iteration directive has the form:

$$\Phi\left(d_{n+1}^\Gamma\right) = d_{n+1}^{\Gamma,k} + \Delta d_{n+1}^{\Gamma,k}. \tag{13}$$

Using a relaxation technique for the update of the interface displacement leads to:

$$d_{n+1}^{\Gamma,k+1} = d_{n+1}^{\Gamma,k} + \Delta d_{n+1}^{\Gamma,k} = d_{n+1}^{\Gamma,k} + \omega^k R_{n+1}^{\Gamma,k} = (1 - \omega^k)d_{n+1}^{\Gamma,k} + \omega^k \tilde{d}_{n+1}^{\Gamma,k}$$

where ω^k is the relaxation factor in iteration k. There exist several possibilities for the choice of the relaxation factor. A constant factor can be used, which is simple, but stability problems are faced for large ω^k and inefficiency and small convergence rates for small ω^k. Choosing a dynamic relaxation factor in every iteration gives improvements w.r.t. efficiency and stability. Therefore, Aitken's Δ^2-method is used to determine the relaxation factor. As a basis, the Aitken factor has to be determined by:

$$\mu_{n+1}^k = \mu_{n+1}^{k-1} + \left(\mu_{n+1}^{k-1} - 1\right) \frac{\left(\Delta d_{n+1}^{\Gamma,k-1} - \Delta d_{n+1}^{\Gamma,k}\right)^T \cdot \Delta d_{n+1}^{\Gamma,k}}{\left(\Delta d_{n+1}^{\Gamma,k-1} - \Delta d_{n+1}^{\Gamma,k}\right)^2} \tag{14}$$

with

$$\Delta d_{n+1}^{\Gamma,k-1} = \tilde{d}_{n+1}^{\Gamma,k-1} - d_{n+1}^{\Gamma,k-1}, \quad \Delta d_{n+1}^{\Gamma,k} = \tilde{d}_{n+1}^{\Gamma,k} - d_{n+1}^{\Gamma,k}$$

and the corresponding relaxation parameter is given by $\omega^k = 1 - \mu_{n+1}^k$.

Quasi-Newton method: The method used within this approach is called a quasi-Newton method, because it is similar to the classical Newton approach in the sense of using history information of the interface residual to minimize the interface residual and determine incremental interface updates, but the Jacobian is not directly calculated within the solution process. Basically, the method consists of two main parts within every iteration: 1. Approximation of the new interface residual based on a linear combination of the previous residuum increments. This leads to a set of coefficient factors. 2. Determination of the new interface displacement increment as a linear combination of those known increments, applying the coefficients determined in the first step.

The algorithm needs a starting procedure, to determine the first residuum and displacement increments. Therefore, a predictor and a relaxation step have to be performed in advance. After this, the quasi-Newton iterations are entered. In every step, the desired residuum of the next iteration is zero, so the desired residuum increment is given by $\Delta \mathbf{R}^k = 0 - \mathbf{R}^{\Gamma,k}$. This is approximated by a linear combination of the known residuum increments from previous iterations, leading to:

$$\Delta \mathbf{R}^k = 0 - \mathbf{R}^{\Gamma,k} \approx \sum_{i=1}^{k-1} \alpha_i^k \Delta \mathbf{R}^{\Gamma,i} = V^k \alpha^k, \tag{15}$$

which gives the linear coefficients α_i^k and V^k being a matrix that stores the residuum increments from past steps. This is in general an overdetermined equation system, which is solved using a least squares method: $\| V^k \alpha^k - \mathbf{R}^{\Gamma,k} \| \to min$. V^k is a non-symmetric matrix and so it can be decomposed into an orthogonal Q^k and a upper-triangular matrix R^k using e.g. Householder transformation. This gives great numerical benefit in the solution of the problem.

The coefficients, determined for the combination of the residuum increments, are now applied to the corresponding relaxed interface displacements:

$$\Delta \tilde{\mathbf{d}}^{\Gamma,k} = \sum_{i=1}^{k-1} \alpha_i^k \Delta \tilde{\mathbf{d}}^{\Gamma,i} = W^k \alpha^k. \tag{16}$$

The correlation between $\mathbf{d}^{\Gamma}$ and $\tilde{\mathbf{d}}^{\Gamma}$ is given by $\mathbf{R}^{\Gamma} = \tilde{\mathbf{d}}^{\Gamma} - \mathbf{d}^{\Gamma}$, and therefore: $\mathbf{d}^{\Gamma} = \tilde{\mathbf{d}}^{\Gamma} - \mathbf{R}^{\Gamma}$. For the corresponding incremental state, this leads to

$$\Delta \mathbf{d}^{\Gamma} = \Delta \tilde{\mathbf{d}}^{\Gamma} - \Delta \mathbf{R} = W^k \alpha^k + \mathbf{R}^{\Gamma,k}, \tag{17}$$

as the new interface displacement increment.

A detailed description of the algorithm is presented in algorithm 1.

Algorithm 1 Quasi-Newton coupling algorithm [11]

1: **for** $t = 0$ to $t = t_{tot}$ **do**
2:     k=1
3:     Predictor step:
4:     $\mathbf{d}_{n+1}^{\Gamma,1} = 5/2\mathbf{d}_n^{\Gamma} - 2\mathbf{d}_{n-1}^{\Gamma} + 1/2\mathbf{d}_{n-2}^{\Gamma}$
5:     $\tilde{\mathbf{d}}_{n+1}^{\Gamma,1} = \mathscr{S} \circ \mathscr{F}(\mathbf{d}_{n+1}^{\Gamma,1})$
6:     $\mathbf{R}_{n+1}^{\Gamma,1} = \tilde{\mathbf{d}}_{n+1}^{\Gamma,1} - \mathbf{d}_{n+1}^{\Gamma,1}$
7:     relaxation step:
8:     $\mathbf{d}_{n+1}^{\Gamma,2} = \mathbf{d}_{n+1}^{\Gamma,1} + \omega\mathbf{R}_{n+1}^{\Gamma,1}$
9:     quasi-Newton iterations:
10:     **while** (!converged) **do**
11:         $\tilde{\mathbf{d}}_{n+1}^{\Gamma,k} = \mathscr{S} \circ \mathscr{F}(\mathbf{d}_{n+1}^{\Gamma,k})$
12:         $\mathbf{R}_{n+1}^{\Gamma,k} = \tilde{\mathbf{d}}_{n+1}^{\Gamma,k} - \mathbf{d}_{n+1}^{\Gamma,k}$
13:         **if** $\left|\mathbf{R}_{n+1}^{\Gamma,k}\right|_{L2} < \epsilon$ **then**
14:             $\mathbf{d}_{n+1}^{\Gamma} = \mathbf{d}_{n+1}^{\Gamma,k}$
15:             converged
16:         **else**
17:             $\Delta\mathbf{R}^{\Gamma,k} = \mathbf{R}_{n+1}^{\Gamma,k} - \mathbf{R}_{n+1}^{\Gamma,k-1}$
18:             $\Delta\tilde{\mathbf{d}}^{\Gamma,k} = \tilde{\mathbf{d}}_{n+1}^{\Gamma,k} - \tilde{\mathbf{d}}_{n+1}^{\Gamma,k-1}$
19:             $V^k = \begin{bmatrix} \Delta\mathbf{R}^{\Gamma,k} & \Delta\mathbf{R}^{\Gamma,k-1} & \ldots & \Delta\mathbf{R}^{\Gamma,1} \end{bmatrix}$
20:             $W^k = \begin{bmatrix} \Delta\tilde{\mathbf{d}}^{\Gamma,k} & \Delta\tilde{\mathbf{d}}^{\Gamma,k-1} & \ldots & \Delta\tilde{\mathbf{d}}^{\Gamma,1} \end{bmatrix}$
21:             Householder transformation: $V^k = Q^k R^k$
22:             $R^k\alpha^k = Q^{k^T}(-\mathbf{R}^{\Gamma,k}) \quad \rightarrow \quad \alpha^k$
23:             $\Delta\mathbf{d}^{\Gamma} = W^k\alpha^k + \mathbf{R}^{\Gamma,k}$
24:             $\mathbf{d}_{n+1}^{\Gamma,k+1} = \mathbf{d}_{n+1}^{\Gamma,k} + \Delta\mathbf{d}^{\Gamma}$
25:         **end if**
26:         $k \leftarrow k + 1$
27:     **end while**
28:     $n \leftarrow n + 1$
29: **end for**

2.5 Optimization

In the optimization module, decisions about the update of the design as well as the convergence of the overall procedure are taken. In general, the optimization problem can be stated as follows:

$$f(\mathbf{s}) \rightarrow \min : g_j(\mathbf{s}) \leq 0 , \, j \in [1, m] \text{ and } h_j(\mathbf{s}) = 0 , \, j \in [m + 1, m + N], \quad (18)$$

where $\underline{\mathbf{s}_l} \leq \mathbf{s}_l \leq \overline{\mathbf{s}_l}$, $\mathbf{s} \in \mathbb{R}^n$ and f, g_j, h_j are the objective function, the inequality and the equality constraints, respectively. They are all functions of $\mathbf{s} = (s_1, \cdots, s_n)^T$, which is the vector of the optimization variables. The existence of constraints changes totally the way to treat the problem. Consequently, constrained and unconstrained problems form two separate classes of optimization problems. Unconstrained gradient based optimization problems, in which the gradient of the objective function ∇f is required, are discussed in the sequel.

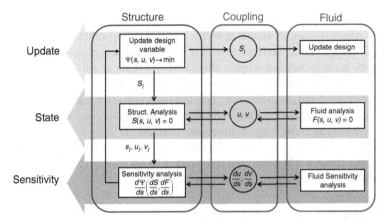

Fig. 5 Three levels of optimization using NAND approach.

Within this project, the physical state (Sect. 2.4) involves a fluid-structure inter-action analysis and the final goal will be to obtain a design which improves the behavior of the structure. The objective function f will be a function of the structural displacement $\mathbf{u}$, the fluid state variable $\mathbf{v}$ and the design variable $\mathbf{s}$, i.e. $f = f(\mathbf{u}, \mathbf{v}, \mathbf{s})$. Design variables of structural shape optimization are the geometric parameters representing the shape of the structure.

Furthermore, the physical state is described by the structural and the fluid par-tial differential equations which have to be satisfied at the optimum. As a result, in the optimization problem, two PDE constraints have to be added. These constraints are not treated as any other constraint, as in the case of Simultaneous Analysis and Design (SAND) but their first order consistency is enforced in each iteration step. This strategy of implementing the optimization problem is known as Nested Anal-ysis and Design (NAND) [1]. For the following, let us assume that the discretized equations describing the two fields are $\mathbf{S(s, u, v)}$, $\mathbf{F(s, u, v)}$, respectively.

NAND consists of three main stages: the optimizer, the analysis of the physical state and the sensitivity analysis (Fig. 5). The analysis of the physical state is already described in Sect. 2.4 whereas the optimizer and the sensitivity analysis are going to be discussed in the sequence. Furthermore, within this work, the physical state is a steady state fluid-structure interaction analysis and thus time is omitted from both fluid and structural equation.

2.5.1 Optimizer

The optimizer is the driving module of the overall optimization procedure. Here, the decision about the size and the direction of the update vector of the structural design is taken based on the gradient information obtained by the sensitivity analysis.

In general, gradient-based methods start with an initial guess for the design variable $\mathbf{s}_0$ and in each iteration the solution is updated as follows:

$$\mathbf{s}_{i+1} = \mathbf{s}_i + \alpha_i \mathbf{D}^i, \tag{19}$$

where $\mathbf{D}^i = \mathbf{D}^i(\nabla f(\mathbf{s}_i); \nabla^2 f(\mathbf{s}_i)) \in \mathbb{R}^n$ is the search direction and $\alpha_i \in \mathbb{R}$ is the step size which is usually calculated with a line-search algorithm. As for the most of engineering application the Hessian matrix is not availabe, simple gradient methods prove to be more robust and reliable alternatives, in particular for a large number of design variables. Standard methods are "Steepest descent" and "Conjugate gradient" for unconstrained and the method of "Feasible directions" for constrained problems. In the steepest descent algorithm, the steepest descent direction (the negative gradient), is assumed as the search direction, therefore, $\mathbf{D} = -\nabla f(\mathbf{s})$. This method specifically is very efficient when the initial guess is comparatively far from the solution but closer to the minimum it converges rather slowly [15]. The description of the algorithm is presented in algorithm 2.

Algorithm 2 Steepest descent algorithm

1: Initialize $i = 0 : s_0$
2: **while** (!converged) **do**
3:     $D^i = -\nabla_{s_i} f$
4:     $s_{i+1} = s_i + \alpha_i D_i$
5:     $i \leftarrow i + 1$
6: **end while**

Conjugate gradient is a method which converges faster than steepest descent, without increasing complexity and computational cost. In this method, the search direction is:

$$\mathbf{D}^i = -\nabla_{s_i} f + \beta_i \cdot \mathbf{D}^{i-1}, \text{ where } \beta_i = \frac{\nabla_{s_i} f^T \cdot \nabla_{s_i} f}{\nabla_{s_{i-1}} f^T \cdot \nabla_{s_{i-1}} f}. \tag{20}$$

In the case that the objective function is quadratic, the method converges in maximum n steps, where n is the number of design variables. This is not the case for non-quadratic functions. For such functions, intelligent techniques which take into account the nonlinearity of the function in order to decide about the restarting of the algorithm are applied [15].

2.5.2 Sensitivity analysis

During sensitivity analysis the gradient of the objective function with respect to the design variables has to be computed. As already described in Sect. 2.5, the objective function is assumed to be $\mathbf{f} = \mathbf{f}(\mathbf{s}, \mathbf{u}, \mathbf{v})$, where $\mathbf{u}$ and $\mathbf{v}$ represent the structural displacements and the fluid forces, respectively.

Applying the chain rule, the gradient of the objective function can be written as:

$$\frac{df}{d\mathbf{s}} = \frac{\partial f}{\partial \mathbf{s}} + \frac{\partial f}{\partial \mathbf{u}} \frac{d\mathbf{u}}{d\mathbf{s}} + \frac{\partial f}{\partial \mathbf{v}} \frac{d\mathbf{v}}{d\mathbf{s}}. \tag{21}$$

In order to compute the gradient of f from the above formula, all the terms in the right hand side have to be computed. The derivatives of the objective function with respect to the state variables ($\frac{\partial f}{\partial \mathbf{u}}$, $\frac{\partial f}{\partial \mathbf{v}}$) can be easily computed since the function f is usually given in analytical form. The derivatives of the state variables with respect to the design variable $\mathbf{s}$ are computed from the sensitivity equations. To obtain this set of equations, first-order consistence of the PDE of the fluid and the structure has to be enforced:

$$
\begin{aligned}
\frac{dS}{d\mathbf{s}} &= \frac{\partial S}{\partial \mathbf{s}} + \frac{\partial S}{\partial \mathbf{u}}\frac{d\mathbf{u}}{d\mathbf{s}} + \frac{\partial S}{\partial \mathbf{v}}\frac{d\mathbf{v}}{d\mathbf{s}} = 0, \\
\frac{dF}{d\mathbf{s}} &= \frac{\partial F}{\partial \mathbf{s}} + \frac{\partial F}{\partial \mathbf{u}}\frac{d\mathbf{u}}{d\mathbf{s}} + \frac{\partial F}{\partial \mathbf{v}}\frac{d\mathbf{v}}{d\mathbf{s}} = 0.
\end{aligned}
\tag{22}
$$

The above equations form a linear system of equations with derivatives of the state variables with respect to the design variables as the unknowns:

$$
\begin{pmatrix} \dfrac{\partial S}{\partial \mathbf{u}} & \dfrac{\partial S}{\partial \mathbf{v}} \\[2mm] \dfrac{\partial F}{\partial \mathbf{u}} & \dfrac{\partial F}{\partial \mathbf{v}} \end{pmatrix} \cdot \begin{pmatrix} \dfrac{d\mathbf{u}}{d\mathbf{s}} \\[2mm] \dfrac{d\mathbf{v}}{d\mathbf{s}} \end{pmatrix} = \begin{pmatrix} -\dfrac{\partial S}{\partial \mathbf{s}} \\[2mm] -\dfrac{\partial F}{\partial \mathbf{s}} \end{pmatrix}.
\tag{23}
$$

The system described by Eq. (23) is solved in an iterative manner, described in Sect. 3.2, and the solution procedure is similar to a fluid structure interaction analysis [2, 35].

3 Solution Strategy and Realization

In this section the strategy followed for the realization of the overall framework is presented. In 3.1 the software environment is introduced and in 3.2 the algorithmic implementation of the shape optimization technique within this environment is explained.

3.1 Computational Framework

A robust computational framework should be able to treat a broad range of applications, from small-sized principle to large industrial examples, from laminar to turbulent, from pure FSI-simulations to optimization of coupled problems. Therefore, a highly modular environment is needed, so that for each specific problem the best-suited approach is possible. To fulfill these requirements, a software realization based on three different codes is chosen. Two of these codes are the single-field solvers for fluid and structure. The third code, called coupling code, is located between the field solvers and operates as a master process, controlling the whole simulation, and working as an interface between fluid and structure field. It has to be noted that all codes within this environment are capable of performing parallel computations and completely object-oriented. These three software are:

Carat: The in-house structure field solver CARAT++ is based on Finite Element method. Different sorts of analysis, such as static and dynamic, linear and nonlinear are available in this code. For the dynamic analysis, different time-integration schemes such as Generalzed-α are used.

OpenFOAM: For the fluid field, a solver based on OpenFOAM is used. OpenFOAM is a freely available set of object-oriented libraries for Finite Volume method. Various numerical schemes for interpolation and integration are available.

CoMA: CoMA (Coupling for Multiphysics Analysis) is an independent software developed specifically for the scope of this work. This program is responsible for the simulation control of surface-coupled simulations and is located as a central process between single-field solvers. It offers a wide variety of features, making it a powerful tool for all kind of surface coupled simulations. The most important features of CoMA are:

- **Parallelization and communication:** In coupled computations, the field solvers typically work in parallel due to high computational effort. This means that the coupled surface of one field is spread over multiple field processes. To handle this, CoMA uses the so-called *process group* concept. The subset of all field processes which take part in the coupled computation is identified and grouped into a process group. The communication with processes can be based on MPI or files, depending on the possibilities offered by the field solvers.
- **Data transfer:** One of the key ingredients of CoMA is to handle data transfer between non-matching surface meshes. For flexibility and due to the parallelization of the field solvers, the surface mesh consists of different partitions, each of which is connected to a process, with the possibility of multiple partitions per process. Both flux and field quantities can be transferred. The data transfer is based on a nearest-neighbor search and linear interpolation. This transfer method is conservative with respect to field quantity fluxes. More details can be found in [21].
- **Coupled simulation control:** CoMA serves as a master process controlling black-box single field solvers in coupled computations. It offers the possibility of different coupling concepts (one-way and two-way coupling, explicit and implicit coupling) and for two-way coupling different coupling algorithms. The key ingredient is the so-called *exchange point* concept. Exchange point is an entity, which transfers a quantity, e.g. force, from a sender process group to a receiver process group. The exchange point can have additional features to influence and adapt the quantity. This can be e.g. a coupling algorithm applying underrelaxation or a convergence proof. Exchange points are arranged into an exchange point group, what resembles one cycle/iteration within a coupled simulation. The chosen coupling algorithm influences the way, how the group is

called, i.e. once per timestep in explicit coupling, multiple times in implicit coupling. This concept offers great flexibility and also new coupling strategies can be easily realized.

3.2 Optimization strategy

As described in Sects. 2.1 and 2.5, the steady-state shape optimization method presented in this work consists of three main steps, the geometry update, physical state evaluation and the sensitivity analysis. Accordingly, in software implementation phase, these steps are introduced to all three modules of the coupled problem, within a loop and in the same order. Therefore, in each design step, all three codes go through this loop and perform these steps one after each other simultaneously. In this section, the structure of the design loop as well as the communication and synchronization of field solvers within the optimization procedure is briefly introduced, which is indeed the challenge in software development of such complex class of analysis.

Optimization loop Each design step starts by updating the shape with the new set of design variables proposed by the optimizer. After performing the structural geometry update, the new geometry is sent to the fluid solver via the coupling code. This new geometry is applied to the fluid by moving its boundaries toward the new design and after that the CFD solver computes a converged solution for the new boundary. Then the fluid-structure interaction analysis is performed by exchanging displacement-pressure values at the interface between the fields until convergence. Resulting pressures and displacements are stored as the "state-variables" of the current time-step, i. Afterwards, the sensitivity analysis is performed for all the design variables (2.5.2). Within the sensitivity loop and after evaluation of sensitivity for each design variable, it is necessary to reset all the fields back to their *state* values. The result of this step is the gradient vector of the objective function with respect to all design variables, which is subsequently given to the optimizer for calculation of the new set of design variables. This is done according to the particular optimization algorithm used in the problem (see Sect. 2.5.1).

The optimization loop is repeated unless the optimizer reports the design convergence. This could be due to reaching a minimum in the objective function or because some technical constraint at the design variables is reached.

Evaluation of sensitivities The aim is to solve Eq. (23) for $\frac{du}{ds}$ and $\frac{dv}{ds}$ which is then used in Eq. (21) in order to calculate $\frac{df}{ds}$. Direct solution of this system of equation is possible just if both fluid and structure derivatives of Eq. (23) are explicitly known. In this project, structural derivatives ($\frac{\partial S}{\partial u}$ and $\frac{\partial S}{\partial s}$) are calculated directly at the structural code, Carat, whereas the fluid solver is treated as a black-box module. Therefore, an iterative mixed strategy is applied, which uses directly the sensitivity equation of the structure and finite-difference evaluation of derivatives for the fluid. The main steps of this method are explained in the following.

At the beginning of each design variable sensitivity analysis, a perturbation of that particular design variable is applied on the design. Then, the coupled sensitivity loop starts. In this loop, first, the structure sensitivity equation is solved using the fluid-derivative information from the previous iteration:

$$\left(\frac{d\mathbf{u}}{d\mathbf{s}}\right)_i = -\left(\frac{\partial S}{\partial \mathbf{u}}\right)_i^{-1} \left[\left(\frac{\partial S}{\partial \mathbf{v}}\right)_i \left(\frac{d\mathbf{v}}{d\mathbf{s}}\right)_{i-1} + \left(\frac{\partial S}{\partial \mathbf{s}}\right)_i\right]. \tag{24}$$

Then, using the resulting structure-derivatives, $\frac{d\mathbf{u}}{d\mathbf{s}}$, a new perturbation on the fluid is calculated:

$$\mathbf{u}_i = \mathbf{u}_0 + \left(\frac{d\mathbf{u}}{d\mathbf{s}}\right)_i \Delta\mathbf{s} = \mathbf{u}_0 + \Delta\mathbf{u}, \tag{25}$$

where u_0 is the state displacement. Response of the fluid to the new perturbation could be written as:

$$\mathbf{v}_i = \mathbf{v}(\mathbf{u}_0 + \Delta\mathbf{u}, \mathbf{s}_0 + \Delta\mathbf{s}). \tag{26}$$

$\mathbf{s}_0$ is the state design variable and $\Delta\mathbf{s}$ is the perturbation of it. This response ($\mathbf{v}_i$) is used to produce the fluid derivative, $\frac{d\mathbf{v}}{d\mathbf{s}}$ by finite differencing:

$$\left(\frac{d\mathbf{v}}{d\mathbf{s}}\right)_i \approx \frac{\mathbf{v}_i - \mathbf{v}_0}{\Delta\mathbf{s}}. \tag{27}$$

$\mathbf{v}_0$ is the state force vector. These steps are repeated until convergence of sensitivity values.

This method is more efficient than external finite difference evaluation of the sensitivity on the coupled problem, since the structural equation is reduced to a linear equation. It should also be mentioned that unlike the evaluation of the state variables, a high accuracy for the calculation of sensitivities is not required since their values are used only for trend information. Therefore, a lower order approximation of the sensitivities is sufficient to perform gradient-based optimization. Thus, by setting a larger tolerance for the sensitivity analysis in this method, one can reduce the computational effort.

4 Verification and Validation

In this section, the solution technique introduced in Sect. 3 is examined by two benchmark examples. The first one, which is a pure CWE problem, is used for validation of the wind simulation strategy. The second example investigates the correctness of the whole FSI solution with the focus on the coupling algorithms and their computational efficiency.

4.1   Numerical simulation of the Silsoe cube

The "Silsoe" cube is a $6m \times 6m \times 6m$ cube constructed by Silsoe research institute in England [31]. The cube is equipped with different sensors for measurements of wind engineering experiments. The flat field around the cube allows the wind to have a close form to the ideal atmospheric boundary layer. Both size of the cube and the wind velocity gives to this experiment the characteristics of a real wind engineering problem. Since accurate measured data from the Silsoe experiment is published, numerical simulation of this experiment is chosen as a "reference" problem for validation of the wind simulation technique. Furthermore, different modeling aspects are studied based on this problem. For instance, it is observed that $k - \omega$ SST is a more proper model than other common RANS turbulence viscosity models w.r.t. accuracy. Furthermore, similarity of the Silsoe experiment ABL data to the inlet consistions for velocity, k and ε confirms the validity of those logarithmic profiles.

Results In Fig. 6 the pressure coefficient on the shown paths of the cube can be seen. These curves correspond to the Silsoe cube experiment and wind tunnel results reported by Hoelscher and Niemann from Windtechnologische Gesellschaft [18], and Castro and Robins [9] and this project. Diversity of the curves shows the complexity of wind engineering simulations. However, it can be seen that in both diagrams, almost on the entire path, the numerical results of this work have the closest value to the Silsoe experiment. It is seen that the obtained pressure coefficients on cube's surfaces are also very close to the values provided in known literature, e.g. "Wind Effects on Structures" [34].

The Silsoe cube problem was solved also using unstructured tetrahedral mesh (instead of cartesian structured mesh) and the difference in the solution was negligible. This shows that this modeling approach is applicable for more complex geometries which cannot be meshed with structured grids. Furthermore, some robustness studies on different modeling parameters such as the mesh size, the time step, Reynolds number etc. were performed. The mentioned achieved accurate results for the wind loads on structures is a required basis for performing wind engineering FSI simulations.

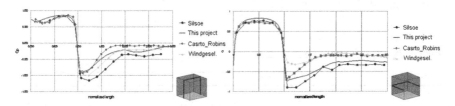

Fig. 6 Pressure Coefficient values on centerlines of the cube [9, 18, 31].

4.2   Coupled Computation in laminar Regime

In this example, a coupled computation is carried out, which scope is twofold: First, it is used to validate the coupled algorithms within the developed environment. Second, the efficiency of different coupling algorithms is examined. The testcases developed within the DFG research group 493 by [38] are chosen, because several groups have performed those simulations and therefore a broad range of verified results exists. Additionally, the testcases cover a broad range of coupling.

The testcase is described as follows: A fixed, rigid cylinder is placed slightly unsymmetric into a wall-bounded channel. An elastic beam is attached at the back of the cylinder. The geometric representation of the computational domain and a full description of the problem setup can be found in [38]. If a fluid flow is imposed onto the system, the flow regime behind the cylinder influences the beam and leads to deformations. Up to a certain Reynolds number, the flow is steady. In larger values of the Reynolds number, the flow gets unsteady and Karmann vortices occur behind the cylinder.

The cantilever beam is modeled using 30 4-node shell elements. The mid-surface of the shell is located in the transverse direction of the domain. The geometric representation of the interface between fluid and structure domain is evaluated by the method described in Sect. 2.4.1. The fluid is modeled based on a block-structured grid with 54000 elements. The values taken as reference are the results published in [38]. Different mesh levels were examined on both fields. The results shown in this contribution are the ones giving the best compromise between accuracy and numerical effort.

Results   Three different cases are examined, called FSI1 to FSI3, in which the inlet velocity, the structure density and the elastic modulus are varied. FSI1 results in a steady, whereas FSI2 and FSI3 in unsteady solutions, showing periodic oscillations of the beam with differing frequency and amplitude. The results are given in Table 1, in which the x- and y-displacement of point **A** located at the beam's tip are given. The numbers in brackets denote the frequency of the oscillations in the case of transient behavior.

The difference in displacements is below 5.5%. Considering the results from other groups differences with similar magnitude occur. Therefore, the verification of the coupled computation is done.

Table 1   Benchmark results FSI1-3.

	FSI1		FSI2		FSI3	
	$u_x(A)$	$u_y(A)$	$u_x(A)$	$u_y(A)$	$u_x(A)$	$u_y(A)$
own	$2.264e^{-05}$	$8.280e^{-04}$	$-1.489e^{-02}\pm$ $1.24e^{-02}[3.16]$	$1.30e^{-03}\pm$ $8.40e^{-02}[1.90]$	$-2.95e^{-03}\pm$ $2.84e^{-03}[11.24]$	$1.55e^{-03}\pm$ $3.66e^{-02}[5.49]$
reference	$2.270e^{-05}$	$8.208e^{-04}$	$-1.485e^{-02}\pm$ $1.27e^{-02}[3.86]$	$1.30e^{-03}\pm$ $8.17e^{-02}[1.93]$	$-2.88e^{-03}\pm$ $2.72e^{-03}[10.93]$	$1.47e^{-03}\pm$ $3.49e^{-02}[5.46]$
difference	$+0.00\%$	-0.88%	$-0.27\%+$ $2.42\%[+18.13\%]$	$+0.00\%-$ $2.82\%[+1.55\%]$	$-2.43\%+$ $4.41\%[-2.83\%]$	$+5.44\%-$ $4.87\%[-0.55\%]$

Table 2 Coupling algorithm comparisons - FSI2-3.

Type	FSI2		FSI3	
	num average iterations	reduction	num average iterations	reduction %
constant	35	+446.9%	no convergence	--
Aitken	6.4	0.0%	11.2	0.0%
QNR	4.68	−26.9%	8.54	−23.8%
QNR-H	3.14	−50.9%	5.34	−52.3%
QNR-HRC	3.26	−49.0%	5.06	−54.8%

Coupling algorithm efficiency In this section, different coupling algorithms introduced in Sect. 2.4.3, are compared with respect to their efficiency. As an example, the cases FSI2 and FSI3 (4.2) are selected, because they cover different density ratios and therefore different coupling strengths. The first 100 timesteps are examined and an average is made over the number of subiterations per timestep. A Newmark type displacement predictor is used giving 2nd order accuracy. As a basis, the fixed-point iteration in combination with Aitken's Δ^2-method is used. It is compared to the quasi-Newton method in 3 different variants: The basic version (QNR), the version taking also history incremental values into account (QNR-H), and the version taking not only history effects into account, but also replacing the constant relaxation step in every timestep (QNR-HRC). For sake of completeness, also constant relaxation method is taken into comparison, showing the inefficiency of this approach. The relaxation parameter is set to 0.2. The results for FSI2 with density ratio of $\frac{\rho_S}{\rho_F} = 10$ and for FSI3 with $\frac{\rho_S}{\rho_F} = 1$ are given in Table 2.

The difference in performance of the mentioned coupling algorithms is clearly noticeable in these tables. As expected, the constant relaxation is extremely inefficient and its computation time is more than 5 times longer compared to Aitken's method. Aitken's method performs well in both cases. The results of the quasi-Newton method with reduced order modeling are very encouraging. Its basic version (QNR) gives a speed-up of around 25 % compared to Aitken's method. But a real boost is achieved by taking history information into account, where the efficiency can be doubled, and a speed-up of over 50 % compared to Aitken's method is reached. Replacing the constant relaxation step by a quasi-Newton step (QNR-HRC), just gives minimal speed-up. It also has to be noted, that the speed-up of the QNR and its derivatives is not only w.r.t. subiterations, but with nearly the same factor in computing time. In comparison with fixed-point methods QNR needs the additional solution of a minimization problem with number of unknowns equal to the number of subiterations plus the number of history steps. These are approximately $5 - 10$ unknowns. This is absolutely negligible in comparison to the unknowns in fluid and structure field. As summary, the quasi-Newton method including history information has a significant performance improvement compared to Aitken's method and serves as an appropriate coupling algorithm for the solution of strongly coupled problems.

5 Numerical example - Shape optimization with FSI

In this section, practical application of the general optimization approach described in Sect. 2 is reviewed by the help of a real size engineering example which combines and relates all the several points described in the previous sections. More precisely, the behavior of a $5m \times 10m$ light-slender membrane roof surrounded by the turbulent wind flow is optimized. Large deformations on the structure are expected and a strong coupling treatment between the fields is required.

In this example, the description of the structural shape is a combination of the parametric and non-parametric shape representations. The structure consists of a membrane part which is described with form-finding and two rigid supporting frames which are described with NURBS curves. The shape of these frames is decisive for the final shape of the overall structure.

5.1 Structural design chain

The main body of the structure is a membrane supported by the rigid frames on sides and by cables at the edges of the structure (Fig. 7). The shape of the membrane is derived from a form-finding procedure with $4/3$ prestress ratio in the principal directions.

First, the initial shape is designed by "Rhinoceros 4.0" which is a CAD software specialized for NURBS curves and surfaces and related operators. For the membrane part, the Rhinoceros plug-in "Rhino-membrane" (developed based on the URS method explained in Sect. 2.2) is used, which finds the membrane geometry by applying form-finding. Rhino and similar software are frequently used by architects for structural design. This initial design is exported to Carat. In Carat, applying form-finding on the structure with prescribed prestresses in the membrane

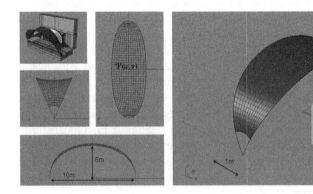

Fig. 7 Geometry of the membrane structure and the wind direction.

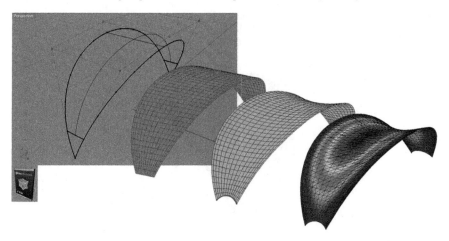

Fig. 8 The design process.

and supporting cables, the initial equilibrium state is found. The geometry of the membrane is sent to ICEM-CFD (Fig. 8).

Design update At the beginning of each design step, having the updated design variables (here, the NURBS parameters) the new curve for the supporting frame is calculated. The nodes on the membrane which are fixed by the frame are mapped to the new curve and then assuming the new given boundaries the updated geometry of the membrane is calculated by form-finding (Fig. 9).

Considering the mentioned steps, the position of finite element nodes on the structure is a function of the design variable **s** through the following function chain:

$$\mathbf{x} = \mathbf{x}(\mathbf{y}(\mathbf{s})), \tag{28}$$

where the **y** operator returns the geometry of the supporting frame, being given the design variables, and **x**, the form finding operator, gives the nodal position of the surface mesh due the given boundary. As explained in 2.3, after few design steps the mesh looses its initial quality, specially in the first and the last rows of membrane elements. Applying "mesh regularization" after each update step remedies this problem and keeps the elements in their best possible shapes (Fig. 10).

5.2   FSI Analysis

Since this optimization example is a steady-state (and not a transient) problem, the physical state is solved also with stationary fluid and static structure assumptions. The fluid-structure interaction is performed over a pseudo time with arbitrary time steps. In each step the load applied by the fluid to the structure is divided to a

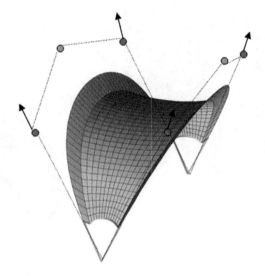

Fig. 9 Updating strategy.

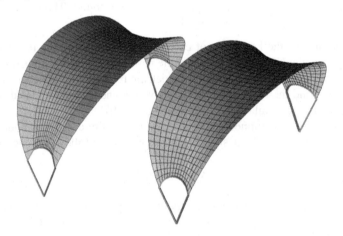

Fig. 10 Regularization of the membrane mesh.

certain number of load steps for the structural static-nonlinear analysis. Similarly, the structural displacement is given to the fluid within some fluid sub-steps. Additionally, after applying the full displacement on the fluid domain, CFD computation is continued for some more sub-steps until the fluid fields are completely converged.

Around the tent, the fluid field has a much finer mesh comparing to the structure. The front edge of the tent is locally refined in order the high gradient flow fields around the separation point to be captured (Fig. 11). The turbulence model, the computational domain and boundary conditions are as described in Sect. (2.4.2). The wind velocity at the height of the tent has an approximate value of $8\frac{m}{s}$. The wind velocity profile as well as the mesh are shown in Fig. 12.

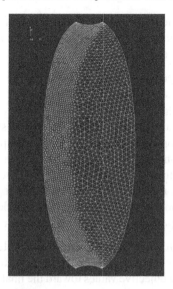

Fig. 11 Refinement at the leading edge.

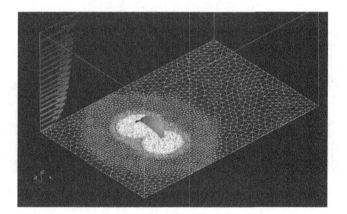

Fig. 12 Computational wind engineering domain and the inlet velocity profile.

5.3 Optimization

The algorithm introduced in Sect. (3.2) is applied to this example. As in many engineering design problems, a multi-objective target function is chosen, i.e. a linear combination of lift and drag forces applied to the whole structure. Since the entire shape is controlled by the NURBS parameters of the frames, the same parameters are selected as design variables of the optimization problem. Therefore, the optimization algorithm minimizes the overall force applied on the structure, by modifying the postion (and the weight) of the control points of the NURBS curve of the supporting frames. Due to the strong coupling and relatively complex doubled-curved shape of the structure, it is not possible to predict the final optimized shape and hence this example is definitely not trivial.

Sensitivity analysis According to Eq. (28), since the objective function, f, is directly a function of nodal postions, its sensitivity w.r.t. the design variables can be obtained by applying the chain rule, $\frac{\partial f}{\partial s} = \frac{\partial f}{\partial x} \frac{\partial x}{\partial y} \frac{\partial y}{\partial s}$ (Fig. 13).

The size of the perturbation for the sensitivity analysis shall be small enough due to the linearity assumptions, but also large enough compared to the element size scale, to be captured in the discretized solution. For this example, the calculated sensitivities with the method described in Sect. (3.2) were compared to external finite difference values, for the sake of verification. Having the same level of accuracy, the suggested method of this work shows a significant improvement in the computational time, compared to external finite differences.

Results The optimization was performed with constant step size, based on steepest decent algorithm and the computation was carried out until convergence. Figure 14 shows the evolution of the objective values toward the final design. An improvement of 23% in the design has been achieved which is certainly a large amount in the field of structural design.

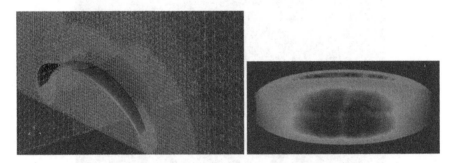

Fig. 13 Distribution of pressure (left) and sensitivity of pressure to one of design variables (right).

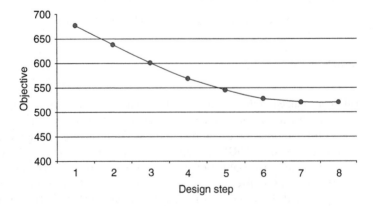

Fig. 14 Evaluation of the objective function.

6 Summary and conclusion

An integrated concept for shape optimal design of light-weight and thin-walled structures considering fluid-induced effects was developed. The Nested Analysis and Design approach (i.e. separate optimizer, state analysis and sensitivity analysis) was followed and a partitioned FSI simulation for the state analysis was embedded. The gained modularity allows for the adaption of the single ingredients to a wide spectrum of applications by choosing appropriate coupling algorithms for the solution of the coupled problem and the sensitivity analysis as well as different strategies to describe the shapes to be optimized. Furthermore, the non-matching grid capability at the coupling interface, which enables the use of problem-specific discretizations in each field, guarantees flexibility to several technical applications.

Within this contribution, the focus was on problems of aeroelasticity in the field of wind engineering. To ensure reliable results, investigations on the correct modeling of single fields and the coupling as well as goal-oriented benchmarking were carried out: In the case of Computational Wind Engineering (CWE) the validation was done with the Silsoe cube example and for verification of all numerical simulations of single fields and in particular the coupled problem, a testcase was developed together with the partners within the DFG research group 493 [38].

With respect to shape optimal design, various methods for the shape definition and the design modifications during the optimization process were introduced and their respective numerical effort and achievable design freedom were pointed out. Moreover, special emphasis was put on the appropriate combination of different approaches for shape description and determination and to establish the resulting closed design cycle. The complete optimization workflow incorporates their connection and the influence on the sensitivity analysis. For the sensitivity analysis of the coupled system a newly developed iterative strategy to compute the coupled sensitivities is applied. To maintain the mesh quality even for large design updates, a new regularization technique based on the principles of numerical form finding was developed and applied.

Finally, the success of the overall solution and optimization strategy was demonstrated with an example of a hybrid structure consisting of supporting frames described by NURBS (as typically used in CAD systems) and a connected membrane roof which was treated by the URS as a stabilized scheme for numerical form finding. The design of this lightweight free-form structure which is subject to turbulent wind flow was analyzed and improved by the developed and implemented methodology.

As an outlook, the presented modular multidisciplinary optimization framework can be enhanced and augmented by more complex methods in the single parts, i.e. the optimizer, the coupled state and sensitivity analysis. These modifications depend on the type of problem to be solved: e.g. in case of shape optimization using an FE-based shape description, an adjoint coupled sensitivity analysis would be beneficial and for problems in Computational Wind Engineering, a still refined modeling of the surrounding wind field is important. Moreover, a continued benchmarking for the different applications is necessary.

Acknowledgements The present research project is part of the DFG Forschergruppe 493 - Fluid-Struktur-Wechselwirkung: Modellierung, Simulation, Optimierung. The authors gratefully acknowledge the financial support for their research work through the German Science Foundation (DFG) - Germany.

References

1. Arora JS (2007) Elements of structural optimization World Scientific Publishing Company
2. Barcelos M, Maute K (2008) Aeroelastic design optimization for laminar and turbulent flows. Computer Methods in Applied Mechanics and Engineering 197:1813–1832
3. Bischoff M, Wall WA, Bletzinger K-U, and Ramm E (2004) Models and finite elements for thin-walled structures. In: Erwin Stein, De Borst R, Hughes TJR (ed) Encyclopedia of Computational Mechanics, pages 59–138, John Wiley & Sons, Ltd., Chichester
4. Bletzinger K-U, Ramm E (1999) A general finite element approach to the form finding of tensile structures by the updated reference strategy. International Journal of Space and Structures 14(2):131–145
5. Bletzinger K-U, Wüchner R, Daoud F, Camprubi N (2005) Computational methods for form finding and optimization of shells and membranes. Computer Methods in Applied Mechanics and Engineering 194:3438–3452
6. Bletzinger K-U, Wüchner R, Kupzok A (2006) Algorithmic treatment of shells and free form-membranes in FSI In: Bungartz HJ, Schäfer M (ed) Fluid-Structure Interaction: Modelling, Simulation, Optimization pages 336–355, Springer
7. Bletzinger K-U, Firl M, Linhard J, Wüchner R (2008) Optimal shapes of mechanically motivated surfaces. Computer Methods in Applied Mechanics and Engineering, doi:10.1016/j.cma.2008.09.009
8. Blocken B, Stathopoulos T, Carmeliet J (2007) CFD simulation of the atmospheric boundary layer: wall function problems. Atmospheric Environment 41:238–252
9. Castro IP, Robins AG (1977) The flow around a surface-mounted cube in uniform, turbulent streams. J. Fluid Mech. 79(2):307
10. Chung J, Hulbert GM (1993) A time integration algorithm for structural dynamics with improved numerical dissipation: the generalized-α method. Journal of Applied Mechanics 60:371–375
11. Degroote J, Bathe K-J and Vierendeels J (2009) Performance of a new partitioned procedure versus a monolithic procedure in fluid-structure interaction. Computers & Structures 87:793–801
12. Etienne S, Pelletier D (2005) A general approach to sensitivity analysis of fluid-structure interactions. Journal of Fluids and Structures 21:169–186
13. Felippa CA, Park KC, Farhat C (2001) Partitioned analysis of coupled mechanical systems. Computer Methods in Applied Mechanics and Engineering 190:3247–3270
14. Franke J, Hellsten A, Schünzen H, Carissimo B (2007) Best practice guideline for the CFD simulation of flows in the urban environment. Cost Action 732, Quality assurance and improvement of microscale meteorological models.
15. Haftka RT, Gürdal Z (1992) Elements of structural optimization. Springer, Dordrecht
16. Haftka RT, Sobieszczanski-Sobieski J, Padula SL (1992) On options for interdisciplinary analysis and design optimization. Structural Optimization 4:65–74
17. Hargreaves DM, Wright NG (2007) On the use of the $k - \varepsilon$ model in commercial CFD software to model the neutral atmospheric boundary layer. Journal of Wind Engineering and Industrial Aerodynamics 95:355–369
18. Hölscher N, Niemann HJ (1998) Towards quality assurance for wind tunnel tests: A comparative testing program of the Windtechnologische Gesellschaft. Journal of Wind Engineering and Industrial Aerodynamics 74-76:599–608

19. Issa RI (1985) Solution of the implicitly discretised fluid flow equations by operator-splitting. Journal of Computational Physics 62:40–65
20. Jasak H, Tukovic Z (2007) Automatic mesh motion for the unstructured finite volume method. Transactions of FAMENA 30:1–18
21. Kupzok A. (2009) Modeling the interaction of wind and membrane structures by numerical simulation. Ph.D. Thesis, Lehrstuhl für Statik der TU München, München
22. Küttler U, Wall W A (2008) Fixed-point fluid-structure interaction solvers with dynamic relaxation. Computational Mechanics 43(1):61–72
23. Küttler U. and Wall W. A. (2009) Vector extrapolation for strong coupling fluid-structure interaction solvers. Journal of Applied Mechanics 76
24. Lübcke H, Schmidt St, Rung T, Thiele F (2001) Comparison of LES and RANS in bluff-body flows. Journal of Wind Engineering and Industrial Aerodynamics 89:1471–1485
25. Lund E, Møller H, Jakobsen LA (2003) Shape design optimization of stationary fluid-structure interaction problems with large displacements and turbulence. Struct Multidisc Optim 25:383–392
26. Maute K, Nikbay M, Farhat C (2003) Sensitivity analysis and design optimization of three-dimensional non-linear aeroelastic systems by the adjoint method. International Journal for Numerical Methods in Engineering 56:911–933
27. Menter FR (1993) Zonal two equation $\kappa - \omega$ turbulence models for aerodynamic flows. AIAA Paper 93-2906
28. Michler C, Van Brummelen EH and De Borst R (2005) An interface Newton-Krylov solver for fluid-structure interaction. International Journal for Numerical Methods in Fluids 47:1189–1195
29. Mohammadi B, Pironneau O (2004) Shape optimization in fluid mechanics. Annual Review of Fluid Mechanics 36:255–279
30. Richards PJ, Hoxey RP (1993) Appropriate boundary conditions for computational wind engineering models using the $k - \varepsilon$ model. Journal of Wind Engineering and Industrial Aerodynamics 46 & 47:145–153
31. Richards PJ, Hoxey RP, Short LJ (2001) Wind pressure on a 6 m cube. Journal of Wind Engineering and Industrial Aerodynamics 89:1553–1564
32. Schmidt S, Thiele F (2002) Comparison of numerical methods applied to flow over wall-mounted cubes. International Journal of Heat and Fluid Flow 23:330–339
33. Sidi A (1991) Efficient implementation of minimal polynomial and reduced rank extrapolation methods. Journal of Computational and Applied Mathematics 36:305–337
34. Simiu E, Scanlan RH (1996) Wind Effects on Structures. 3rd edn. Wiley-Interscience
35. Sobieszcanski-Sobieski J (1990) Sensitivity of complex, internally coupled systems. AIAA Journal 28(1):153–160
36. Soto O, Löhner R (2001) CFD shape optimization using an incomplete-gradient adjoint formulation. International Journal for Numerical Methods in Engineering 51:735–753
37. Suga K, Craft TJ, Iacovides H (2006) An analytical wall-function for turbulent flows and heat transfer over rough walls International Journal of Heat and Fluid Flow 27:852–866
38. Turek S and Hron J (2006) Proposal for numerical benchmarking of fluid-structure interaction between an elastic object and laminar incompressible flow. Lecture Notes in Computational Science and Engineering 52:371–385
39. Vierendeels J, Lanoye L, Degroote J, Verdonck P (2007) Implicit coupling of partitioned fluid-structure interaction problems with reduced order models. Computers & Structures 85:970–976
40. Weller HG, Tabor G, Jasak H, Fureby C (1998) A tensorial approach to CFD using object oriented techniques. Computers in Physics 12(6):620–631
41. Wilcox DC (2006) Turbulence Modeling for CFD. 3rd edn. DCW Industries
42. Wüchner R, Bletzinger K-U (2005) Stress-adapted numerical form finding of pre-stressed surfaces by the updated reference strategy. International Journal for Numerical Methods in Engineering 64(2):143–166
43. Wüchner R, Kupzok A, Bletzinger K-U (2007) A framework for stabilized partitioned analysis of thin membrane-wind interaction. International Journal for Numerical Methods in Fluids 54:945–963

Experimental Benchmark: Self-Excited Fluid-Structure Interaction Test Cases

J. Pereira Gomes and H. Lienhart

Abstract The swivelling motion of a flexible structure immersed in a flow can become self-excited as a result of different fluid-structure interaction mechanisms. The accurate simulation of these mechanisms still constitutes a challenge with respect to mathematical modelling, numerical discretization, solution techniques, and implementation as software tools on modern computer architectures. Thus, to support the development of numerical codes for fluid structure interaction computations, in the present work an experimental investigation on the two-dimensional self-excited periodic swivelling motion of flexible structures in both laminar and turbulent uniform flows was performed. The investigated structural model consisted of a stainless-steel flexible sheet attached to a cylindrical front body. At the trailing edge of the flexible sheet, a rectangular mass was considered. The entire structure model was free to rotate around an axle located in the central point of the front body. During the experimental investigation, the general character of the elastic-dynamic response of the structure model was studied first. The tests in laminar flows were performed in a polyglycol syrup (dynamic viscosity: 1.64×10^{-4} m^2/s) for a Reynolds number smaller than 270, whereas the tests in turbulent flows were conducted in water for Reynolds numbers up to 44000. In both cases, the maximum incoming velocity tested was about 2 m/s. Subsequently, three specific test cases were selected and characterized in more detail as far as the flow velocity field and structure mechanical behavior are concerned. Thus, the present contribution presents the detailed results obtained at 1.07 m/s and at 1.45 m/s in laminar and at 0.68 m/s in turbulent flows. It also compares the experimental data with numerical results obtained for the same conditions using different simulating approaches. They revealed very good agreement in some of the fluid-structure interaction modes whereas in others deficiencies were observed that need to be analyzed in more detail.

J.P. Gomes and H. Lienhart
Institute of Fluid Mechanics and Erlangen Graduate School in Advanced Optical Technologies, University of Erlangen-Nürnberg Cauerstr. 4, 91058 Erlangen, Germany
e-mail: jorge.gomes@lstm.uni-erlangen.de, hermann.lienhart@lstm.uni-erlangen.de

H.-J. Bungartz et al. (eds.), *Fluid Structure Interaction II*, Lecture Notes
in Computational Science and Engineering 73, DOI 10.1007/978-3-642-14206-2_14,
© Springer-Verlag Berlin Heidelberg 2010

1 Introduction

The excitation, and posterior amplification, of a structure and fluid system oscillation can be described, in a first approach to the problem, as follows. If a structure deforms, its orientation to the flow changes and a change on the fluid forces acting upon the surface of the structure results. A new set of fluid forces determines a new structure deformation. As soon as this coupled mechanism is initiated, as a result of any initial flow or movement instability, the damping imposed by the fluid can become negative as a result of different mechanisms by which energy is transfered from the fluid to the structure. Whether the vibration is damped out or amplified is just a matter of the sign of the net damping coefficient. For lightly damped structures, the fluid damping becomes dominant and the coupled fluid and structure movement may become self-excited. Depending on whether the fluctuation of the flow plays a significant role in the initial excitation process or not, the excitation is either called flow-induced or movement-induced excitation (MIE). In the case of flow-induced excitation, one can further distinguish between the extraneously-induced and instability-induced excitation (EIE and IIE) [5, 6].

In the case of EIE, the fluctuations are produced by an extraneous source. Therefore the exciting frequency is independent of the structure movement and controls the frequency of the resulting oscillation. Excitation mechanisms of this kind induce the structure to undergo forced-oscillations. In opposition to EIE, IIE is caused by an instability of the flow which gives rise to flow fluctuations if a certain threshold value of flow velocity is reached. As a rule, this instability is intrinsic to the flow created by the structure and exists even in those cases in which structure oscillations do not occur. An example of flow instability is the alternate vortex shedding formation in the wake behind a bluff body. Depending on the control and amplification mechanism affecting the instability, these fluctuations and the forces they generate can become well correlated and close to a dominant frequency of the mechanical oscillator, so they can lead to large-amplitude movements of the oscillating system. Thus, this type of excitation is expected to occur in a finite small range of flow velocities in which the flow fluctuation is in resonance with the first or higher harmonics of the natural frequencies of the structure [7]. This corresponds to the condition

$$\frac{U}{f_N d} \approx \frac{1}{n S_t}, \tag{1}$$

where U is the flow velocity, f_N and d are the natural frequency and the characteristic length of the structure and S_t is the Strouhal number.

In the resonance range, the main feature of fluid-elastic and fluid-resonant control is the amplification of the exciting force (because of the negative damping) and a "locking-in" of the frequency of the fluid to that of the structure frequency. The velocity range of locking-in and its extension is very sensitive to the instability amplification and phase condition, which is intimately associated with the compatibility between the flow and structure kinematics. It is also to be expected that in the resonance range no change in the dynamics of the structure occurs because it is reasonable to assume that the net effect of the fluid forces in-phase with the structure acceleration is in average zero, i.e., the fluid stiffness balances the added mass.

Excitation of the type MIE is characterized by fluctuating forces that arise from the movement of the structure. Whenever the structure is accelerated in the fluid, an unsteady flow is induced that alters the fluid forces acting upon the structure. If this alteration in the fluid load leads to a negative damping or to a transfer of energy to the moving structure, a MIE process starts. In this process, the forces that are responsible for the excitation are inherently linked to the structure movement and disappear if the structure comes to rest. The sensitivity of the oscillating system to MIE can be described by the equation of movement

$$\left(m + A'\right) \ddot{y}\left(t\right) + \left(B + B'\right) \dot{y}\left(t\right) + C y\left(t\right) = 0, \tag{2}$$

where m, B and C are the mass, damping and stiffness of the structure, respectively. A' and B' are known as added, or fluid coefficients and account for the effects of the fluid. The criterion for dynamic instability with respect to infinitesimal disturbances can be stated in terms of the net damping of the system as

$$2\zeta m \omega_N + B' \leq 0 . \tag{3}$$

Regarding the behavior of the added mass, it can be seen from Eq. (2) that A' is not zero at the onset of MIE. Thus, in opposition to IIE, an alteration of the dynamics of the structure in the case of MIE is expected.

Regardless the excitation mechanism, the self-excitation onset and resulting oscillation movements are very difficult to predict whenever the structure has a complex geometry or multiple degrees-of-freedom (such as flexible structures). From the numerical research view point, the challenges with respect to mathematical modeling, numerical discretization, solution techniques, and their implementation as software tools on modern computer architectures are still huge, in particular if accuracy, flexibility and simulation efficiency are considered. Most of the actual software packages for computational fluid dynamics (CFD) and computational structure dynamics (CSD) are already on a high level and allow, to some extend, the simulations of certain classes of fluid-structure interaction (FSI) problems. However, although much research has been invested in the development of models and coupling strategies for numerical simulations and in the creation of coupling algorithms between CFD and CSD solvers, many of the key questions regarding accuracy, robustness, flexibility and efficiency have not been yet completely answered. An overview of the present development of numerical solution strategies and their applications is given, for instance, in [9] and [4].

The recent developments in new numerical methodologies triggered the present work to perform an experimental research in the field of FSI investigating the instability and the resulting FSI induced swivelling motion of complex flexible structures immersed in uniform laminar and turbulent flows. The object of the research were two-dimensional structure models which combine the elastic behavior of a flexible sheet in an axial flow with the dynamic of a rigid cylinder in cross-flow. The selection of the experiments and the detailed measurements performed during the present work were used to compile a reliable and extensive experimental data base on reference test cases. The data base created addressed the need for well-defined experimental benchmarks to be used as a diagnostic and validation tool for numerical models for FSI simulations. The data base on these reference test cases included the time-phase resolved characterization of the flow velocity field and the mechanical behavior of the structure such as its deflection, principal deflection modes, periodic motion amplitude and frequency. The present contribution presents three selected benchmarks extracted from the present work as well as comparisons between the experimental results and numerical simulations.

2 Definition of the test cases

The project requirements for reproducibility and periodicity of the resulting flow and structure motion, both in the laminar and turbulent regime, imposed stringent restrictions on the design of the experiments. Thus, the definition of the test case took into account six principal requirements: (i) reproducibility of the resulting motion, (ii) well controlled working and boundary conditions, (iii) two-dimensionality of the structure deflection, (iv) moderate structure motion frequency, (v) significant deformation of the structure, and (vi) linear material properties.

2.1 Structure definition

The structure found to serve the requirements of the project, see Fig. 1, consisted of a 0.04 mm thick stainless-steel flexible sheet attached to an aluminum cylindrical front body. At the rear end of the sheet, a rectangular stainless-steel mass was located. Both the rear mass and the front body could be considered rigid. An extra degree of freedom was taken into account as the structure was free to rotate around an axle located in the center of the cylindrical front body.

The densities of the different materials used in the construction of the model as well as all parameters of the different test cases discussed in the present paper are

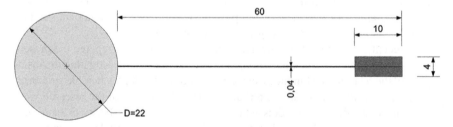

Fig. 1 Structure model geometry (all dimensions in mm).

Table 1 Parameters of the test cases.

Variable	1st test	2nd test	3rd test
	Laminar	Laminar	Turbulent
Reynolds number	≈ 140	≈ 190	≈ 15000
$\rho_{cylinder}$	$2828 \, kg/m^3$	$2828 \, kg/m^3$	$2828 \, kg/m^3$
$\rho_{flexible\ sheet}$	$7855 \, kg/m^3$	$7855 \, kg/m^3$	$7855 \, kg/m^3$
$\rho_{rear\ mass}$	$7800 \, kg/m^3$	$7800 \, kg/m^3$	$7800 \, kg/m^3$
$E_{flexible\ sheet}$	$2 \times 10^{11} \, N/m^2$	$2 \times 10^{11} \, N/m^2$	$2 \times 10^{11} \, N/m^2$
ρ_{fluid}	$1050 \, kg/m^3$	$1050 \, kg/m^3$	$998 \, kg/m^3$
ν_{fluid}	$1.64 \times 10^{-4} \, m^2/s$	$1.64 \times 10^{-4} \, m^2/s$	$0.97 \times 10^{-6} \, m^2/s$
gravity	$9.81 \, m/s^2$ in x	$9.81 \, m/s^2$ in x	$9.81 \, m/s^2$ in x
incoming velocity	$1.07 \, m/s$	$1.45 \, m/s$	$0.68 \, m/s$

Table 2 Natural frequencies of the structure model.

mode	0	1st	2nd	3rd	4th
frequency	1.94 Hz	5.89 Hz	27.43 Hz	92.73 Hz	234.41 Hz

summarized in Table 1. For the range of forces acting on the structure during the tests, the mechanical behavior of the flexible stainless-steel sheet could be considered to be linear and Young's modulus was measured to be $200 \, kN/mm^2$ within this range of strain. As it was already discussed, in fluid-structure interaction problems the natural frequencies of the structure play an important role in the overall characteristics of the elastic-dynamic response of the coupled system. To support this statement, one can argue that the domain of relevance of each individual instability-induced exciting mechanism is directly dependent on the natural frequencies of the structure. Hence, for a better understanding and analysis of the results, the definition of the model also include its natural frequencies. The natural frequencies were computed in a finite element code and are listed in Table 2. The mode identified as the zero mode corresponded to the rigid body motion mode, i.e. the structure swiveled around its free axle without changing its original straight shape.

2.2 Flow definition

The tests were conducted in a vertical, closed-loop tunnel capable of operating with liquids of different viscosity. The spatial dimensions of the tunnel test section are given in Fig. 2. It had an overall length of 338 mm and a cross-sectional area of 180 mm × 240 mm. The structure was mounted 55 mm downstream of the inlet plane.

Opting for a vertical tunnel, the gravity force was aligned with the x-axis and so it did not introduce any asymmetry. Special attention was given to the model support on the test section. Low-friction bearings were used for this specific task to guarantee a frictionless rotational degree of freedom of the front cylinder. The supporting

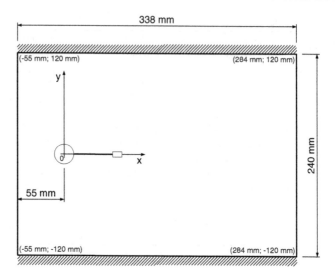

Fig. 2 Experimental domain.

system also considered a non-contacting magnetic position sensor to measure the angular position of the structure front cylinder. The output of this sensor was used during the tests to measure the time-dependent angle of the structure front body and to determine the beginning of a new model swivelling motion period.

For the tests in the laminar regime, a polyethylene glycol PG-12000 (polyglycol) syrup was used. The measurements were performed at 25°C within an uncertainty of 0.5°C. For this temperature range, the properties of the syrup could be considered constant and the kinematic viscosity was measured as 1.64×10^{-4} m^2/s. For the same conditions, the density was equal to 1050 kg/m^3. After an analysis of the elastic-dynamic behavior of the structure for an incoming flow velocities up to 2 m/s, it was decided to perform the laminar reference test case at two different velocities, the first at 1.07 m/s and the second at 1.45 m/s. The tests inlet velocity profiles were measured in the absence of the model at the location $x = -55$ mm (see Fig. 3) in the center of the spanwise direction.

On the other hand, the tests in turbulent flows were conducted in water at the temperature of $22°C^{+0°C}_{-1°C}$ ($\rho_w = 998$ kg/m^3; $\nu_w = 0.97 \times 10^{-6}$ m^2/s). For this regime, only one incoming velocity was selected for the reference tests, namely 0.68 m/s. Figure 4 shows the measured inlet velocity profile at 0.68 m/s. It was also measured in the absence of the model at the location $x = -55$ mm in the center of the spanwise direction.

For both flow regimes, the flow angularity fluctuation was measured to be less than 0.5° and the RMS of the velocity magnitude variation was less than 1%. As a synopsis, Table 1 gives a complete overview of the values relevant for the three referent test cases discussed in this paper. The Reynolds number of the tests was defined based on the diameter of the structure front cylinder.

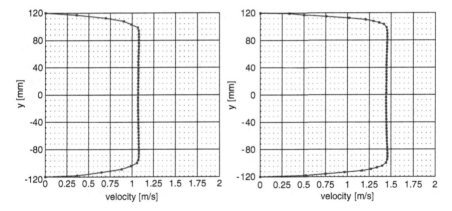

Fig. 3 Measured inlet velocity profiles for the first (left) and second (right) laminar reference test case.

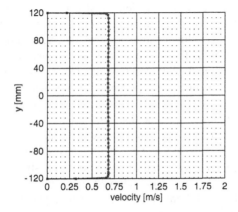

Fig. 4 Measured inlet velocity profile for the turbulent reference test case.

3 Measurement techniques

3.1 Flow velocity measurements

The task of sampling the flow velocity of the fluid surrounding the flexible structure was addressed to a two-component, two-dimensional PIV system. The PIV system adopted for the measurements consisted of two 1280 pixel × 1024 pixel synchronized cameras and a pulsed double-head 120 mJ laser.

The cameras were used to acquire two simultaneous time-dependent pairs of PIV images per measurement. For that purpose, they are mounted parallel to the rotating axle of the flexible structure to visualize the flow in a plane perpendicular to it. To assure the correct position of the two adjacent images, a special support was designed to hold both cameras and to permit the adjustment of each individual camera or of both cameras simultaneously. The PIV images acquired by the two cameras were then imported into a MATLAB-based post-processing software to stitch the correspondent pairs of images before being cross-correlated. Opting for the solution of two synchronized cameras, it was possible to achieve an extended measuring area while keeping the spatial resolution as low as 2.1 mm × 2.1 mm for a 16 pixel × 16 pixel interrogation area. In relation to Fig. 2, the flow velocity field measuring area extended from −35 mm up to 225 mm in the x-direction and from −85 mm up to 85 mm in the y-direction.

Preliminary tests with the present system showed that reliable and continuous data acquisition from the two synchronized cameras was possible for long time periods using a maximum frequency of 1 Hz. Concerning the velocity accuracy, the system proved to have a measurement uncertainty of less than 1.5 % for the undisturbed flow.

The laser sheet for flow illumination is positioned perpendicular to the rotating axis of the flexible structure at the center of the structure in the spanwise direction. The presence of the swivelling flexible structure in the measurement plane imposes extra problems. The most important of them is related to the fact that the flexible structure is a swivelling opaque body which creates a time-dependent dark shadow region in the fluid. This behavior not only reduces the measuring area to almost only one side of the flexible structure but also makes the masking of the PIV images in post-processing difficult to perform. To cope with this problem, multiple light sources were used to illuminate the structure from both sides and, therefore, to cover the entire flow field avoiding the shade behind the model. As seeding particles, 10 μm mean diameter hollow glass spheres were chosen to be used in water. They provided a good match of density and enough scattering signal over all the measuring area. During the laminar tests, in the polyglycol syrup, silver-coated hollow glass spheres with the same diameter were adopted as seeding particles. They produce higher signal levels in high light-absorbing media compared with non-coated hollow glass spheres. The major drawback of the silver-coated glass spheres is related to their density; the relative density of this kind of particles is about 1.4. Nevertheless, this drawback was acceptable because of the high viscosity of the fluid and the velocity of the flow during the tests.

3.2 Structure deflection measurements

For structure deflection measurements, the PIV system was modified to provide it with structure deflection analysis capabilities. The idea behind this setup was to use the PIV system to acquire images from the swivelling structure and to adopt specially developed software to analyze and reconstruct the time-dependent deflection of the structure. The major advantage of this approach was that the same measuring system as used for the velocity field measurements could be employed.

In order to generate coherent results, the relative position of the laser source was maintained unchanged to illuminate the time-dependent deflection of the flexible structure at the same plane as for the velocity field measurements. The cameras were now mounted in a side-by-side arrangement in such a way as to acquire simultaneous images of the flexible structure illuminated by the laser sheet from each side of the model.

The quantitative analysis was performed in MATLAB workspace by a script developed for the specific task. The software analyzed and compared the images illuminated from both sides of the model and reconstructed the time-dependent image of the light sheet reflected by the structure. To achieve that purpose, it mapped the pixel value in the gray-scale of the entire image and detected the line resulting from the intersection of the laser sheet and the structure in addition to the edges of the rear mass. Then the same code, including camera calibration routines, reconstructed the real scale and perspective-corrected position of that line. With the information on the position of the flexible sheet and of the time-phase detector module, the algorithm finally computed all the relevant data for the structure movement such as time-phase resolved angle of the front body, structure deflection shape and trailing edge coordinates. Based on these data, the modes present in the structure were identified and characterized. Taking into account the accuracy of the measurement technique together with the resolution of the time-phase reconstruction, one could expect an uncertainty of the structure deflection measurements of the order of 0.3 mm.

3.3 Resolution of the results in the time-phase space

When investigating fluid-structure interaction problems, two additional difficulties appear when it comes to resolving the measured data in time-phase space. First, the periodicity of the structure motion is sensitive and therefore there are cycle-to-cycle variations of the period time. Second, the velocity of the structure motion is not predefined (as it is, for example, in crank shaft-driven set-ups), which makes it impossible to reconstruct the time-phase resolved data from position resolved measurements. For these reasons, a different approach was adopted. Instead of triggering the data acquisition by the experiment, the measuring system was operated at a constant acquisition rate and both events, the acquisition of a measurement and the start of a new movement cycle, were recorded based upon an absolute clock. Using this time information, the data were reorganized in a post-processing program in order to provide the time-phase resolved values. The hardware module designed to perform the event monitoring was based on an FPGA (Field Programmable Gate Array) and a 1 MHz internal clock. In this way, it was able to record up to 250 events per second with an accuracy of $2\,\mu s$.

Two different kinds of events were recorded: the measurements (t_j) and the beginning of the flexible structure motion period (t_i'). Depending on the type of measurements performed (flow velocity measurements or structure deflection measurements), the measurement events were detected using the first laser pulse trigger signal or the camera trigger signal, respectively. After deciding which angular position of the structure front body corresponded to the beginning of the motion period, this position was detected by a magnetic angular position sensor and the small magnet attached to the axle of the model. This sensor was selected to perform the task based on two criteria: non-contacting position angular sensor and direction resolved output signal. An additional reason, namely the high sensitivity around a predefined angular position, also contributed to the selection of this sensor.

From the recorded events time information, a software computed the measurement time-phase angle tpa_j within the structure motion period as

$$tpa_j = \frac{t_j - t_i'}{T_i} \times 360, \qquad (4)$$

where t_i' and T_i corresponded to the starting instant and the period of the swivelling cycle in which the measurements t_j took place.

Finally, the time-phase resolved measurements were reorganized in the time-phase space in a reference structure motion period, as time-phase angle, between $0°$ and $360°$ at a resolution of $2.5°$. Further details can be found elsewhere [2, 3].

4 Results

To assist the understanding of the results, they are sub-divided according to the flow regime. Thus, in Sect. 4.1 the results obtained in the laminar regime are presented and in Sect. 4.2 the results in turbulent flows. For each regime, the structure model was first tested at different incoming flow velocities up to 2 m/s. These results defined the general character of the dynamic response of the structure model as a function of the incoming flow velocity and showed the different swivelling-modes exhibited by the structure. In the second stage, detailed measurements were conducted at selected velocities to characterize each combined flow and structure swivelling-mode. Section 4.1 includes detailed measurements obtained in laminar flow at 1.07 m/s and 1.45 m/s. In turbulent flows, detailed measurements were performed only for one approaching flow velocity, 0.68 m/s. These measurements are presented in Sect. 4.2.

4.1 Results in laminar flows

Figure 5 shows the dynamic response of the structure in laminar flow. The Reynolds number, based on the diameter of the front cylindrical body, reached the maximum value of 270 at 2 m/s. At very low flow velocities, it was not possible to identify any kind of motion. On increasing the flow speed, it was observed that the minimum velocity needed to excite the movement of the structure varied slightly from test to test. Nevertheless, in all cases it was possible to achieve a periodic cyclic swivelling

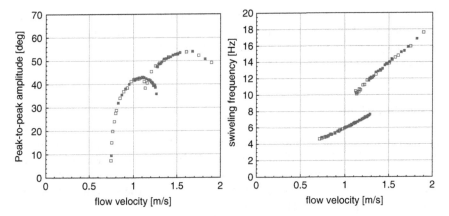

Fig. 5 Structure front body peak-to-peak amplitude (left) and structure swivelling frequency (right) versus incoming flow velocity (solid squares correspond to measurements acquired while increasing and open squares while decreasing the flow velocity).

motion for velocities slightly smaller than 1 m/s. It should be mentioned that as soon as the structure started to swivel, its motion frequency coincided with the line in Fig. 5 independently of the velocity value at which the movement started. From the instant that the structure started to swivel and for all the range of velocities tested, the resulting motion proved to be symmetric and periodic. The RMS value of the cycle-to-cycle structure motion period was measured to be less than 1%.

The most obvious aspect revealed by Fig. 5 is the existence of two distinctive structure swivelling-modes separated by a pronounced, well-defined hysteretic region. For both swivelling-modes, the frequency of the resulting motion increased linearly with the velocity of the incoming flow. While the frequency increased monotonically with the flow velocity, the amplitude of the structure motion showed a maximum value for each swivelling motion mode. The first swivelling-mode, registered for incoming flow velocities up to approximately 1.1 m/s, was characterized by the fact that the deflection of the structure model was strongly governed by the first bending mode. In connection with this, the movement of the rear mass was in concordance with the movement of the front body. The second swivelling-mode, observed for incoming velocities higher than 1.3 m/s, was characterized by the fact that the rear mass motion is in opposition to the movement of the structure front body. At the same time, higher bending modes were present in the deflection of the structure. The conclusion about the deflection modes exhibited by the structure was supported by visualizations performed during the tests at different flow speeds. Within the transition region, from 1.1 m/s to 1.3 m/s, the structure presented a hysteretic behavior where both swivelling-modes could be observed depending on the previous frequency of the structure. In both cases, a delay of the movement of the rear mass in relation to the front body was identified. This delay is a consequence of the flexibility of the sheet and it is a function of the mechanical properties of the structure. After identifying the two different self-excited swivelling-modes of the flow-structure system, each was characterized in more detail. To this end, two flow velocities, 1.07 m/s and 1.45 m/s, were selected as representative of the two swivelling-modes. Both values are located close to the velocity of the maximum structure amplitude excitation (Fig. 5(a)).

Results at 1.07 m/s

In the following figures, the characterization of the structure movement for an incoming flow velocity of 1.07 m/s is shown. At this velocity, the Reynolds number is approximately 140, based on a front body diameter of 22 mm and a kinematic viscosity of the polyglycol syrup of 1.64×10^{-4} m^2/s. The swivelling frequency observed was equal to 6.38 Hz.

Figure 6 shows the evolution of the angle of the structure front body within the swivelling motion averaged period and Figs. 7 and 8 present the successive positions of the structure flexible sheet and the coordinates of the trailing edge during one period. The time-phase resolution in Fig. 7 was set to 30° whereas that used for Fig. 8 was 5°.

As far as the flow field surrounding of the structure model is concerned, Fig. 9 compiles the time-phase resolved combined flow field and structure deflection at eight instants of the reference swivelling period. The successive results indicate that the movement of the trailing edge is in-phase, but delayed, with respect to the movement of the front body.

This delay could be quantified by comparing the time-phase resolved angle of the front body with the y-coordinate of the structure trailing edge. Thus, at 1.07 m/s the delay of the trailing edge with respect to the front body movement was computed to be approximately 60° of time-phase.

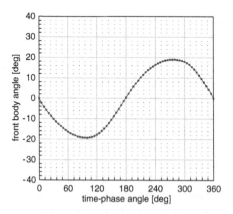

Fig. 6 Time-phase resolved front body angle within a period of motion at 1.07 m/s ($Re \approx 140$).

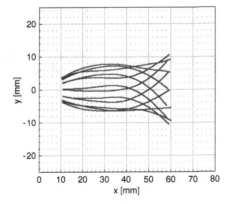

Fig. 7 Flexible sheet deflection within a period of motion at 1.07 m/s ($Re \approx 140$).

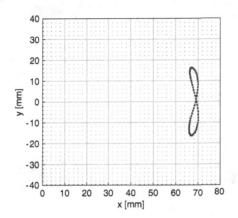

Fig. 8 Trailing edge coordinates within a period of motion at 1.07 m/s ($Re \approx 140$).

Results at 1.45 m/s

A similar set of measurements were performed for an incoming flow velocity of 1.45 m/s, which corresponds to a Reynolds number close to 190. At 1.45 m/s, the structure exhibited a more complex and faster swivelling motion. The resulting motion frequency was measured to be 13.58 Hz and the maximum front body angular amplitude was ±22°. In Fig. 10 the angle of the front body within the averaged period of motion is displayed. The time-phase delay of the trailing edge excursion in relation to the front body movement increased to about 210°.

Figures 11 and 12 show the time-phase resolved position of the structure and of the structure model trailing edge within the averaged swivelling motion. Now the collection of the deformations display a pronounced nodal region, indicating the existence of higher bending modes in the structure deflection. In Figs. 11 and 12, time-phase angle resolutions of 30° and 5° were used. Regarding the unsteady flow

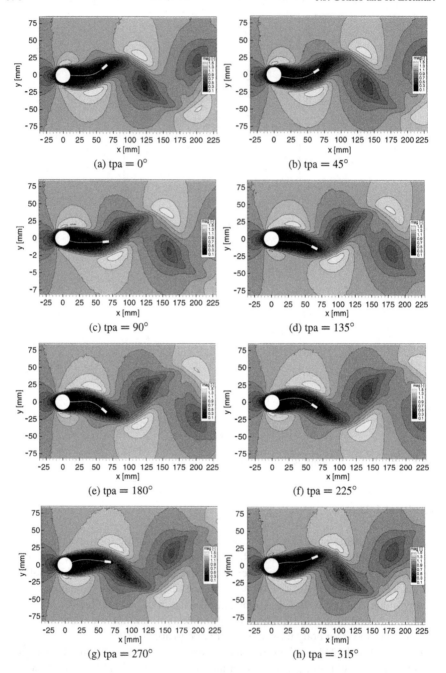

Fig. 9 Time-phase resolved combined flow field/structure deflection measurements at eight different instants of the swivelling motion period at 1.07 m/s ($Re \approx 140$).

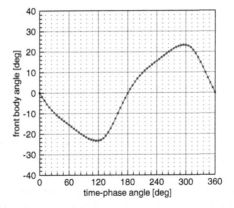

Fig. 10 Time-phase resolved front body angle within a period of motion at 1.45 m/s ($Re \approx 190$).

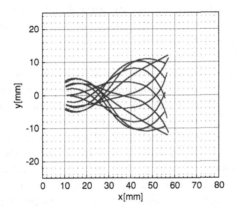

Fig. 11 Flexible sheet deflection within a period of motion at 1.45 m/s ($Re \approx 190$).

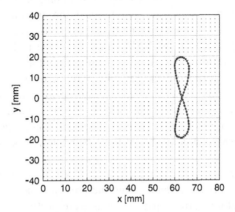

Fig. 12 Trailing edge coordinates within a period of motion at 1.45 m/s ($Re \approx 190$).

field results, Fig. 13 presents the velocity field around the structure for eight, 45° equally distant, time-phase angles within the reference movement cycle.

4.2 Results in turbulent flows

In the turbulent regime, using water as the working fluid, the structure proved to have the same well-defined multi swivelling-mode behavior as observed in laminar flows. Figure 14 presents the dynamic response of the structure versus the incoming water flow velocity up to 2 m/s ($Re \approx 44000$).

Now the structure could be excited to a periodic swivelling motion at very low flow velocities. Visualizations showed that this first excited mode corresponded to the rigid body motion mode, i.e., the structure swiveled in the fluid around its free rotating axle without changing its original and straight shape. Because it corresponded to the rigid body motion, this mode was named the zero swivelling-mode. As an example of this zero mode, Fig. 15 shows the structure behavior at approximately 0.19 m/s in water.

A swivelling-mode transition was registered for a flow velocity close to 0.4 m/s. The transition between modes is abrupt and it was not possible to observe either any evolution of the structure motion during the transition or any hysteretic behavior. In the new swivelling-mode, the structure deflection was dominated by the first bending mode. The behavior of the structure in this mode is in all respects similar to the first swivelling-mode observed in laminar flows. For both swivelling-modes, the amplitude of the structure movement was limited and exhibited a local maximum value. The structure movement frequency increased approximately linearly with the velocity of the incoming flow in both modes. The only exception occurred in the first mode at approximately 0.6 m/s where a change in slope was registered because of a lock-in occurrence. On further increasing the incoming flow velocity, an unusual behavior was observed: as soon as the amplitude of the structure started to decrease, after reaching the local maximum, the motion characteristics degraded very rapidly. This effect was supported by the RMS value presented in Fig. 14. The coupled movement became non-periodic and non-symmetric and led to a rapid destruction of the structure. Therefore, no measurements could be obtained for flow velocities higher than 0.9 m/s. This sequence of facts occurred when the structure swivelling frequency was showing the first signs of transition to a new, second swivelling-mode. In the range in which the structure movement is periodic and reproducible, up to 0.9 m/s, the RMS value of the cycle-to-cycle motion period remained lower than 1%.

Except for the trivial rigid body mode, the only self-exciting swivelling-mode that could be characterized in detail using the present structure configuration was the first one. Therefore, further investigations were performed in a water flow at 0.68 m/s. The decision regarding that velocity followed the same criteria as used for the laminar investigations.

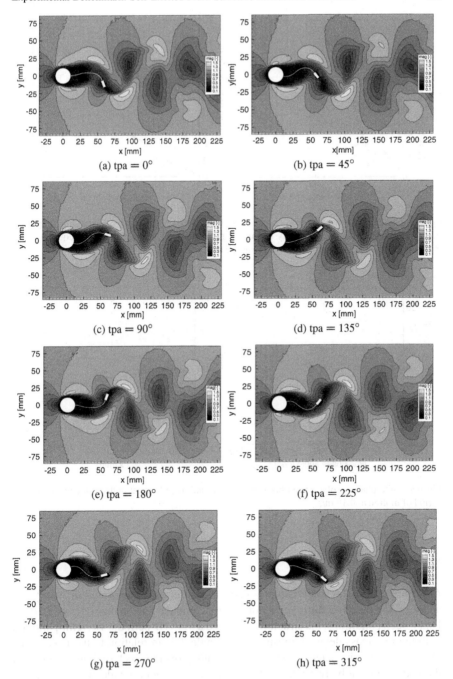

(a) tpa = 0°

(b) tpa = 45°

(c) tpa = 90°

(d) tpa = 135°

(e) tpa = 180°

(f) tpa = 225°

(g) tpa = 270°

(h) tpa = 315°

Fig. 13 Time-phase resolved combined flow field/structure deflection measurements at eight different instants of the swivelling motion period at 1.45 m/s ($Re \approx 190$).

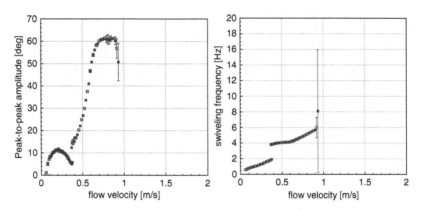

Fig. 14 Structure front body peak-to-peak amplitude (left) and structure swivelling frequency (right) versus incoming flow velocity (solid squares correspond to measurements acquired while increasing and open squares while decreasing the flow velocity).

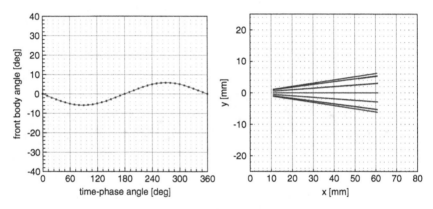

Fig. 15 Time-phase resolved front body angle (left) and flexible sheet deflection (right) within a period of motion at 0.19 m/s.

Results at 0.68 m/s

At 0.68 m/s, the Reynolds number of the measurements performed in water was about 15000, based on the diameter of the front cylinder and on the water properties. Under such conditions, the structure exhibited a 4.45 Hz periodic swivelling motion with the rear mass delayed, in time-phase angle, 95° in relation to the front body. Once again, the characterization of the resulting movement is based on the angle of the structure front body, trailing edge position and structure deflection. Figures 16 to 18 show the time-phase resolved evolution of these quantities within the averaged period of motion. The same resolutions as used to generate the laminar results were used in the following figures.

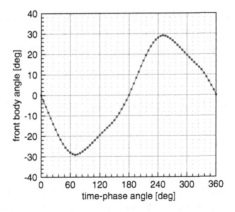

Fig. 16 Time-phase resolved front body angle within a period of motion at 0.68 m/s ($Re \approx$ 15000).

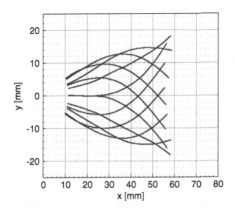

Fig. 17 Flexible sheet deflection within a period of motion at 0.68 m/s ($Re \approx$ 15000).

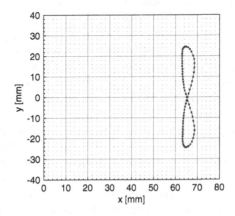

Fig. 18 Trailing edge coordinates within a period of motion at 0.68 m/s ($Re \approx$ 15000).

Figure 19 shows the flow velocity field results obtained in water at 0.68 m/s for eight successive time-phase angles measured within a period of the structure motion. The time-phase angles presented correspond to those adopted in Sect. 4.1.

4.3 Discussion of the experimental results

The analysis of the structure model dynamic response proved the existence of different structure swivelling-modes. In the laminar regime, it was possible to observe two swivelling-modes depending on the approaching flow velocity. For both modes, the structure movement frequency increased linearly whereas the front body amplitude presented a local maximum. The maximum excitation occurred at 1.1 m/s and 1.6 m/s for the first and second excitation mode, respectively. The corresponding movement frequencies for these two instants of maximum excitation were measured to be around 6.5 Hz and 15 Hz.

The first mode was excited for the first time at 0.8 m/s. For this flow velocity, the corresponding Strouhal number ($St \approx 0.175$) and the first natural frequency of the structure ($N_1 = 5.9$ Hz) showed a strong interconnection between the excited movement and the classical von Karman vortex shedding triggered by the structure front cylinder. Despite a small delay, the trailing edge movement could be considered in phase with the angular movement of the front body. Concerning the deflection of the structure, this mode was characterized almost exclusively by the existence of the first bending mode. At 1.07 m/s, the structure vibrated around its first natural frequency; more precisely, the coupled fluid and structure unsteady motion was registered to occur at 6.38 Hz associated with a maximum excursion of the front body and trailing edge of 19° and 16 mm, respectively.

The transition to the second, more complex mode was observed between 1.12 m/s and 1.3 m/s and it showed a strong hysteretic behavior. The second swivelling-mode was characterized by a frequency much lower than the second natural frequency of the structure ($N_2 = 27.4$ Hz), indicating a self-exciting mechanism of a different type than the first one. The trailing edge was now almost in phase opposition in relation to the front body position and the structure deflection was mainly characterized by the second structure bending mode and higher. The presence of the second bending mode justified the pronounced node observed in the structure deformation. At 1.45 m/s, the front cylinder reached a maximum deflection of 26° and the trailing edge excursion was limited to 19 mm.

In the turbulent regime, one more mode was observed for very small inlet flow velocities. The lowest mode to be excited in water was the rigid body mode (referred to as the zero swivelling-mode). This mode started to be observed at a very small approaching velocity and it was characterized by small structure deflections and movement frequencies. The maximum rigid body excitation was registered for 0.2 m/s at the same time the structure swivelled at about 1.1 Hz. The transition to the first self-excitation mode was registered at about 0.4 m/s and no hysteresis was observed. In this mode, the maximum excitation of the structure was achieved

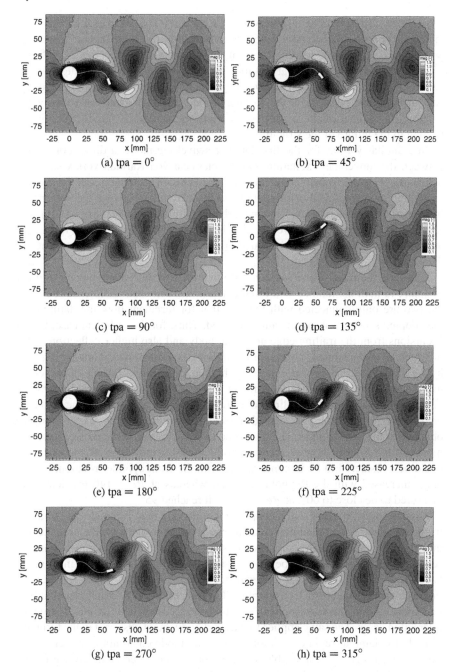

Fig. 19 Time-phase resolved combined flow field/structure deflection measurements at eight different instants of the swivelling motion period at 0.68 m/s ($Re \approx 15000$).

for 0.8 m/s corresponding to a movement frequency of 5 Hz. Detailed measurements at 0.68 m/s registered a 4.45 Hz self-exciting resulting motion associated with a maximum excursion of the front body and the trailing edge of 29° and 25 mm, respectively. Finally, at 0.9 m/s, the structure start to give indications of a swivelling-mode transition. However, beyond this value it was not possible to register the dynamic behavior of the structure model. On increasing the velocity, the resulting movement of the model became unstable and non-reproducible, finally leading to the failure of the structure.

In a similar way as for the first laminar swivelling-mode, a direct connection between the movement excitation and the classical von Karman vortex shedding behind the structure front cylinder was proved to exist. This is supported for the first swivelling-mode observed in turbulent flows by the relation between the first natural frequency of the structure (N_1) and the turbulent Strouhal number ($St \approx 0.21$). The same direct relation was observed for the rigid body swivelling-mode (or rigid body mode) on comparing the Strouhal number and the rigid body natural frequency of the structure ($N_0 = 0.19\,\text{Hz}$).

Based on the results, one can conclude that the first swivelling-mode is similar in nature in both the laminar and turbulent regimes. The main differences between the two are only connected with the fact that for turbulent flow the damping of the coupled system was significantly reduced. Thus, for the turbulent case, higher excursions from the trailing edge and front body and also higher deflection of the flexible part of the structure were observed. Because of the lower damping imposed by the fluid, the structure was exposed to higher accelerations during its swivelling movement in the turbulent tests. Another difference appears when comparing the movement of the structure trailing edge. At higher Reynolds number, the area covered by the "figure-of-eight" shaped trajectory was considerably bigger than that obtained in the laminar tests. This is related to the delay registered between the rear mass and the front body movement. In both cases, the movement of the rear mass could be considered to be in concordance with the front cylinder rotation, but the delay increased with the Reynolds number: whereas at $Re = 140$ the delay was measured to be close to 60°, at $Re = 15000$ it reached 95°.

Comparing both self-excited coupled motions, one may conclude that both are triggered by the vortex shedding created around the front cylinder. Because the natural frequencies of the structure are constant and the Strouhal number is not so sensitive to the Reynolds number in the range 140-15000, the resulting movements have the same response as far as frequency of the movement versus approaching flow velocity is concerned. Considering all evidences, it can be concluded that both first swivelling-modes correspond to instability-induced excited (IIE) fluid-structure interaction cases. The same applies to the rigid body mode observed in the turbulent tests. In the laminar regime, a similar rigid body mode could not to be registered. The excitation process responsible for the second mode observed in laminar flow is more difficult to examine. However, the results indicate strongly that this mode can be attributed to movement-induced excitation (MIE).

5   Comparison of numerical and experimental results

Within the DFG Research Unit 493 (Fluid-Structure Interaction: Modelling, Simulation, Optimisation), two sets of numerical simulation data were produced. More
information about these two numerical studies including details on the numerical
schemes, modelling and coupling approaches adopted can be found in [8] and [1].
Both data sets covered the two swivelling-modes of the laminar test cases and are
discussed in the following subsections.

5.1   First swivelling-mode in laminar flows

At the inlet velocity of 1.07 m/s, a self excited cyclic swivelling of the structure
exposed to the uniform and constant flow was observed during the experiments
in which the movement of the rear mass was in concordance with the rotation
of the front cylinder. The numerical simulations reproduced the same kind of
movement for this first swivelling-mode. The experimentally determined swivelling frequency was 6.38 Hz while the simulations yielded 6.72 Hz ([8]) and 7.47 Hz
([1]). Therefore, both the numerical values were somewhat high but still in fairly
good agreement.

   A comparison of the time-phase resolved front cylinder angle for one swivelling
period is given in Fig. 20. The qualitative time traces turned out to be very similar, in
special for [8]. Quantitatively, the maximum values showed differences. The simulated value was about 26% greater than the measured value. For [1], the peak values
better matched the experimental results but some non-symmetry in the positive and
negative excursion were displayed.

   Figure 21 shows the trailing edge xy-coordinates for one swivelling period.
In this graph the two numerical results almost coincide. The qualitative behavior

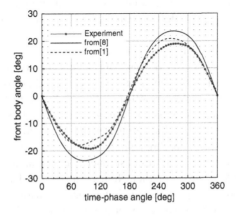

Fig. 20   Comparison of the front body angle for one period at 1.07 m/s.

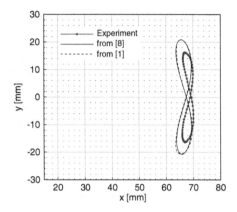

Fig. 21 Comparison of the x-y trailing edge coordinates for one period at 1.07 m/s.

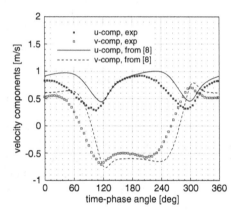

Fig. 22 Comparison of the flow velocity at point 1 for one period at 1.07 m/s.

was again well captured but the absolute values compared to the experiment differed noticeably. Here again, the simulated excursions were larger than the measured ones.

For comparing the flow field two monitor points were chosen. The first one (point 1) was positioned shortly downstream behind the structure on the symmetry plane at the coordinates (82 mm, 0 mm) and the second one (point 2) at the same x-coordinate but off-axis at (82 mm, 40 mm). Especially the first monitor point represented a very selective choice. Its position in the center of the structure near wake resulted in the fact that all the fluctuating velocities caused by the passage of the rear mass and the vortices emerging behind the swivelling structure were of direct impact. The second monitor point was positioned at the boundary of the wake region and turned out to be less hard to predict. Numerical velocity data was available only in [8] and the u-, and v-components are plotted together with the experimental data in Fig. 22 and Fig. 23 for the two monitor points, respectively. The numerical and

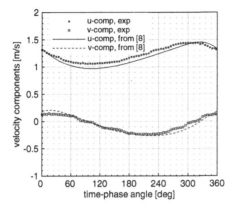

Fig. 23 Comparison of the flow velocity at point 2 for one period at 1.07 m/s.

experimental results matched very well at the second monitor point. At the first monitor point, the velocity time traces were very similar in shape and amplitude but a phase shift between the two results was displayed. A cross check with the position data of the trailing edge proved that this phase shift was not present in the trailing edge movement and, therefore, could not be the cause for the shift in the velocity data.

5.2   Second swivelling-mode in laminar flows

At the inlet velocity of 1.45 m/s both numerical approaches resulted in a movement of the same character than that observed during the experiments; the rear mass now moved in the opposite direction with reference to the front cylinder. The frequency of the self excited motion was measured to be 13.58 Hz and the simulations resulted in values of 14.42 Hz ([8]) and 16.78 Hz ([1]), respectively. Hence, a comparably good agreement between the measured and computed frequencies was obtained as observed for the first swivelling-mode. Again the computed frequencies were higher than the measured values being [8] closer to the experiments than [1].

In Fig. 24, the time-phase resolved front body angle for one period is plotted. A comparison between Figs. 20 and 24 clearly shows the differences in the dynamic response of the structure for the two swivelling-modes. In the second swivelling-mode, the two numerical data sets resulted in distinctively different time traces. While the data from [8] almost coincided with the experimental one, indicating that both the amplitude and the pronounced non-linear characteristic were accurately reproduced by the simulation, the data from [1] exhibited a strong non-symmetry. An examination to figure out the reasons for that somewhat odd behaviour did not lead to convincing conclusions yet and had to be left over to future investigations. Figure 25 displays the traces of the trailing edge for one period at 1.45 m/s.

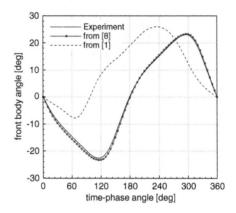

Fig. 24 Comparison of the front body angle for one period at 1.45 m/s.

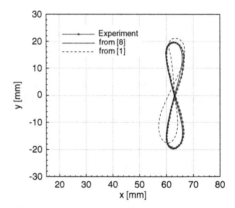

Fig. 25 Comparison of the x-y trailing edge coordinates for one period at 1.45 m/s.

Comparing the two different numerical and the experimental results yielded in a similar evaluation than for the front body angle but the differences were less pronounced. The data from [8] and the experimental results virtually collapsed while the data from [1] was noticeably off and showed some amount of non-symmetry. But compared with Fig. 21, both numerical results were in closer agreement to the experiment for the second swivelling-mode.

As was to be expected from the very good agreement of the numerical results from [8] with the experimental data in the behaviour of the swivelling motion of the structure the same good match was found in the flow velocities. In Figs. 26 and 27 the u-, and v-components are plotted for the two monitor points. At the second monitor point, positioned off-axis, the graphs almost collapsed. For the more critical monitor point positioned on the symmetry axis the agreement is still good, although a slightly too large value for the u-component was predicted. The phase-shift observed in the first swivelling-mode was no longer present in the second.

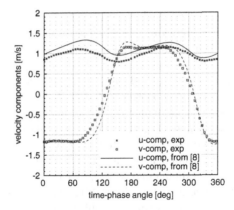

Fig. 26 Comparison of the flow velocity at point 1 for one period at 1.45 m/s.

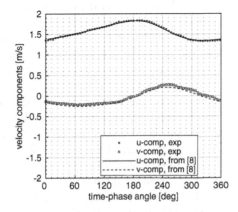

Fig. 27 Comparison of the flow velocity at point 2 for one period at 1.45 m/s.

5.3 *Conclusions of the comparison of numerical and experimental results*

The comparison of the outcome of the numerical and experimental studies described above showed that for both tested flow velocities the models reached a very reproducible steady-state oscillation within a short time (a few cycles). The comparison of the results performed on the basis of the time-phase resolved front body angle, the trailing edge coordinates, and the flow velocities at two monitor points proved that all the trends observed in the experiments were correctly captured by the numerical approaches except for some non-symmetric front body oscillation predicted in [1] for the second mode. For both self-excited swivelling-modes, the outcomes of the measurements and the simulations were, in general, in good agreement. The numerical simulations always resulted in slightly higher oscillation frequencies compared with the corresponding measurements and also the amplitudes tended to be on the larger side.

Nevertheless, there is one point which deserves further attention and which shall be shortly discussed in the following, namely that for the first swivelling-mode the simulation results matched the measured values much less closely compared with the second mode. Whereas for the second mode the data for the simulations and the experiments almost collapsed, for the first mode significant discrepancies were observed. There are two facts that made a difference between the experimental and the numerical models. First, the numerical model had perfectly two-dimensional boundary conditions, while in the experiment the span of the model was limited, and hence there were test section side walls with boundary layers and gaps between the model and the side walls. Second, the rotational degree of freedom was considered frictionless in the numerical simulations, whereas the bearings in the experiment had extremely small but non-zero friction. Both differences introduced higher damping into the experimental set-up compared with the computations, which in turn would result in the tendencies observed. Nevertheless, the differences between the experimental and the numerical outcomes for the first swivelling-mode can hardly be attributed to these minor differences in the damping only.

A possible explanation for the different degrees of agreement between the measurements and numerical results for the two swivelling-modes can be related to the different fluid-structure excitation mechanisms identified. In the first mode of oscillation, the relevant frequency for the excitation process [Instability-Induced Excitation (IIE); [6]] is the first natural frequency of the structure (the structure alone, i.e. surrounded by vacuum). This gives rise to the assumption that the dynamic response of the system is very sensitive to the damping coefficient of the structure associated with its first natural frequency, which is small because of the material properties of the flexible sheet. Therefore, the reaction of the system is more sensitive to deviations in the damping of the structure which affect the experimental results but which are not considered in the numerical model. On the other hand, in the second mode of vibration, the relevant frequency for the exciting process [Movement-Induced Excitation (MIE); [6]] is the natural frequency of the coupled system structure and surrounding fluid. For this fully coupled mode, the added damping introduced by the viscous fluid exceeds the damping of the structure by orders of magnitude. This additional damping term is included in the numerical model because it takes into account the properties of the liquid, specially the viscosity. Hence, for the second swivelling-mode, and because of the presence of the added damping, the dynamic response of the system becomes less sensitive to deviations of the structural damping. These arguments can justify the differences observed between the measurements and the numerical results for the first swivelling-mode while at the same time the agreement between the results is almost perfect for the second swivelling-mode.

Acknowledgements The present research project is part of the DFG Forschergruppe 493 - Fluid-Struktur-Wechselwirkung: Modellierung, Simulation, Optimierung. The authors gratefully acknowledge the financial support for their research work through the German Science Foundation (DFG) - Germany - and Fundação para a Ciência e a Tecnologia (FCT) - Portugal. In addition the authors acknowledge the funding of the Erlangen Graduate School in Advanced Optical Technologies (SAOT) by DFG in the framework of the German excellence initiative.

References

1. Geller S., Bettah M., Krafczyk M., Kollmannsberger S., Scholz D., Düster A., Rank E. (2010) An explicit model for three-dimensional fluid-structure interaction using LBM and p-FEM. In Bungartz H. J., Schäfer M. (eds.) Fluid-Structure Interaction, Part II, Lecture Notes Comput. Sci. & Eng., LNCSE. Springer
2. Gomes J. P., Lienhart H. (2006) Time-phase resolved PIV/DMI measurements on two-dimensional fluid-structure interaction problems. International symposium 13th Int Symp on Applications of Laser Techniques to Fluid Mechanics. Calouste Gulbenkian Fundation, Lisbon, 16. - 29. June
3. Gomes J. P., Lienhart H. (2006) Experimental Study on a Fluid–Structure Interaction Reference Test Case. In: Bungartz H J, Schäfer M (eds.) Fluid–Structure Interaction, Lecture Notes Comput. Sci. & Eng., LNCSE. Springer
4. Hartmann S., Meister A., Schäfer M., Turek S. (eds.) (2009) Fluid-structure interaction: Theory, numerics, applications. Kassel University Press
5. Naudascher E., Rockwell D. (1980) Oscillator-model approach to the identification and assessment of flow-induced vibrations in a system. Journal Hydraulics Research 18:59-82
6. Naudascher E., Rockwell D. (1980) Flow-induced Vibrations, IAHR Hydrodynamic Structures Design Manual, Vol.7. Balkena
7. Sarpkaya T. (1978) Fluid forces on oscillating cylinders. Journal of the Waterway Port Coastal and Ocean Division WW4:275-290
8. Schäfer M., Sternel D.C., Becker G., Pironkov P. (2010) Efficient Numerical Simulation and Optimization of Fluid-Structure Interaction. In: Bungartz H. J., Schäfer M. (eds.) Fluid-Structure Interaction, Part II, Lecture Notes Comput. Sci. & Eng., LNCSE. Springer
9. Souli M., Hamdouni A. (eds.) (2007) Fluid structure interaction: Industrial and Academic Applications. European Journal of Computational Mechanics 16:303-548

Numerical Benchmarking of Fluid-Structure Interaction: A Comparison of Different Discretization and Solution Approaches

S. Turek, J. Hron, M. Razzaq, H. Wobker, and M. Schäfer

Abstract Comparative benchmark results for different solution methods for fluid-structure interaction problems are given which have been developed as collaborative project in the DFG Research Unit 493. The configuration consists of a laminar incompressible channel flow around an elastic object. Based on this benchmark configuration the numerical behavior of different approaches is analyzed exemplarily. The methods considered range from decoupled approaches which combine Lattice Boltzmann methods with hp-FEM techniques, up to strongly coupled and even fully monolithic approaches which treat the fluid and structure simultaneously.

1 Introduction

In [6] specific benchmark problems have been defined in order to have a well-founded basis for validating and also comparing different numerical methods and code implementations for fluid-structure interaction (FSI) problems. In the present paper numerical results obtained with different FSI solution approaches are presented and discussed. The methods involve different discretization schemes and coupling mechanisms.

The FSI benchmark is based on the widely accepted *flow around cylinder* configuration developed in [7] for incompressible laminar flow and on the setup in [8]. Similar to these configurations we consider the fluid to be incompressible and in the laminar regime. The structure is allowed to be compressible, and its deformations are significant. The overall setup of the interaction problem is such that the

S. Turek, J. Hron, M. Razzaq, and H. Wobker
Institut für Angewandte Mathematik, LS III, TU Dortmund, Vogelpothsweg 87,
44227 Dortmund, Germany
e-mail: stefan.turek@math.tu-dortmund.de

M. Schäfer
Institut für Numerische Berechnungsverfahren im Maschinenbau, TU Darmstadt,
Dolivostr. 15, 64293 Darmstadt, Germany
e-mail: schaefer@fnb.tu-darmstadt.de

H.-J. Bungartz et al. (eds.), *Fluid Structure Interaction II*, Lecture Notes
in Computational Science and Engineering 73, DOI 10.1007/978-3-642-14206-2_15,
© Springer-Verlag Berlin Heidelberg 2010

solid object with an elastic part is submerged in a channel flow. Two cases with a steady and an unsteady solution, respectively, are considered. Results for characteristic physical quantities are provided, facilitating comparisons without the use of complicated graphical evaluations. The provided benchmark computations are the result of collaborative simulations which have been part of the research projects inside of the DFG Research Unit 493 and which are presented as separate contributions to this volume (which also contains a more detailed description of the different used FSI codes). The parameter settings and results can be downloaded from http://featflow.de/en/benchmarks/cfdbenchmarking/fsi_benchmark.html.

2 Benchmark configuration

We recall the benchmark configuration defined in [6]. It concerns the flow of an *incompressible Newtonian fluid* interacting with an *elastic solid*. We denote by Ω_t^f and Ω_t^s the domains occupied by the fluid and the solid, resp., at the time $t \geq 0$. Let $\Gamma_t^0 = \bar{\Omega}_t^f \cap \bar{\Omega}_t^s$ be the part of the boundary where the elastic solid interacts with the fluid.

2.1 Fluid properties

The fluid is considered to be *Newtonian, incompressible* and its state is described by the velocity and pressure fields v^f, p^f. The balance equations are

$$\rho^f \frac{\partial v^f}{\partial t} + \rho^f (\nabla v^f) v^f = \operatorname{div} \sigma^f, \qquad \text{in } \Omega_t^f. \tag{1}$$
$$\operatorname{div} v^f = 0$$

The material constitutive equation is

$$\sigma^f = -p^f I + \rho^f v^f (\nabla v^f + \nabla v^{f\mathrm{T}}). \tag{2}$$

The constant density of the fluid is ρ^f and the viscosity is denoted by v^f.

2.2 Structure properties

The structure is assumed to be *elastic* and *compressible*. Its deformation is described by the displacement u^s, with velocity field $v^s = \frac{\partial u^s}{\partial t}$. The balance equations are

$$\rho^s \frac{\partial v^s}{\partial t} + \rho^s (\nabla v^s) v^s = \text{div}(\sigma^s) + \rho^s g \qquad \text{in } \Omega_t^s. \tag{3}$$

Written in the more common Lagrangian description, i.e. with respect to some fixed reference (initial) state Ω^s, we have

$$\rho^s \frac{\partial^2 u^s}{\partial t^2} = \text{div}(J \sigma^s F^{-T}) + \rho^s g \qquad \text{in } \Omega^s, \tag{4}$$

where $F = I + \nabla u^s$ is the deformation gradient tensor. For further details see for example [1].

The material is specified by the Cauchy stress tensor σ^s or by the 2nd Piola-Kirchhoff stress tensor $S^s = J F^{-1} \sigma^s F^{-T}$ via the *St. Venant-Kirchhoff* constitutive law

$$\sigma^s = \frac{1}{J} F (\lambda^s (\text{tr } E) I + 2\mu^s E) F^T, \tag{5}$$

$$S^s = \lambda^s (\text{tr } E) I + 2\mu^s E, \tag{6}$$

where $E = \frac{1}{2}(F^T F - I)$ is the Green-St. Venant strain tensor.

The density of the structure in the undeformed configuration is ρ^s. The elasticity of the material is characterized by the Poisson ratio v^s ($v^s < 0.5$ for a compressible structure) and by the Young modulus E^s. The alternative characterization is described by the Lame coefficients λ^s and μ^s (the shear modulus):

$$v^s = \frac{\lambda^s}{2(\lambda^s + \mu^s)} \qquad\qquad E^s = \frac{\mu^s (3\lambda^s + 2\mu^s)}{(\lambda^s + \mu^s)}, \tag{7}$$

$$\mu^s = \frac{E^s}{2(1 + v^s)} \qquad\qquad \lambda^s = \frac{v^s E^s}{(1 + v^s)(1 - 2v^s)}. \tag{8}$$

2.3 Interaction conditions

The boundary conditions on the fluid-solid interface are assumed to be

$$\begin{aligned} \sigma^f n &= \sigma^s n \\ v^f &= v^s \end{aligned} \qquad \text{on } \Gamma_t^0, \tag{9}$$

where n is a unit normal vector to the interface Γ_t^0. This implies the no-slip condition for the flow, and that the forces on the interface are in balance.

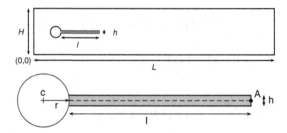

Fig. 1 Computational domain and details of the structure part.

2.4 Domain definition

The problem domain, which is based on the 2D version of the well-known CFD benchmark in [7], is illustrated in Fig. 1.

The geometry parameters are given as follows (all values in meters):

- The domain has length $L = 2.5$ and height $H = 0.41$.
- The circle center is positioned at $C = (0.2, 0.2)$ (measured from the left bottom corner of the channel) and the radius is $r = 0.05$.
- The elastic structure bar has length $l = 0.35$ and height $h = 0.02$, the right bottom corner is positioned at $(0.6, 0.19)$, and the left end is fully attached to the fixed cylinder.
- The control point is $A(t)$, attached to the structure and moving in time with $A(0) = (0.6, 0.2)$.

The setting is intentionally non-symmetric (see [7]) to prevent the dependence of the onset of any possible oscillation on the precision of the computation.

2.5 Boundary conditions

The following boundary conditions are prescribed:

- A parabolic velocity profile is prescribed at the left channel inflow

$$v^f(0, y) = 1.5\bar{U} \frac{y(H - y)}{(H/2)^2} = 1.5\bar{U} \frac{4.0}{0.1681} y(0.41 - y), \qquad (10)$$

such that the mean inflow velocity is $\bar{U}$ and the maximum of the inflow velocity profile is $1.5\bar{U}$.
- The outflow condition can be chosen by the user, for example *stress free* or *do nothing* conditions. The outflow condition effectively prescribes some reference value for the pressure variable p. While this value could be arbitrarily set in the incompressible case, in the case of compressible structure this will have influence

Table 1 Parameter settings
for the FSI benchmarks.

	FSI1	FSI3
ρ^s [10^3kg/m^3]	1	1
ν^s	0.4	0.4
μ^s [10^6kg/ms^2]	0.5	2.0
ρ^f [10^3kg/m^3]	1	1
ν^f [10^{-3}m^2/s]	1	1
$\bar{U}$ [m/s]	0.2	2

on the stress and consequently the deformation of the solid. In this proposal, we set the reference pressure at the outflow to have *zero mean value*.
- The *no-slip* condition is prescribed for the fluid on the other boundary parts, i.e. top and bottom wall, circle and fluid-structure interface Γ_t^0.

2.6 Initial conditions

The suggested starting procedure for the non-steady tests was to use a smooth increase of the velocity profile in time as

$$v^f(t, 0, y) = \begin{cases} v^f(0, y)[1 - \cos(\pi t/2)]/2 & \text{if } t < 2.0, \\ v^f(0, y) & \text{otherwise}, \end{cases} \qquad (11)$$

where $v^f(0, y)$ is the velocity profile given in Eq. (10).

2.7 Physical parameters

We consider physical parameters for two different test cases as indicated in Table 1. Defining the Reynolds number by Re $= 2r\bar{U}/v^f$ the two cases correspond to Re=20 and Re=200. FSI1 is resulting in a steady state solution, while FSI3 results in a periodic solution (the numbering refers to the original benchmark definition in [6]).

3 Quantities for comparison

Comparisons will be done for *fully developed flow*, and particularly for *one full period of the oscillation* with respect to the position of the point $A(t)$. The quantities of interest are:

1. The displacements $u_1(t)$ and $u_2(t)$ in x- and y-direction of the point $A(t)$ at the end of the beam structure (see Fig. 1).

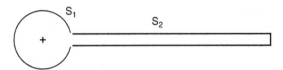

Fig. 2 Integration path $S = S_1 \cup S_2$ for the force calculation.

2. Forces exerted by the fluid on the *whole* submerged body, i.e. lift and drag forces
   acting on the cylinder and the beam structure together

$$(F_D, F_L)^{\mathrm{T}} = \int_S \sigma^f n \, dS = \int_{S_1} \sigma^f n \, dS + \int_{S_2} \sigma^f n \, dS,$$

where $S = S_1 \cup S_2$ (see Fig. 2) denotes the part of the circle being in contact
with the fluid and n is the outer unit normal vector to the integration path with
respect to the fluid domain.

Time dependent values are represented by mean value, amplitude, and frequency.
The mean value and the amplitude are computed from the last period of the oscillations by taking the maximum and minimum values, the mean value is taken as
average of the max/min values, and the amplitude is the difference of the max/min
from the mean:

$$\text{mean} = \frac{1}{2}(\text{max} + \text{min}),$$

$$\text{amplitude} = \frac{1}{2}(\text{max} - \text{min}).$$

The frequency of the oscillations are computed either as $1/T$ with the period time T
or by using Fourier analysis on the periodic data and taking the lowest significant
frequency present in the spectrum.

4 FSI solution approaches

In the following, a short description of the different approaches, which have been
applied in the subsequent benchmark calculations, will be provided. A more detailed
description of the underlying methodology and the corresponding software tools is
given as separate contributions to this volume.

Method 1 (Schäfer):

This method involves an implicit partitioned solution approach (see [4, 5] for
details). It is realized based on the finite-volume multigrid flow solver FASTEST

involving an ALE formulation (hexahedral control volumes, second-order in space and time), the finite-element structural solver FEAP (bilinear brick elements, second-order Newmark time discretization), and the coupling interface MpCCI. For each time step the implicit solution procedure consists in the application of different nested iteration processes for linearization, pressure-velocity coupling, and linear system solving. These are linked by an iterative fluid-structure coupling procedure with structural underrelaxation and displacement prediction.

Method 2 (Rannacher):

Two different solvers have been used in this project, both based on monolithic variational formulations of the FSI problem: (a) one based on a unified Eulerian framework for describing the fluid as well as the structure deformation ("interface capturing") which allows for large structural deformations and topology changes of the flow domain, and (b) for comparison one based on the standard ALE approach ("interface fitting"). The discretization is fully implicit and uses in case (a) the second-order fractional-step-θ scheme and in case (b) the stabilized Crank-Nicolson scheme in time. The spatial discretization is by a conforming finite element Galerkin method on quadrilateral meshes with mixed global and zonal refinement using in case (a) the "equal-order" Q_1^c/Q_1^c Stokes-element for velocity and pressure with pressure and transport stabilization by "local projection", and in case (b) the inf-sup-stable Q_2^c/P_1^d Stokes-element. For the resulting nonlinear algebraic problems in each time step a Newton-like iteration is used (combined with pseudo-time stepping in the stationary test case). The linear subproblems are solved in fully coupled form in case (a) by a multigrid methods with GMRES acceleration, block-ILU smoothing and canonical grid transfers, and in case (b) by a direct solver ("UMFPACK") which allows for up to 10^6 unknowns in acceptable time.

Method 3 (Turek/Hron):

The applied FSI solver is a fully implicit, monolithic ALE-FEM approach. The discretization is based on the classical (nonparametric) Q2/P1 finite element pair, together with the Crank-Nicolson scheme for time stepping. The resulting nonlinear systems are solved by a discrete Newton method, while the linear subsystems are treated via Krylov-multigrid solvers with smoothing operators of local pressure Schur complement/Vanka-like type with canonical grid transfer routines.

Method 4 (Breuer):

The method relies on a partitioned solution approach. For the fluid flow a finite-volume scheme (FASTEST-3D) is used to discretize the (filtered) Navier–Stokes equations for an incompressible fluid. The discretization is done on a curvilinear, blockstructured body-fitted grid with colocated variable arrangement by applying standard schemes. Linear interpolation of the flow variables to the cell faces and a midpoint rule approximation for the integrals is used to obtain a second-order accurate central scheme. In order to account for displacement or deformation of the structure the Arbitrary Lagrangian-Eulerian formulation is applied. A new partitioned coupling method based on the predictor-corrector scheme often used for LES is prefered. A strongly coupled but nevertheless still explicit time-stepping algorithm results, which is very efficient in the LES–FSI context. For the prediction of the deformation and displacement of the structure, the forces are transferred to computational structure dynamics code, here the finite-element solver Carat provided by the Chair of Structural Analysis of TU Munich. The response of the structure, i.e. the displacements are transferred to the fluid solver. The coupling interface CoMA also developed at TUM is used for this purpose. This interface is based on the Message-Passing-Interface (MPI) and thus runs in parallel to the fluid and structure solver. Presently, the grid adjustment on the fluid side is performed based on a transfinite interpolation.

Method 5 (Krafczyk/Rank):

This method involves an explicit partitioned solution approach. It is realized with the Lattice-Boltzmann flow solver VirtualFluids (VF) and the structural p-FEM solver AdhoC. VirtualFluids is based on an Eulerian grid (cubic, hierarchical, graded FD grids, second-order in space and time). AdhoC discretizes the equations of structural dynamics in space with a Bubnov-Galerkin method utilizing hierarchic ansatz functions while the time domain is discretized with a second-order finite difference scheme (Newmark or generalized alpha). The coupling interface MshPI manages the data transfer on an interface mesh common to all involved solvers. For each time step, the partitioned solution procedure consists of nested substeps. VirtualFluids sends loads via the interface mesh to AdhoC which generates displacements. These, in turn, are then mapped to the fluid grid. The exchange of tractions and displacements is performed only once per structural time step. As the time step for the flow solver is typically smaller than the structural one, VF proceeds for several time steps with an interpolated geometry. It is interesting to note that this setup proved to be stable without the need to perform interfield iterations.

Method 6 (Wall):

For this benchmark scenario we have used our rather "classic" approach selected from the different FSI approaches that we have developed in recent years for solving incompressible fluid flow coupled with large structural deformations. A main ingredient is a strongly coupled, iterative staggered scheme based on [2,9]. The fluid field uses an ALE formulation, stabilized, quadratic Q2Q2 elements and BDF2 time-discretiyation. The fluid mesh is deformed based on linear-elastic material behavior.

The structure is discretized by bi-linear quadrilateral elements with EAS formulation and the generalized-α method for time discretization. Each field is solved implicitly and an iterative procedure over the fields using Aitken relaxation ensures convergence for the interface conditions at the new time step level $n + 1$.

Method 7 (Bletzinger):

The coupled problem is solved by a partitioned approach. Three independent software components are combined: in-house codes CARAT++ and CoMA (Computer Aided Research Analysis Tool, Coupling for Multiphysics Analysis) for structural analysis, coupling control and data transfer between non-matching grids, as well as OpenFOAM, an open source finite volume solver. The single field solvers use individual, at the interface non-matching, grids. Implicit coupling schemes based on fixed-point iterations with Aitken relaxation or a quasi-Newton method are used. Different to all other groups the structure has been modelled as a shell (with mid-surface in x-z-plane). Two different shell theories have been applied: a "classical", 5-parameter, Reissner-Mindlin shell model neglecting normal stresses in thickness direction as well as a 7-parameter, 3D solid shell model with a straight thickness director as only kinematic assumption. The difference becomes obvious in FSI1 where a longitudinal Poisson effect due to cross thickness normal stress is evident for the horizontal deformation $u_1 = 1.85 \times 10^{-5}$ (classical shell) or $u_1 = 2.26 \times 10^{-5}$ (solid shell), respectively. All other data match well and show good results in all benchmarks. It is concluded, that shell finite element formulations can be used effectively in FSI analyses and give correct results for structures up to a moderate thickness.

5 Numerical results

The results of the benchmark computations are summarized in Tables 2 and 3 (units are omitted). Indicated are the displacements $u_1(A)$ and $u_2(A)$ in x- and y-direction of the point A as well as the drag and lift forces F_D and F_L. For the unsteady case also the frequencies f_1 and f_2 obtained for the displacements $u_1(A)$ and $u_2(A)$, respectively, are given. The number in the first column refers to the methods given in the previous section. The column "Unknowns" refers to the total number (in space), i.e., the sum of unknowns for all velocity components, pressure, and displacement components.

As a first result for the FSI1 benchmark, which leads to stationary displacement of the attached elastic beam, it is obvious that all applied methods and codes can approximate the same results, at least with decreasing mesh width.

For FSI3, the evaluation of the results is a little bit more difficult: First of all, all schemes show the tendency to converge towards the (more or less) same solution values, at least for increasing mesh level. Although the applied FSI techniques

Table 2 Results for steady benchmark FSI1.

	Unknowns	$u_1(A)$ $[\times 10^{-5}]$	$u_2(A)$ $[\times 10^{-4}]$	F_D	F_L
1	82722	–	–	14.2770	0.77200
	322338	–	–	14.2890	0.76900
2a	11250	2.4800	7.7800	–	–
2b	19488	2.2821	8.1957	14.2382	0.76481
	29512	2.2793	8.2201	14.2263	0.76420
	51016	2.2733	8.1867	14.2408	0.76400
	93992	2.2710	8.1702	14.2500	0.76392
	179912	2.2700	8.1609	14.2561	0.76389
	351720	2.2695	8.1556	14.2603	0.76388
3	19488	2.2871	8.1930	14.2736	0.76175
	76672	2.2774	8.2042	14.2918	0.76305
	304128	2.2732	8.2071	14.2948	0.76356
	1211392	2.2716	8.2081	14.2949	0.76370
	4835328	2.2708	8.2085	14.2945	0.76374
	19320832	2.2705	8.2088	14.2943	0.76375
5	884736	2.1270	11.0800	14.3179	0.85491
	3538944	2.1990	8.3370	14.3127	0.75138
	14155776	2.2160	8.2010	14.3815	0.75170
6	7059	2.5396	8.8691	14.2800	0.73690
	19991	2.2630	8.2935	14.2970	0.76687
	77643	2.2676	8.2347	14.2940	0.76545
	164262	2.2680	8.2310	14.2940	0.76487
7	217500	2.2640	8.2800	14.3510	0.76351

are very different w.r.t. discretization, solver and coupling mechanisms, the FSI3 benchmark setting proves to be a very valuable tool for numerical FSI benchmarking, leading to grid independent results for the prescribed geometrical and parameter settings.

However, also clear differences between the different approaches with regard to accuracy are visible. Particularly for the drag and lift values, which lead to differences of up to order 50%, and also for the displacement values which are in the range of 10% errors. A more detailed evaluation and also more rigorous comparisons w.r.t. the ratio 'accuracy vs. efficiency' are therefore planned for the future.

6 Summary

We have presented 2D benchmark results for different numerical approaches for fluid-structure interaction problems. These benchmarks have been developed as a collaborative project in the DFG Research Unit 493. The configurations have been carefully chosen and validated via extensive numerical tests (see also [3, 6])

Table 3 Results for unsteady benchmark FSI3.

	Unknowns	Δt	$u_1(A)\,[\times 10^{-3}]$	$u_2(A)\,[\times 10^{-3}]$	F_D	F_L	f_1	f_2
1	61318	1.0e−3	−2.54 ± 2.41	1.45 ± 32.80	450.3 ± 23.51	-0.10 ± 143.0	10.90	5.13
	237286	2.0e−3	−2.88 ± 2.73	1.53 ± 34.94	458.6 ± 27.18	2.08 ± 153.1	10.60	5.30
	237286	1.0e−3	−2.87 ± 2.73	1.54 ± 34.94	458.6 ± 27.31	2.00 ± 153.3	10.34	5.91
	237286	5.0e−4	−2.86 ± 2.72	1.53 ± 34.90	458.6 ± 27.27	2.01 ± 153.4	12.16	6.08
	941158	1.0e−3	−2.91 ± 2.77	1.47 ± 35.26	459.9 ± 27.92	1.84 ± 157.7	11.63	4.98
2a	11250	5.0e−3	−2.48 ± 2.24	1.27 ± 36.50	–	–	10.10	5.10
2b	7176	5.0e−3	−2.44 ± 2.32	1.02 ± 31.82	473.5 ± 56.97	8.08 ± 283.8	11.07	5.29
	7176	2.0e−3	−2.48 ± 2.39	0.92 ± 32.81	471.3 ± 62.28	6.11 ± 298.6	10.73	5.35
	7176	1.0e−3	−2.58 ± 2.49	0.94 ± 33.19	470.4 ± 64.02	4.65 ± 300.3	10.69	5.36
	27744	5.0e−3	−2.43 ± 2.27	1.41 ± 31.73	483.7 ± 22.31	2.21 ± 149.0	10.53	5.37
	27744	2.0e−3	−2.63 ± 2.61	1.46 ± 33.46	483.3 ± 24.48	2.08 ± 161.2	10.66	5.43
	27744	1.0e−3	−2.80 ± 2.64	1.45 ± 34.12	483.0 ± 25.67	2.21 ± 165.3	10.75	5.41
	42024	2.5e−3	−2.40 ± 2.26	1.39 ± 31.71	448.7 ± 21.16	1.84 ± 141.3	10.72	5.42
	42024	1.0e−3	−2.53 ± 2.38	1.40 ± 32.49	449.7 ± 22.24	1.61 ± 142.8	10.77	5.44
	42024	5.0e−4	−2.57 ± 2.42	1.42 ± 32.81	450.1 ± 22.49	1.49 ± 143.7	10.79	5.42
	72696	2.5e−3	−2.64 ± 2.48	1.38 ± 33.25	451.1 ± 24.57	2.04 ± 150.6	10.73	5.38
	72696	1.0e−3	−2.79 ± 2.62	1.28 ± 34.61	452.0 ± 25.78	1.91 ± 152.7	10.78	5.42
	72696	5.0e−4	−2.84 ± 2.67	1.28 ± 34.61	452.4 ± 26.19	2.36 ± 152.7	10.84	5.42
3	19488	1.0e−3	−3.02 ± 2.83	1.41 ± 35.47	458.2 ± 28.32	2.41 ± 145.6	10.75	5.37
	19488	5.0e−4	−3.02 ± 2.85	1.42 ± 35.63	458.7 ± 28.78	2.23 ± 146.0	10.75	5.37
	19488	2.5e−4	−3.02 ± 2.85	1.32 ± 35.73	458.7 ± 28.80	2.23 ± 146.0	10.74	5.33
	76672	1.0e−3	−2.78 ± 2.62	1.44 ± 34.36	459.1 ± 26.63	2.41 ± 151.3	10.93	5.46
	76672	5.0e−4	−2.78 ± 2.62	1.44 ± 34.35	459.1 ± 26.62	2.39 ± 150.7	10.92	5.46
	76672	2.5e−4	−2.77 ± 2.61	1.43 ± 34.43	459.1 ± 26.50	2.36 ± 149.9	10.93	5.46
	304128	1.0e−3	−2.86 ± 2.70	1.45 ± 34.93	460.2 ± 27.65	2.47 ± 154.9	10.95	5.47
	304128	5.0e−4	−2.86 ± 2.70	1.45 ± 34.90	460.2 ± 27.47	2.37 ± 153.8	10.92	5.46
	304128	2.5e−4	−2.88 ± 2.72	1.47 ± 34.99	460.5 ± 27.74	2.50 ± 153.9	10.93	5.46
4	81120	9.0e−5	−5.18 ± 5.04	1.12 ± 45.10	477.0 ± 48.00	7.00 ± 223.0	10.14	4.99
	324480	2.0e−5	−4.54 ± 4.34	1.50 ± 42.50	467.5 ± 39.50	16.20 ± 188.7	10.12	5.05
5	2480814	5.1e−5	−2.88 ± 2.71	1.48 ± 35.10	463.0 ± 31.30	1.81 ± 154.0	11.00	5.50
6	7059	5.0e−4	−1.60 ± 1.60	1.50 ± 25.90	525.0 ± 22.50	-0.55 ± 106.0	10.90	5.45
	27147	5.0e−4	−2.00 ± 1.89	1.45 ± 29.00	434.0 ± 17.50	2.53 ± 88.6	10.60	5.30
7	271740	5.0e−4	−3.04 ± 2.87	1.55 ± 36.63	474.9 ± 28.12	3.86 ± 165.9	10.99	5.51

with various CFD codes so that, as a main result, characteristic flow quantities can be provided which allow a quantitative validation and comparison of different numerical methods and software tools. As an extension, corresponding 3D simulations are planned as well as the embedding into outer optimization tools (see http://jucri.jyu.fi/?q=node/14 for a first attempt towards optimal control on the basis of the presented FSI1 configuration).

Acknowledgements The described benchmarks were developed in collaboration with G. Becker, M. Heck, S. Yigit, M. Krafczyk, J. Tölke, S. Geller, H.-J. Bungartz, M. Brenk, R. Rannacher, T. Dunne, T. Wick, W. Wall, A. Gerstenberger, P. Gamnitzer, E. Rank, A. Düster, S. Kollmannsberger, D. Scholz, M. Breuer, M. Münsch, G. De Nayer, H. Lienhart, J. Gomes, K.-U. Bletzinger, A. Kupzok, and R. Wüchner.
This work has been supported by German Research Association (DFG), Reasearch Unit 493.

References

1. P. G. Ciarlet. *Mathematical Elasticity. Volume I, Three-Dimensional Elasticity*, volume 20 of *Studies in Mathematics and its Applications*. Elsevier Science Publishers B.V., Amsterdam, 1988.
2. Ulrich Küttler and Wolfgang A. Wall. Fixed-point fluid-structure interaction solvers with dynamic relaxation. *Computational Mechanics*, 43(1):61–72, 2008.
3. M. Razzaq, S. Turek, J. Hron, and J. F. Acker. Numerical simulation and benchmarking of fluid-structure interaction with application to hemodynamics. In *Fundamental Trends in Fluid-Structure Interaction*. World Scientific Publishing Co. Pte Ltd, 2010.
4. M. Schäfer, M. Heck, and M. Schäfer (eds.) Fluid-Structure Interaction: Modelling Simulation Optimization 53 pp. 171-194. Springer Berlin Heidelberg Yigit, S.: An implicit partitoned method for the numerical simulation of fluid-structure interaction. In: H. J. Bungartz. 2006.
5. D.C. Sternel, M. Schäfer, M. Heck, and S. Yigit. Efficiency and accuracy of fluid-structure interaction simulations using an implicit partitioned approach. *Computational Mechanics*, 43(1):103–113, 2008.
6. S. Turek and J. Hron. Proposal for numerical benchmarking of fluid-structure interaction between an elastic object and laminar incompressible flow. In H.-J. Bungartz and M. Schäfer, editors, *Fluid-Structure Interaction: Modelling, Simulation, Optimisation*, LNCSE-53. Springer, 2006.
7. S. Turek and M. Schäfer. Benchmark computations of laminar flow around cylinder. In E.H. Hirschel, editor, *Flow Simulation with High-Performance Computers II*, volume 52 of *Notes on Numerical Fluid Mechanics*. Vieweg, 1996. co. F. Durst, E. Krause, R. Rannacher.
8. W. A. Wall and E. Ramm. Fluid-structure interaction based upon a stabilized (ALE) finite element method. In S. Idelsohn, E. Oñate, and E. Dvorkin, editors, *4th World Congress on Computational Mechanics: New Trends and Applications*, Barcelona, 1998. CIMNE.
9. Wolfgang A. Wall, Daniel P. Mok, and Ekkehard Ramm. Partitioned analysis approach of the transient coupled response of viscous fluids and flexible structures. In W. Wunderlich, editor, *Solids, Structures and Coupled Problems in Engineering, Proc. ECCM '99*, Munich, August/September 1999.

Editorial Policy

1. Volumes in the following three categories will be published in LNCSE:

i)   Research monographs
ii) Tutorials
iii) Conference proceedings

Those considering a book which might be suitable for the series are strongly advised to contact the publisher or the series editors at an early stage.

2. Categories i) and ii). Tutorials are lecture notes typically arising via summer schools or similar events, which are used to teach graduate students. These categories will be emphasized by Lecture Notes in Computational Science and Engineering. **Submissions by interdisciplinary teams of authors are encouraged.** The goal is to report new developments – quickly, informally, and in a way that will make them accessible to non-specialists. In the evaluation of submissions timeliness of the work is an important criterion. Texts should be well-rounded, well-written and reasonably self-contained. In most cases the work will contain results of others as well as those of the author(s). In each case the author(s) should provide sufficient motivation, examples, and applications. In this respect, Ph.D. theses will usually be deemed unsuitable for the Lecture Notes series. Proposals for volumes in these categories should be submitted either to one of the series editors or to Springer-Verlag, Heidelberg, and will be refereed. A provisional judgement on the acceptability of a project can be based on partial information about the work: a detailed outline describing the contents of each chapter, the estimated length, a bibliography, and one or two sample chapters – or a first draft. A final decision whether to accept will rest on an evaluation of the completed work which should include

– at least 100 pages of text;
– a table of contents;
– an informative introduction perhaps with some historical remarks which should be accessible to readers unfamiliar with the topic treated;
– a subject index.

3. Category iii). Conference proceedings will be considered for publication provided that they are both of exceptional interest and devoted to a single topic. One (or more) expert participants will act as the scientific editor(s) of the volume. They select the papers which are suitable for inclusion and have them individually refereed as for a journal. Papers not closely related to the central topic are to be excluded. Organizers should contact the Editor for CSE at Springer at the planning stage, see *Addresses* below.

In exceptional cases some other multi-author-volumes may be considered in this category.

4. Only works in English will be considered. For evaluation purposes, manuscripts may be submitted in print or electronic form, in the latter case, preferably as pdf- or zipped ps-files. Authors are requested to use the LaTeX style files available from Springer at http://www. springer.com/authors/book+authors?SGWID=0-154102-12-417900-0.

For categories ii) and iii) we strongly recommend that all contributions in a volume be written in the same LaTeX version, preferably LaTeX2e. Electronic material can be included if appropriate. Please contact the publisher.

Careful preparation of the manuscripts will help keep production time short besides ensuring satisfactory appearance of the finished book in print and online.

5. The following terms and conditions hold. Categories i), ii) and iii):

Authors receive 50 free copies of their book. No royalty is paid.
Volume editors receive a total of 50 free copies of their volume to be shared with authors, but no royalties.

Authors and volume editors are entitled to a discount of 33.3 % on the price of Springer books purchased for their personal use, if ordering directly from Springer.

6. Commitment to publish is made by letter of intent rather than by signing a formal contract. Springer-Verlag secures the copyright for each volume.

Addresses:

Timothy J. Barth
NASA Ames Research Center
NAS Division
Moffett Field, CA 94035, USA
barth@nas.nasa.gov

Michael Griebel
Institut für Numerische Simulation
der Universität Bonn
Wegelerstr. 6
53115 Bonn, Germany
griebel@ins.uni-bonn.de

David E. Keyes
Mathematical and Computer Sciences
and Engineering
King Abdullah University of Science
and Technology
P.O. Box 55455
Jeddah 21534, Saudi Arabia
david.keyes@kaust.edu.sa

and

Department of Applied Physics
and Applied Mathematics
Columbia University
500 W. 120 th Street
New York, NY 10027, USA
kd2112@columbia.edu

Risto M. Nieminen
Department of Applied Physics
Aalto University School of Science
and Technology
00076 Aalto, Finland
risto.nieminen@tkk.fi

Dirk Roose
Department of Computer Science
Katholieke Universiteit Leuven
Celestijnenlaan 200A
3001 Leuven-Heverlee, Belgium
dirk.roose@cs.kuleuven.be

Tamar Schlick
Department of Chemistry
and Courant Institute
of Mathematical Sciences
New York University
251 Mercer Street
New York, NY 10012, USA
schlick@nyu.edu

Editor for Computational Science
and Engineering at Springer:
Martin Peters
Springer-Verlag
Mathematics Editorial IV
Tiergartenstrasse 17
69121 Heidelberg, Germany
martin.peters@springer.com

Lecture Notes
in Computational Science
and Engineering

24. T. Schlick, H.H. Gan (eds.), *Computational Methods for Macromolecules: Challenges and Applications.*

25. T.J. Barth, H. Deconinck (eds.), *Error Estimation and Adaptive Discretization Methods in Computational Fluid Dynamics.*

26. M. Griebel, M.A. Schweitzer (eds.), *Meshfree Methods for Partial Differential Equations.*

27. S. Müller, *Adaptive Multiscale Schemes for Conservation Laws.*

28. C. Carstensen, S. Funken, W. Hackbusch, R.H.W. Hoppe, P. Monk (eds.), *Computational Electromagnetics.*

29. M.A. Schweitzer, *A Parallel Multilevel Partition of Unity Method for Elliptic Partial Differential Equations.*

30. T. Biegler, O. Ghattas, M. Heinkenschloss, B. van Bloemen Waanders (eds.), *Large-Scale PDE-Constrained Optimization.*

31. M. Ainsworth, P. Davies, D. Duncan, P. Martin, B. Rynne (eds.), *Topics in Computational Wave Propagation.* Direct and Inverse Problems.

32. H. Emmerich, B. Nestler, M. Schreckenberg (eds.), *Interface and Transport Dynamics.* Computational Modelling.

33. H.P. Langtangen, A. Tveito (eds.), *Advanced Topics in Computational Partial Differential Equations.* Numerical Methods and Diffpack Programming.

34. V. John, *Large Eddy Simulation of Turbulent Incompressible Flows.* Analytical and Numerical Results for a Class of LES Models.

35. E. Bänsch (ed.), *Challenges in Scientific Computing - CISC 2002.*

36. B.N. Khoromskij, G. Wittum, *Numerical Solution of Elliptic Differential Equations by Reduction to the Interface.*

37. A. Iske, *Multiresolution Methods in Scattered Data Modelling.*

38. S.-I. Niculescu, K. Gu (eds.), *Advances in Time-Delay Systems.*

39. S. Attinger, P. Koumoutsakos (eds.), *Multiscale Modelling and Simulation.*

40. R. Kornhuber, R. Hoppe, J. Périaux, O. Pironneau, O. Wildlund, J. Xu (eds.), *Domain Decomposition Methods in Science and Engineering.*

41. T. Plewa, T. Linde, V.G. Weirs (eds.), *Adaptive Mesh Refinement – Theory and Applications.*

42. A. Schmidt, K.G. Siebert, *Design of Adaptive Finite Element Software.* The Finite Element Toolbox ALBERTA.

43. M. Griebel, M.A. Schweitzer (eds.), *Meshfree Methods for Partial Differential Equations II.*

44. B. Engquist, P. Lötstedt, O. Runborg (eds.), *Multiscale Methods in Science and Engineering.*

45. P. Benner, V. Mehrmann, D.C. Sorensen (eds.), *Dimension Reduction of Large-Scale Systems.*

46. D. Kressner, *Numerical Methods for General and Structured Eigenvalue Problems.*

47. A. Boriçi, A. Frommer, B. Joó, A. Kennedy, B. Pendleton (eds.), *QCD and Numerical Analysis III.*

48. F. Graziani (ed.), *Computational Methods in Transport.*

49. B. Leimkuhler, C. Chipot, R. Elber, A. Laaksonen, A. Mark, T. Schlick, C. Schütte, R. Skeel (eds.), *New Algorithms for Macromolecular Simulation.*

50. M. Bücker, G. Corliss, P. Hovland, U. Naumann, B. Norris (eds.), *Automatic Differentiation: Applications, Theory, and Implementations.*

51. A.M. Bruaset, A. Tveito (eds.), *Numerical Solution of Partial Differential Equations on Parallel Computers.*

52. K.H. Hoffmann, A. Meyer (eds.), *Parallel Algorithms and Cluster Computing.*

53. H.-J. Bungartz, M. Schäfer (eds.), *Fluid-Structure Interaction.*

54. J. Behrens, *Adaptive Atmospheric Modeling.*

55. O. Widlund, D. Keyes (eds.), *Domain Decomposition Methods in Science and Engineering XVI.*

56. S. Kassinos, C. Langer, G. Iaccarino, P. Moin (eds.), *Complex Effects in Large Eddy Simulations.*

57. M. Griebel, M.A Schweitzer (eds.), *Meshfree Methods for Partial Differential Equations III.*

58. A.N. Gorban, B. Kégl, D.C. Wunsch, A. Zinovyev (eds.), *Principal Manifolds for Data Visualization and Dimension Reduction.*

59. H. Ammari (ed.), *Modeling and Computations in Electromagnetics: A Volume Dedicated to Jean-Claude Nédélec.*

60. U. Langer, M. Discacciati, D. Keyes, O. Widlund, W. Zulehner (eds.), *Domain Decomposition Methods in Science and Engineering XVII.*

61. T. Mathew, *Domain Decomposition Methods for the Numerical Solution of Partial Differential Equations.*

62. F. Graziani (ed.), *Computational Methods in Transport: Verification and Validation.*

63. M. Bebendorf, *Hierarchical Matrices.* A Means to Efficiently Solve Elliptic Boundary Value Problems.

64. C.H. Bischof, H.M. Bücker, P. Hovland, U. Naumann, J. Utke (eds.), *Advances in Automatic Differentiation.*

65. M. Griebel, M.A. Schweitzer (eds.), *Meshfree Methods for Partial Differential Equations IV.*

66. B. Engquist, P. Lötstedt, O. Runborg (eds.), *Multiscale Modeling and Simulation in Science.*

67. I.H. Tuncer, Ü. Gülcat, D.R. Emerson, K. Matsuno (eds.), *Parallel Computational Fluid Dynamics 2007.*

68. S. Yip, T. Diaz de la Rubia (eds.), *Scientific Modeling and Simulations.*

69. A. Hegarty, N. Kopteva, E. O'Riordan, M. Stynes (eds.), *BAIL 2008 – Boundary and Interior Layers.*

70. M. Bercovier, M.J. Gander, R. Kornhuber, O. Widlund (eds.), *Domain Decomposition Methods in Science and Engineering XVIII.*

71. B. Koren, C. Vuik (eds.), *Advanced Computational Methods in Science and Engineering.*

72. M. Peters (ed.), *Computational Fluid Dynamics for Sport Simulation.*

73. H.-J. Bungartz, M. Mehl, M. Schäfer (eds.), *Fluid Structure Interaction II - Modelling, Simulation, Optimization.*

For further information on these books please have a look at our mathematics catalogue at the following URL: www.springer.com/series/3527

Monographs in Computational Science and Engineering

1. J. Sundnes, G.T. Lines, X. Cai, B.F. Nielsen, K.-A. Mardal, A. Tveito, *Computing the Electrical Activity in the Heart.*

For further information on this book, please have a look at our mathematics catalogue at the following URL: www.springer.com/series/7417

Texts in Computational Science and Engineering

1. H. P. Langtangen, *Computational Partial Differential Equations.* Numerical Methods and Diffpack Programming. 2nd Edition

2. A. Quarteroni, F. Saleri, P. Gervasio, *Scientific Computing with MATLAB and Octave.* 3rd Edition

3. H. P. Langtangen, *Python Scripting for Computational Science.* 3rd Edition

4. H. Gardner, G. Manduchi, *Design Patterns for e-Science.*

5. M. Griebel, S. Knapek, G. Zumbusch, *Numerical Simulation in Molecular Dynamics.*

6. H. P. Langtangen, *A Primer on Scientific Programming with Python.*

7. A. Tveito, H. P. Langtangen, B. F. Nielsen, X. Cai, *Elements of Scientific Computing.*

For further information on these books please have a look at our mathematics catalogue at the following URL: www.springer.com/series/5151